21世纪高等学校规划教材 | 计算机应用

Android应用程序
开发与案例分析

杨国燕　聂佳志　编著

清华大学出版社
北京

内 容 简 介

本书从初学者的角度出发,通过通俗易懂的语言、丰富多彩的案例分析、关键代码的分析,详细介绍了 Android 平台基础知识以及进行项目开发应该掌握的基本应用技术。全书共分 12 章,内容包括 Android 集成开发环境搭建、Android 项目的组成及开发流程、常用基本组件的使用、后台服务开发、数据存储技术、组件之间的通信技术、多媒体、网络通信技术、图形和图像处理及项目案例分析等。

书中注重应用实例开发,由浅入深、循序渐进地将理论知识和实例紧密结合,以加深读者对 Android 系统基础知识和基本应用的理解。本书既可作为高等院校信息技术的教材,也可供相关工程技术人员和其他自学者参考。

本书封面贴有清华大学出版社防伪标签,无标签者不得销售。
版权所有,侵权必究。侵权举报电话: 010-62782989　13701121933

图书在版编目(CIP)数据

Android 应用程序开发与案例分析/杨国燕,聂佳志编著. —北京: 清华大学出版社,2016(2018.2 重印)
21 世纪高等学校规划教材·计算机应用
ISBN 978-7-302-42272-3

Ⅰ. ①A… Ⅱ. ①杨… ②聂… Ⅲ. ①移动终端-应用程序-程序设计-高等学校-教材
Ⅳ. ①TN929.53

中国版本图书馆 CIP 数据核字(2015)第 283745 号

责任编辑: 郑寅堃　王冰飞
封面设计: 傅瑞学
责任校对: 白　蕾
责任印制: 宋　林

出版发行: 清华大学出版社
　　　　网　　址: http://www.tup.com.cn, http://www.wqbook.com
　　　　地　　址: 北京清华大学学研大厦 A 座　　　邮　编: 100084
　　　　社 总 机: 010-62770175　　　　　　　　　　邮　购: 010-62786544
　　　　投稿与读者服务: 010-62776969, c-service@tup.tsinghua.edu.cn
　　　　质量反馈: 010-62772015, zhiliang@tup.tsinghua.edu.cn
　　　　课件下载: http://www.tup.com.cn,010-62795954
印 装 者: 北京中献拓方科技发展有限公司
经　　销: 全国新华书店
开　　本: 185mm×260mm　　印　张: 22.5　　　　　字　数: 546 千字
版　　次: 2016 年 2 月第 1 版　　　　　　　　　　印　次: 2018 年 2 月第 3 次印刷
印　　数: 2501~2700
定　　价: 49.00 元

产品编号: 061734-01

出版说明

随着我国改革开放的进一步深化,高等教育也得到了快速发展,各地高校紧密结合地方经济建设发展需要,科学运用市场调节机制,加大了使用信息科学等现代科学技术提升、改造传统学科专业的投入力度,通过教育改革合理调整和配置了教育资源,优化了传统学科专业,积极为地方经济建设输送人才,为我国经济社会的快速、健康和可持续发展以及高等教育自身的改革发展做出了巨大贡献。但是,高等教育质量还需要进一步提高以适应经济社会发展的需要,不少高校的专业设置和结构不尽合理,教师队伍整体素质亟待提高,人才培养模式、教学内容和方法需要进一步转变,学生的实践能力和创新精神亟待加强。

教育部一直十分重视高等教育质量工作。2007年1月,教育部下发了《关于实施高等学校本科教学质量与教学改革工程的意见》,计划实施"高等学校本科教学质量与教学改革工程(简称'质量工程')",通过专业结构调整、课程教材建设、实践教学改革、教学团队建设等多项内容,进一步深化高等学校教学改革,提高人才培养的能力和水平,更好地满足经济社会发展对高素质人才的需要。在贯彻和落实教育部"质量工程"的过程中,各地高校发挥师资力量强、办学经验丰富、教学资源充裕等优势,对其特色专业及特色课程(群)加以规划、整理和总结,更新教学内容、改革课程体系,建设了一大批内容新、体系新、方法新、手段新的特色课程。在此基础上,经教育部相关教学指导委员会专家的指导和建议,清华大学出版社在多个领域精选各高校的特色课程,分别规划出版系列教材,以配合"质量工程"的实施,满足各高校教学质量和教学改革的需要。

为了深入贯彻落实教育部《关于加强高等学校本科教学工作,提高教学质量的若干意见》精神,紧密配合教育部已经启动的"高等学校教学质量与教学改革工程精品课程建设工作",在有关专家、教授的倡议和有关部门的大力支持下,我们组织并成立了"清华大学出版社教材编审委员会"(以下简称"编委会"),旨在配合教育部制定精品课程教材的出版规划,讨论并实施精品课程教材的编写与出版工作。"编委会"成员皆来自全国各类高等学校教学与科研第一线的骨干教师,其中许多教师为各校相关院、系主管教学的院长或系主任。

按照教育部的要求,"编委会"一致认为,精品课程的建设工作从开始就要坚持高标准、严要求,处于一个比较高的起点上;精品课程教材应该能够反映各高校教学改革与课程建设的需要,要有特色风格、有创新性(新体系、新内容、新手段、新思路,教材的内容体系有较高的科学创新、技术创新和理念创新的含量)、先进性(对原有的学科体系有实质性的改革和发展,顺应并符合21世纪教学发展的规律,代表并引领课程发展的趋势和方向)、示范性(教材所体现的课程体系具有较广泛的辐射性和示范性)和一定的前瞻性。教材由个人申报或各校推荐(通过所在高校的"编委会"成员推荐),经"编委会"认真评审,最后由清华大学出版社审定出版。

目前,针对计算机类和电子信息类相关专业成立了两个"编委会",即"清华大学出版社计算机教材编审委员会"和"清华大学出版社电子信息教材编审委员会"。推出的特色精品教材包括:

(1) 21世纪高等学校规划教材·计算机应用——高等学校各类专业,特别是非计算机专业的计算机应用类教材。

(2) 21世纪高等学校规划教材·计算机科学与技术——高等学校计算机相关专业的教材。

(3) 21世纪高等学校规划教材·电子信息——高等学校电子信息相关专业的教材。

(4) 21世纪高等学校规划教材·软件工程——高等学校软件工程相关专业的教材。

(5) 21世纪高等学校规划教材·信息管理与信息系统。

(6) 21世纪高等学校规划教材·财经管理与应用。

(7) 21世纪高等学校规划教材·电子商务。

(8) 21世纪高等学校规划教材·物联网。

清华大学出版社经过三十多年的努力,在教材尤其是计算机和电子信息类专业教材出版方面树立了权威品牌,为我国的高等教育事业做出了重要贡献。清华版教材形成了技术准确、内容严谨的独特风格,这种风格将延续并反映在特色精品教材的建设中。

清华大学出版社教材编审委员会
联系人:魏江江
E-mail:weijj@tup.tsinghua.edu.cn

前 言

　　Android 是谷歌(Google)公司发布的一款开源移动设备操作系统,它基于 Linux 平台,是目前世界上最流行的移动设备操作系统之一。Android 是一个完全免费的操作系统平台,开发应用项目的费用也较以前大幅降低,并且还开放了应用程序的开发工具,从而使 Android 平台有了丰富的应用程序,吸引了无数软件开发者投身其中。目前很多高校也开设了 Android 应用程序开发课程,本书旨在满足于高等院校教学或初学者入门学习 Android 程序开发的需要,使读者轻松、愉快地进入移动应用软件开发大门。

　　本书基于最新的 Android SDK5.0 版本编写源代码,使读者能及时跟上 Android 应用程序开发最新技术的发展。书中注重应用实例开发,由浅入深、循序渐进地将理论知识和实例紧密结合进行介绍、剖析和实现,以加深读者对 Android 系统基础知识和基本应用的理解,帮助读者系统、全面地掌握 Android 程序设计的基本思想和基本应用技术,快速提高开发技能,为进一步深入学习 Android 应用开发打下坚实的基础。

　　全书共有 12 章,各章的具体知识点介绍如下。

　　第 1 章　Android 概述,介绍智能手机及智能手机操作系统的发展、Android 操作系统的发展史和系统特征、Android 平台的技术架构和 Android 应用程序的构成。

　　第 2 章　搭建 Android 开发环境,掌握安装、配置 Android 开发环境的步骤和注意事项,理解 Android SDK 和 ADT-Bundle 环境的使用,熟悉在应用程序开发过程中可能使用到的开发工具。

　　第 3 章　学习 Android 界面开发常用控件,包括 EditText、Button、ImageButton、RadioButton、CheckBox、Spinner、ListView 和 ProgressBar 等,读者熟悉这些控件的功能和用法,将可以设计出优秀的图形界面。

　　第 4 章　学习 Android 界面布局与菜单处理,六大布局方式分别是线性布局、帧布局、表格布局、相对布局、绝对布局、网格布局,常见的菜单处理包括选项菜单、子菜单和快捷菜单。

　　第 5 章　介绍 Android 生命周期,以 Activity 为例说明 Android 系统如何管理程序组件的生命周期。

　　第 6 章　Android 组件之间的通信,学习 Intent 的各种属性、Intent 过滤器和广播消息机制,了解 Android 系统的组件通信原理,掌握利用 Intent 启动其他组件的方法。

　　第 7 章　学习后台服务,Service 用于后台完成用户指定的操作,是 Android 的四大组件之一,掌握 Service 的启动方式和基础、本地服务应用,了解 Service 的生命周期。

　　第 8 章　学习数据存储与访问,Android 平台中实现数据存储的方式有 5 种,分别是使用 Shared Preferences 存储数据、文件存储数据、SQLite 数据库存储数据、使用 Content Provider 存储数据和网络存储数据。

　　第 9 章　学习多媒体技术,Android 提供了常见媒体的编码、解码机制,可以调用

Android 提供的现有 API，非常容易地集成音频、视频和图片等多媒体文件到应用程序中。

第 10 章　学习 Android 网络通信技术基础和 HTTP 通信应用、WebKit 应用及 Socket 通信。

第 11 章　学习图形和图像，Android 处理图形的能力非常强大，掌握图片浏览器的应用和访问图片、2D 绘图、图像特效的应用及了解内存优化。

第 12 章　以"理财系统"作为示例，综合运用所学的知识和技巧，从需求分析、界面设计、模块设计和程序开发等几个方面，详细介绍 Android 应用程序的设计思路与开发方法。

本书由杨国燕和聂佳志负责主要编写工作，其中杨国燕编写第 1~6 章和第 12 章，聂佳志编写第 7~11 章，全书由杨国燕完成整体结构的设计。因作者水平有限，书中难免存在不足和疏漏，欢迎大家批评指正，衷心希望各位读者提出宝贵的意见和建议。

编　者

2015 年 12 月

目 录

第1章 Android 概述 …………………………………………………………………… 1

1.1 智能手机的发展 …………………………………………………………………… 1
    1.1.1 智能手机的特点 …………………………………………………………… 1
    1.1.2 智能手机的未来发展趋势 ………………………………………………… 2
1.2 智能手机操作系统简介 …………………………………………………………… 3
    1.2.1 智能手机操作系统的发展 ………………………………………………… 4
    1.2.2 智能手机操作系统的分类 ………………………………………………… 4
1.3 Android 操作系统简介 …………………………………………………………… 8
    1.3.1 开放手机联盟 ……………………………………………………………… 8
    1.3.2 Android 发展史 …………………………………………………………… 10
    1.3.3 Android 系统特征 ………………………………………………………… 16
1.4 Android 平台的技术架构 ………………………………………………………… 17
1.5 Android 应用程序的构成 ………………………………………………………… 19
习题 …………………………………………………………………………………… 20

第2章 Android 开发环境与开发工具 ……………………………………………… 21

2.1 安装 Android 开发环境 …………………………………………………………… 21
    2.1.1 JDK 下载及安装 …………………………………………………………… 21
    2.1.2 ADT-Bundle for Windows 下载及安装 ………………………………… 24
2.2 使用 Android SDK 开发 Android 应用 …………………………………………… 27
    2.2.1 Android SDK 目录结构 …………………………………………………… 27
    2.2.2 Android SDK 中的示例 …………………………………………………… 28
2.3 Android 常用的开发工具 ………………………………………………………… 32
2.4 Android 程序目录结构 …………………………………………………………… 38
    2.4.1 创建第一个 Android 应用程序 …………………………………………… 38
    2.4.2 Android 程序结构 ………………………………………………………… 42
习题 …………………………………………………………………………………… 48

第3章 Android 界面开发常用控件 ………………………………………………… 49

3.1 用户界面基础 ……………………………………………………………………… 49
    3.1.1 手机用户界面应解决的问题 ……………………………………………… 49
    3.1.2 Android 平台中的 View 类 ……………………………………………… 50

3.2 TextView 控件 ……………………………………………………… 52
 3.2.1 TextView 控件常见的属性和方法 ……………………………… 52
 3.2.2 TextView 控件实例 ……………………………………………… 53
3.3 EditText 控件 ………………………………………………………… 54
 3.3.1 EditText 控件常见的属性和方法 ………………………………… 54
 3.3.2 EditText 控件实例 ……………………………………………… 55
3.4 Button 控件 …………………………………………………………… 57
 3.4.1 Button 控件常见的属性和方法 …………………………………… 57
 3.4.2 Button 控件实例 ………………………………………………… 58
3.5 ImageButton 控件 …………………………………………………… 60
 3.5.1 ImageButton 控件常见的属性和方法 …………………………… 60
 3.5.2 ImageButton 控件实例 ………………………………………… 61
3.6 RadioButton 控件 …………………………………………………… 63
 3.6.1 RadioButton 控件常见的方法 …………………………………… 64
 3.6.2 RadioButton 控件实例 ………………………………………… 64
3.7 CheckBox 控件 ……………………………………………………… 66
 3.7.1 CheckBox 控件常见的方法 ……………………………………… 66
 3.7.2 CheckBox 控件实例 …………………………………………… 67
3.8 Toast ………………………………………………………………… 70
 3.8.1 Toast 常量和常见的方法 ………………………………………… 70
 3.8.2 Toast 实例 ……………………………………………………… 70
3.9 Spinner 控件 ………………………………………………………… 72
 3.9.1 Spinner 控件常见的属性和方法 ………………………………… 72
 3.9.2 Spinner 控件实例 ……………………………………………… 73
3.10 ListView 控件 ……………………………………………………… 74
 3.10.1 ListView 控件常见的属性和方法 ……………………………… 75
 3.10.2 ListView 控件实例 …………………………………………… 76
3.11 ProgressBar 控件 …………………………………………………… 77
 3.11.1 ProgressBar 常见方法 ………………………………………… 78
 3.11.2 ProgressBar 控件实例 ………………………………………… 78
习题 ………………………………………………………………………… 80

第 4 章 Android 界面布局与菜单处理 ……………………………………… 81

4.1 界面布局概述 ………………………………………………………… 81
4.2 线性布局 ……………………………………………………………… 82
 4.2.1 LinearLayout 类简介 …………………………………………… 82
 4.2.2 线性布局实例 …………………………………………………… 83
4.3 帧布局 ………………………………………………………………… 85
4.4 表格布局 ……………………………………………………………… 87

| 4.4.1　TableLayout 类简介 …………………………………………………………… 87
| 4.4.2　表格布局实例 ………………………………………………………………… 88
|　4.5　相对布局 ……………………………………………………………………………………… 90
| 4.5.1　RelativeLayout 类简介 ………………………………………………………… 90
| 4.5.2　相对布局实例 ………………………………………………………………… 91
|　4.6　绝对布局 ……………………………………………………………………………………… 93
| 4.6.1　AbsoluteLayout 类简介 ………………………………………………………… 93
| 4.6.2　绝对布局实例 ………………………………………………………………… 93
|　4.7　网格布局 ……………………………………………………………………………………… 94
|　4.8　菜单 …………………………………………………………………………………………… 97
| 4.8.1　菜单资源 ……………………………………………………………………… 97
| 4.8.2　选项菜单 ……………………………………………………………………… 98
| 4.8.3　子菜单 ………………………………………………………………………… 102
| 4.8.4　快捷菜单 ……………………………………………………………………… 104
|　习题 ………………………………………………………………………………………………… 107

第 5 章　Android 生命周期 ………………………………………………………………………… 108

　5.1　Android 应用程序组件 ………………………………………………………………………… 108
　5.2　Android 程序生命周期 ………………………………………………………………………… 109
　5.3　Activity 生命周期 ……………………………………………………………………………… 111
　5.4　程序调试 ……………………………………………………………………………………… 119
　　　 5.4.1　LogCat ………………………………………………………………………… 119
　　　 5.4.2　DevTools ……………………………………………………………………… 121
　习题 ………………………………………………………………………………………………… 127

第 6 章　Android 组件之间的通信 ………………………………………………………………… 128

　6.1　Intent 简介 …………………………………………………………………………………… 128
　　　 6.1.1　Intent 的 action 属性 …………………………………………………………… 129
　　　 6.1.2　Intent 的 data 属性 …………………………………………………………… 129
　　　 6.1.3　Intent 的 type 属性 …………………………………………………………… 130
　　　 6.1.4　Intent 的 category 属性 ………………………………………………………… 130
　　　 6.1.5　Intent 的 extras 属性 …………………………………………………………… 130
　　　 6.1.6　Intent 的 component 属性 ……………………………………………………… 131
　6.2　系统标准 ActivityAction 应用 ………………………………………………………………… 131
　　　 6.2.1　Activity 的启动 ………………………………………………………………… 131
　　　 6.2.2　获取 Activity 返回值 …………………………………………………………… 134
　6.3　Intent 过滤器 ………………………………………………………………………………… 140
　　　 6.3.1　注册 Intent 过滤器 ……………………………………………………………… 140
　　　 6.3.2　Intent 解析 …………………………………………………………………… 141

6.4 广播消息实例 143
习题 145

第7章 后台服务 146

7.1 Service 介绍 146
　　7.1.1 Service 启动方式 146
　　7.1.2 Service 基础 147
7.2 本地服务 147
　　7.2.1 不需要与组件交互本地服务 148
　　7.2.2 本地服务结合广播接收器 153
　　7.2.3 与组件交互本地服务 159
　　7.2.4 Service 与 Thread 的区别 164
7.3 管理 Service 的生命周期 165
习题 169

第8章 数据存储与访问 170

8.1 SharedPreferences 170
　　8.1.1 SharedPreferences 简介 170
　　8.1.2 存储应用程序数据实例 173
　　8.1.3 读取其他应用程序数据实例 176
8.2 文件存储 179
　　8.2.1 文件存储简介（内部存储）... 179
　　8.2.2 文件存储应用实例 180
　　8.2.3 SD Card 存储简介 184
　　8.2.4 SD 卡存储应用实例 185
8.3 SQLite 数据库存储 191
　　8.3.1 SQLite 数据库简介 191
　　8.3.2 创建 SQLite 数据库方式 193
　　8.3.3 SQLite 数据库操作 196
　　8.3.4 SQLite 数据库管理 199
　　8.3.5 SQLite 数据库应用案例 201
8.4 数据共享 207
　　8.4.1 ContentProvider 简介 207
　　8.4.2 URI、UriMatcher 和 ContentUris 简介 208
　　8.4.3 创建 ContentProvider 210
　　8.4.4 ContentResolver 操作数据 211
　　8.4.5 ContentProvider 应用实例 211
习题 213

第 9 章　多媒体 ··· 214

9.1　音频播放 ··· 214
9.1.1　MediaPlayer 的介绍 ······································· 214
9.1.2　MediaPlayer 播放音频 ······································ 216
9.2　视频播放 ··· 219
9.2.1　自带播放器播放视频 ······································ 220
9.2.2　VideoView 播放视频 ······································ 222
9.2.3　MediaPlayer 结合 SurfaceView 播放视频 ························ 224
9.3　音频录制 ··· 229
9.4　视频录制 ··· 234
9.5　TTS 的使用 ··· 238
习题 ··· 242

第 10 章　Android 网络通信技术 ·································· 243

10.1　Android 网络通信技术基础 ····································· 243
10.1.1　无线网络技术 ·· 243
10.1.2　Android 网络基础 ······································ 243
10.1.3　Android 中的蓝牙 ······································ 245
10.1.4　Android 中的 Wi-Fi ····································· 249
10.2　HTTP 通信 ··· 252
10.2.1　HttpURLConnection 接口 ································· 253
10.2.2　HttpClient 接口 ·· 262
10.3　WebKit 应用 ·· 270
10.3.1　WebKit 概述 ·· 270
10.3.2　WebView 浏览网页 ····································· 270
10.3.3　WebView 加载 HTML 代码 ································ 272
10.3.4　WebView 与 JavaScript ··································· 274
10.4　Socket 通信 ··· 278
10.4.1　Socket 传输模式 ······································· 278
10.4.2　Socket 编程原理 ······································· 279
习题 ··· 281

第 11 章　图形和图像 ·· 282

11.1　图片浏览器 ··· 282
11.1.1　Gallery ·· 282
11.1.2　ImageSwither ··· 286
11.2　访问图片 ··· 290
11.2.1　Drawable ·· 290

11.2.2 Bitmap 和 BitmapFactory 293
11.3 内存优化 295
11.3.1 Drawable 与 Bitmap 占用内存比较 295
11.3.2 防止内存溢出 298
11.4 2D 绘图 303
11.4.1 View 类 303
11.4.2 SurfaceView 类 303
11.4.3 Paint 类 304
11.4.4 Canvas 类 306
11.4.5 绘制几何图形 308
11.4.6 绘制文本 311
11.4.7 绘制路径 314
11.5 为图像添加特效 316
11.5.1 旋转图像实例 316
11.5.2 缩放图像实例 319
11.5.3 倾斜图像实例 321
11.5.4 平移图像实例 323
11.5.5 使用 BitmapShader 渲染图像实例 326
习题 328

第 12 章 综合示例设计与开发 329

12.1 需求分析 329
12.2 程序设计 330
12.2.1 系统功能模块设计 330
12.2.2 系统流程设计 331
12.2.3 数据库设计 331
12.3 程序开发 335
12.3.1 工程结构 335
12.3.2 数据库操作类 335
12.3.3 界面设计类 337
12.3.4 辅助工具类 338
12.3.5 主控制类 339
12.3.6 用户界面 342
习题 344

参考文献 345

第1章 Android概述

智能手机操作系统是一种运算能力及功能比传统功能手机更强的操作系统。目前应用最广泛的是 Android(安卓)操作系统。Android 是一个以 Linux 为基础的开源移动设备操作系统,主要用于智能手机和平板电脑。通过本章的学习可以让读者对智能手机及操作系统的发展有一个初步的了解,同时掌握 Android 平台的起源、发展、特征和体系结构等方面的内容。

本章主要学习内容:
- 了解智能手机及智能手机操作系统的发展;
- 了解 Android 操作系统发展史和系统特征;
- 掌握 Android 平台的技术架构;
- 掌握 Android 应用程序的构成。

1.1 智能手机的发展

智能手机是一种具有相对独立、开放式操作系统的手机,可以由用户自由安装应用软件、游戏等第三方服务商提供的程序,通过此类程序来不断对手机的功能进行扩充,并可以通过移动通信网络来实现无线网络接入。随着移动通信技术的飞速发展和移动多媒体时代的到来,手机作为人们必备的移动通信工具,已从简单的通话工具向智能化发展,演变成一个移动的个人信息收集和处理平台。借助操作系统和丰富的应用软件,智能手机成了一台移动终端。

1.1.1 智能手机的特点

(1) 具备普通手机的全部功能,能够进行正常的通话,收发短信等手机应用。

(2) 具备无线接入互联网的能力,即需要支持 GSM 网络下的 GPRS 或者 CDMA 网络下的 CDMA 1X 或者 4G 网络。

(3) 具备 PDA 的功能,包括 PIM(个人信息管理)、日程记事、任务安排、多媒体应用和浏览网页。

(4) 具备一个具有开放性的操作系统,在这个操作系统平台上,可以安装更多的应用程序,从而使智能手机的功能可以得到无限的扩充。

(5) 具有人性化的一面,可以根据个人需要扩展机器的功能。

(6) 功能强大,扩展性能强,第三方软件支持多。

1.1.2 智能手机的未来发展趋势

1. 便携性与舒适性将是智能手机进一步发展的首要目标

基于统计学的仿真模型和实践证明,手机的重量与体积是存在一个相对平衡的"舒适区"范围,最近几年,各个厂家在手机屏幕和体积方面都做了积极的尝试,最大的移动终端的屏幕做到了笔记本级别的 12 英寸,重量达到 800 克,但很快就发现这种重量和体积是不符合人类的"舒适区"的,许多公司在试探人类智能手机体积舒适区的极限。

2. 基于硬件的新界面与交互,是未来智能手机的创新方向

基于电容屏的多点触摸技术,给用户带来了更好的体验,这也是苹果手机在电阻屏已经普及的时代能够异军突起的关键因素,交互技术的创新给智能手机带来了新的发展空间,而下一步智能手机如何突破多点触摸技术是智能手机创新的关键。目前三星 Galaxy 的眼球追踪技术可以追踪你的目光范围,你看到哪里它都知道,用于分析读者的关注点,自动适配大小,例如看电影时自动调节屏幕,看书以及上网自动翻页等。未来智能手机可以用眼球操作、分析情绪、表情操作、智能手机甚至可以读唇,而基于陀螺仪及传感器智能手机可以感知用户的动作,实现接听电话、挂断电话、锁屏、解锁、自动转换情景模式、打开地图、打开摄像头和自动拍照等。

3. 以智能手机为服务器的可穿戴电子器官延伸

尽管可穿戴设备目前尚处于概念阶段,但可穿戴设备与智能手机的融合一定是智能手机下一步发展的重要领域。重要的不是穿戴什么,根据让渡价值理论,可穿戴产品不要超过目前人们生活中常用的穿戴设备,那么可穿戴的东西也不过就是眼镜、手表、戒指、手环、项链等,重要的是这种可穿戴的电子器官,不仅仅是手机的延伸,更是基于手机为服务器的终端,这些可穿戴设备,可以全方位采集数据,可以记录个人的一切生活,进而形成数据库,使智能终端更好地适配个性化的需求,这些电子器官,可以实时记录血压、心跳、卡路里消耗、行为习惯和社交环境等,更可以实现大脑的延伸,可穿戴电子设备使智能手机不仅仅能够读懂语音、读懂手势,不仅仅是读唇,而是能够"读心"。例如,根据个人数据库,智能手机可以主动提醒、主动推送、主动建议、主动记录,当通过谷歌眼镜看到一段文字、广告、路牌等,就可以瞬间翻译并读出等,看到一个曾经见过的朋友,就可以马上提示在哪里、什么时候认识的等。

4. 云计算、云存储成为未来手机的标配

移动终端一直受到效率与能耗的矛盾的限制,尽管硬件技术以超越摩尔定律的速度发展,电池技术也突飞猛进,但是移动终端的计算能力与存储能力毕竟要受到便携性和续航能力的限制,在本地实现全部的计算与存储很显然制约了智能手机的进一步发展。随着 4G 的成熟 5G 的来临,甚至更加高效、高速的移动数据网络技术的出现,进入了泛网络时代云计算、云存储的春天才真正来临,未来的手机本地进行的运算与存储将越来越少,基于高速网络的云计算云存储成为下一代手机的标配。当智能手机把大量庞杂的运算交给了云,那么智能手机就将变得更加便捷、拥有更长的续航能力。而云计算、云存储也能更好地实现多终端共享、多屏幕共享。

5. 人工智能成为智能手机革命性发展的方向

智能手机的普及是大数据时代的基础,而大数据最终的发展将导致人工智能的出现,目前智能手机是自己具有一定计算能力和通信能力的工具,未来智能手机将向拥有人工智能的智慧手机发展,成为一个帮助人们进行分析、预测、判断和决策的工具,成为个人参谋智囊。系统决策目前已经有成熟的模型,但是决策模型的输出结果正确与否很大程度上取决于系统参数与阈值的设定,过去系统参数都是基于统计学的统计结果,而在大数据时代,统计学的采样正在失去意义,大数据时代加上高性能的计算能力,是对全样本量的统计,而不是采样的统计,当数据足够大、足够广的时候,高性能计算机就将具有人工智能。相信在未来的几年,手机可以帮助人们判断生活中的种种决策,包括求职、婚姻、购物、社交、投资和博弈等所有的分析判断。

6. 智能手机未来将不是手机,而是个人的数据中心

随着智能手机运算能力的提升,云计算、云存储的发展,未来智能手机将成为个人的数据中心,智能手机将全面替代一般的PC,当客户到办公室时,只要把手机放到基座上,就自动连接到键盘、鼠标及21英寸的屏幕上,这时客户仿佛是在操作一台高性能PC,而实际上是没有主机的;当客户回到家里把智能手机放到家里的基座上,马上自动连接到家里的键盘、鼠标、屏幕上,就仿佛是把单位的计算机带到家里,除了可以继续工作,查阅资料之外,也可以实现家庭计算机的娱乐、教育、资信和社交等全部功能。

在机场、餐厅、咖啡厅等移动场景下,可以直接使用手机的屏幕直接处理工作,成为了笔记本电脑;如果放到电视的基座上,智能手机就成为机顶盒,可以播放视频和电视节目;当客户驾车的时候,把手机放到车载底座上,智能手机就成为导航仪、行车记录仪,甚至成为车辆的中控台……

未来个人的所有工作、资料、数据都存储在智能手机里,而这些数据是实时与云端同步的,即便是手机损坏丢失,都可以瞬间无损恢复,智能手机真正成为个人的大数据中心,实现一机在手走遍天下。

1.2 智能手机操作系统简介

智能手机操作系统是在嵌入式操作系统基础之上发展而来的专为手机设计的操作系统,除了具备嵌入式操作系统的功能(如进程管理、文件系统、网络协议栈等)外,还需有针对电池供电系统的电源管理部分、与用户交互的输入输出部分、对上层应用提供调用接口的嵌入式图形用户界面服务、针对多媒体应用提供底层编解码服务、Java运行环境和针对移动通信服务的无线通信核心功能及智能手机的上层应用等。

1.2.1 智能手机操作系统的发展

流行的智能手机操作系统有 Symbian OS、Android OS、Windows Phone、iOS 和 Blackberry 等。按照源代码、内核和应用环境等的开放程度划分，智能手机操作系统可分为开放型平台（基于 Linux 内核）和封闭型平台（基于 UNIX 和 Windows 内核）两大类。1996 年，微软发布了 Windows CE 操作系统，微软开始进入手机操作系统。2001 年 6 月，塞班公司发布了 Symbian S60 操作系统，作为 S60 的开山之作，把智能手机提高了一个概念，塞班系统以其庞大的客户群和终端占有率称霸世界智能手机中低端市场。2007 年 6 月，苹果公司的 iOS 登上了历史的舞台，手指触控的概念开始进入人们的生活，iOS 将创新的移动电话、可触摸宽屏、网页浏览、手机游戏和手机地图等几种功能完美地融合为一体。2008 年 9 月，当苹果和诺基亚两个公司还沉溺于彼此的争斗之时，Android OS，这个由 Google 研发团队设计的小机器人悄然出现在世人面前，良好的用户体验和开放性的设计，让 Android OS 很快地打入了智能手机市场。现在 Android OS 和 iOS 系统不仅仅在智能手机市场份额中维持领先，而且这种优势仍在不断增加。而 Windows Phone 8 与 Windows 系统绑定的优势不容忽视。如果微软公司在手机性能和第三方软件及开发上做出提升和让步，也是市场份额的有力竞争者。

1.2.2 智能手机操作系统的分类

目前，手机上的操作系统主要包括以下几种，分别是 Android、iPhone OS、Windows Mobile、Windows Phone 8、Symbian、黑莓、PalmOS 和 Linux。

Android 也是一种手机操作系统，是谷歌（Google）公司发布的基于 Linux 的开源手机平台，该平台由操作系统、中间件和应用软件组成，是第一个可以完全定制、免费、开放的手机平台。Android 是一个完全免费的手机平台，使用 Android 并不需要授权费，而且因为 Android 平台有丰富的应用程序，也大幅度降低了应用程序的开发费用，可以节约 15%～20%的手机制造成本。Android 底层使用开源的 Linux 操作系统，同时开放了应用程序开发工具，使所有程序开发人员都在统一、开放的开发平台上进行开发，保证了 Android 应用程序的可移植性。Android 平台使用 Java 语言进行开发，支持 SQLite 数据库、2D/3D 图形加速、多媒体播放和摄像头等硬件设备，并内置了丰富的应用程序，如电子邮件客户端、闹钟、Web 浏览器、计时器、通讯录和 MP3 播放器等，如图 1-1 所示。

iPhone OS 是由苹果公司开发的操作系统，以开放源代码的操作系统 Darwin 为基础，主要是供苹果公司生产的 iPhone、iPod touch、iPad 以及 Apple TV 使用。iOS 的系统架构分为 4 个层次，分别是核心操作系统层、核心服务层、媒体层和可轻触层。为了便于 iPhone 应用程序开发，苹果公司提供了 iPhone SDK，为 iPhone 应用程序进行开发、测试、运行和调试提供工具。多点触摸操作是 iPhone OS 的用户界面基础，也是 iPhone OS 区别与其他手机操作系统的特性之一。此外，iPhone OS 还通过支持内置加速器，允许系统界面根据屏幕的方向而改变方向。iPhone OS 自带大量的应用程序，包括 SMS 简讯、日历、照片、相机、YouTube、股市、地图、天气、时间、计算机、备忘录、系统设定、iTunes 和通讯录等，如图 1-2 所示。

图 1-1　Android 界面

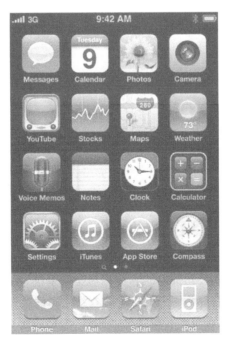

图 1-2　iPhone OS 界面

Windows Mobile 是微软公司推出的移动设备操作系统,捆绑了一系列针对移动设备而开发的应用软件,这些软件构建在 Microsoft Win32 API 基础之上,可以播放音视频文件、浏览网页、MSN 聊天和收发电子邮件。由于该操作系统对硬件配置要求较高,一般需要使用高主频的嵌入式处理器,从而出现了耗电量大、电池续航时间短和硬件成本高等情况。Windows Mobile 系列操作系统包括 Pocket PC、Smartphone 和 Portable Media Center。Smartphone 提供的功能侧重点在联系方面,主要支持的功能有电话、电子邮件、联系人和即时消息等。Pocket PC 的功能侧重个人事务处理和简单的娱乐,主要支持的功能有日程安排、移动版 Office 和多媒体播放功能等。Portable Media Center 提供的功能侧重点在移动多媒体功能,主要支持音频播放和视频播放等,如图 1-3 所示。Windows Mobile 的最新版本是 2010 年 2 月 2 日发布的 6.5.3 版,因为 Windows Phone 8 的出现,Windows Mobile 正逐渐走出历史舞台。

Windows Phone8 是微软 2012 年 6 月发行的新一代移动操作系统,具有独特的"方格子"用户界面,增加了多点触控和动力感应功能,并集成了 Xbox Live 游戏和 Zune 音乐功能。Windows Phone 8 的用户界面非常简洁,黑色背景下的亮蓝色方形图标,显得十分清晰、醒目,如图 1-4 所示。界面上直接提供了 Xbox Live 游戏和社交网站的入口,可见 Windows Phone 8 对游戏功能和社交功能的重视。虽然 Windows Mobile 和 Windows Phone 8 都是微软的手机操作系统,但这两个系统上的应用软件并不互相兼容,因此为 Windows Mobile 开发的软件并不能直接在 Windows Phone 8 上使用。

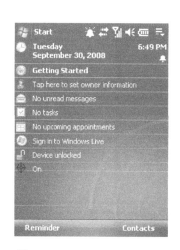
图 1-3 Windows Mobile 界面

图 1-4 Windows Phone 8 界面

黑莓系统是加拿大 RIM 公司推出的一种移动操作系统,主要在黑莓手机上使用,其特色是支持电子邮件推送功能,邮件服务器主动将收到的邮件推送到用户的手持设备上,而不需要用户频繁地连接网络查看是否有新邮件。同时,黑莓系统提供手提电话、文字短信、互联网传真、网页浏览及其他无线信息服务功能,如图 1-5 所示。黑莓系统主要针对商务应用,具有很高的安全性和可靠性。

图 1-5　黑莓系统界面

Symbian 是一个的实时多任务 32 位操作系统，提供了开发使用的函数库、用户界面、通用工具和参考示例，如图 1-6 所示。Symbian 最初由塞班公司开发和维护，后被诺基亚收购。Symbian 操作系统具有功耗低、内存占用少等特点，适合手机等移动设备使用，具有灵活的应用界面框架，并提供公开的 API 文档，不但使开发人员可以快速地掌握关键技术，还可以使手机制造商推出不同界面的产品。早期，Symbian 系统并不对外开放核心代码，核心代码仅提供给手机制造商和其他合作伙伴。后期随着 Android 和 iPhone OS 市场占有率的不断提升，诺基亚曾经一度将 Symbian 系统开源，希望借此改变 Symbian 系统的命运。因为 Symbian 系统在架构、用户体验和应用数量等方面的不足，诺基亚最终决定放弃 Symbian 系统，与微软合作将 Windows Phone 作为诺基亚的主要操作系统，而 Symbian 将在一到两年内被 Windows Phone 所取代。

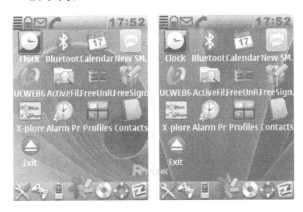

图 1-6　Symbian 界面

PalmOS 是 32 位的嵌入式操作系统，主要在移动终端上使用。PalmOS 由 3Com 公司的 Palm Computing 部门开发，拥有较多的第三方软件。PalmOS 在设计时考虑到了移动设备的内存相对较小，所以操作系统本身所占的内存极小，基于 PalmOS 编写的应用程序所占的空间也很小。PalmOS 的操作界面采用触控式，基本所有的控制选项都排列在屏幕上，仅使用手写笔就可以完成所有操作。PalmOS 向用户免费提供了开发工具,允许用户利用该工具编写或修改相关软件,使支持 PalmOS 的应用程序丰富多彩,如图 1-7 所示。PalmOS 在其他方面还存在一些不足,例如自身不具有录音和 MP3 播放功能,如果需要使用这些功能,还需要加入第三方软件或硬件设备方可实现。

Linux 手机操作系统是由计算机 Linux 操作系统演变而来的。Linux 进入移动终端操作系统近以来,就以其开放源代码的优势吸引了越来越多的终端厂商和运营商的关注。因为 Linux 开放源代码的特性,能够大幅度地降低手机的软件成本,而且有利于独立软件开发商开发出硬件利用效率高、功能更强大的应用软件,也便于行业用户开发安全、可靠的应用系统。同时也满足了手机制造商根据实际情况有针对性地开发 Linux 手机操作系统的要求,又吸引了众多软件开发商对内容应用软件的开发,丰富了第三方应用,如图 1-8 所示。然而,Linux 操作系统有其先天的不足。首先,入门难度高、熟悉其开发环境的工程师少、集成开发环境较差;其次,由于微软操作系统源代码的不公开,基于 Linux 的产品与个人计算机的连接性较差;最后,尽管目前从事 Linux 操作系统开发的公司数量较多,但真正具有很强开发实力的公司却很少,而且这些公司之间是相互独立的开发,很难实现更大的技术突破。

图 1-7　PalmOS 界面

图 1-8　Linux 界面

1.3　Android 操作系统简介

Android 是 Google 开发的基于 Linux 平台的开源手机操作系统(中文名为安卓),它涵盖移动信息设备工作所需要的全部软件,包括操作系统、用户界面和应用程序,正在逐渐成为目前移动信息设备应用程序开发的最主要的平台,而且必将成为今后移动信息设备应用程序开发的主流工具。

1.3.1　开放手机联盟

说到 Android 的发展史,首先要介绍一下 Android 平台的推动者开放手机联盟(Open Handset Alliance,OHA)。OHA 是美国谷歌公司于 2007 年发起的一个全球性的联盟组织,目标是研发用于移动设备的新技术,用以大幅削减移动设备开发与推广成本。同时通过联盟各个合作方的努力,建立移动通信领域建立新的协作环境,促进创新移动设备的开发,使消费者的用户体验远远超过今天的移动平台所能享受到的。

OHA成立时由34个成员组织构成,包括电信运营商、半导体芯片商、手机硬件制造商、软件厂商和商品化公司五类,涵盖移动终端产业链各个环节。目前,OHA的成员组织数目已经增加到82个。谷歌通过与运营商、设备制造商、开发商和其他有关各方结成深层次的合作伙伴关系,借助建立标准化、开放式的移动电话软件平台,在移动产业内形成一个开放式的生态系统。

在OHA的组织成员中,电信运营商主要有中国移动通信、KDDI(日本)、NTT DoCoMo(日本)、Sprint Nextel(美国)、T-Mobile(美国)、Telecom(意大利)、中国联通、SoftBank(日本)、Telefonica(西班牙)和Vodafone(英国),如图1-9所示。

图1-9 电信运营商

OHA中的半导体芯片商有Audience(美国)、AKM(日本)、ARM(英国)、Atheros Communications(美国)、Broadcom(美国)、Intel(美国)、Marvell(美国)、nVIDIA(美国)、Qualcomm(美国)、SiRF(美国)、Synaptics(美国)、ST-Ericsson(意大利、法国和瑞典)和Texas Instruments(美国),如图1-10所示。

图1-10 半导体芯片商

OHA中的手机硬件制造商有Acer(中国台湾)、华硕(中国台湾)、Garmin(中国台湾)、宏达电(中国台湾)、LG(韩国)、三星(韩国)、华为(中国)、摩托罗拉(美国)、索尼爱立信(日本和瑞典)和东芝(日本),如图1-11所示。

图1-11 手机硬件制造商

OHA 中的软件厂商有 Ascender Corp（美国）、eBay（美国）、谷歌（美国）、LivingImage（日本）、NuanceCommunications（美国）、Myraid（瑞士）、Omron（日本）、PacketVideo（美国）、SkyPop（美国）、Svox（瑞士）和 SONiVOX（美国），如图 1-12 所示。

图 1-12　软件厂商

OHA 中的商品化公司有 Aplix Corporation（日本）、Noser Engineering（瑞士）、Borqs（中国）、TAT-The Astonishing（瑞典）、Teleca AB（瑞典）和 Wind River（美国），如图 1-13 所示。

图 1-13　商品化公司

1.3.2　Android 发展史

2007 年 11 月 5 日，开放手机联盟成立，由电信运营商、半导体芯片商、手机硬件制造商、软件厂商和商品化公司在内的 34 个组织构成，推动 Android 平台的研发和推广，如图 1-14 所示。

2007 年 11 月 12 日，发布了 Android SDK 预览版，这是第一个对外公布的 Android SDK，为发布正式版收集用户反馈。

2008 年 4 月 17 日，谷歌举办总共 1000 万美金的 Android 开发者竞赛，奖励最有创意的 Android 程序开发者，使 Android 平台在短时间积累了大量优秀的应用程序。涌现出像 cab4me（出租车呼叫）、BioWallet（生物特征识别）和 CompareEverywhere（实时商品查询）极具创意的应用程序，如图 1-15 所示。

图 1-14　开放手机联盟徽标

2008 年 8 月 28 日，谷歌开通了 Android Market，供 Android 手机下载需要使用的应用程序。程序开发人员可以将自己设计的 Android 软件上传到 Android Market，并决定软件是否收取费用。但在 Android Market 上销售软件需要向谷歌支付 25 美元的注册费，并在每次交易中将 30% 的利润支付给运营商，如图 1-16 所示。

2008 年 9 月 23 日，发布 Android 1.0 版，这是第一个稳定的版本。1.0 版的 SDK 中分别提供了基于 Windows、Mac 和 Linux 操作系统的集成开发环境，包含完整高效的 Android

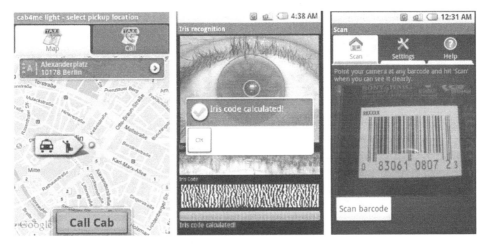

图 1-15　Android 开发者竞赛作品

图 1-16　Android Market

模拟器和开发工具,详尽的说明文档和开发示例。程序开发人员可以快速掌握 Android 应用程序的开发方法,同时也降低了开发手机应用程序的门槛。

2008 年 10 月 21 日,谷歌公布了 Android 平台的源代码。Android 作为开放源代码的手机平台,任何人或机构都可以免费使用 Android,并对它做出改进。开放源代码的 Android 有利于创新,能够为用户提供更好的体验。同时也意味着任何厂商都可以推出基于 Android 的手机,且不用支付任何的许可费用。Android 的源代码可以到谷歌的官方网站下载,地址是 http://source.android.com,如图 1-17 所示。

2008 年 10 月 22 日,第一款 Android 手机 T-Mobile G1(HTC Dream)在美国上市,由中国台湾的宏达电(HTC)公司制造,如图 1-18 所示。在硬件方面,内置 528MHz 的 Qualcomm MSM 7201A 处理器,有 192MB RAM 和 256MB ROM 的内存空间,提供侧面滑动的全键盘,支持 Wi-Fi 功能和内置 GPS 模块,支持最大 8GB 容量的 micro SD 存储卡扩展

图 1-17　Android 源代码的下载网站

容量,支持 GSM/UMTS/GPRS/EDGE/HSDPA 网络,在软件方面,集成了众多的应用功能,包括谷歌的地图功能、YouTube 视频功能、全方位的导航定位以及 360°查看浏览目标位置的功能。

2009 年 2 月,Android 1.1 正式发布。修正了 1.0 版本存在的缺陷,如设备休眠状态的稳定性问题、邮件冻结问题、POP3 链接失败问题和 IMAP 协议的密码引用问题等。同时,增加了新的特性,例如当用户搜索地图或详细查看时,允许用户对地图进行评论;允许用户保存彩信的附件;为了更加便于使用拨号盘,拨号盘可以显示或隐藏在通话菜单中等。

2009 年 2 月 17 日,第二款 Android 手机 T-Mobile G2(HTC Magic)正式发售,仍然由中国台湾的宏达电公司制造,如图 1-19 所示。T-Mobile G2 的硬件配置与 T-Mobile G1 基本相同,不同之处主要在于将内存空间从 256MB 提高到 512MB,并略微增加了电池的容量,用以提升整机的使用时间。T-Mobile G1 放弃了影响手机尺寸的滑动全键盘设计,使 T-Mobile G2 的体积(113mm×55mm×13.65mm)比 T-Mobile G1(117.7mm×55.7mm×17.1mm)明显减小。

图 1-18　T-Mobile G1

图 1-19　T-Mobile G2

2009年4月15日，Android 1.5 正式发布。此版本提升了性能表现，提高了摄像头的启动速度和拍摄速度，提高了GPS位置的获取速度，使浏览器的滚动更为平滑，提高了获取Gmail中对话列表的速度等。在新特性方面，支持了软键盘、中文显示和中文输入功能，并可以将视屏录制的内容直接上传到YouTube。

2009年6月24日，中国台湾的宏达电发布了第三款Android手机HTC Hero，如图1-20所示。在硬件方面，使用Qualcomm MSM 7200A处理器，500万像素摄像头，提供3.5mm的耳机插孔。在软件方面，首次支持Adobe Flash和支持多点触控技术，最突出的改进是使用了HTC Sense界面，使HTC Hero的界面异常美观、绚丽。

2009年10月28日，发布Android 2.0(Eclair)。此版本引入了大量的新特性，如数字变焦、多点触摸和多个账户邮箱等，并在账户同步和蓝牙通信等方面上增加了新的API，开发者可使这些新API令手机与各种联系源进行同步，并实现点对点连接和游戏功能。新版SDK改进了图形架构性能，可以更好地利用硬件加速，改进了虚拟键盘，使操作更为便利。

2010年1月6日，谷歌公司初次发布了自主品牌的Android手机Google Nexus One，如图1-21所示。使用SnapDragon 1GHz处理器，3.7英寸AMOLED电容屏，由中国台湾的宏达电代工生产。Nexus One搭载纯净的Android 2.1，由于谷歌出品的手机系统上没有额外的限制和多余的功能，这款手机在较长的时间内都是Android软件理想的开发和测试平台。

图1-20　HTC Hero　　　　　　　　图1-21　Google Nexus One

2010年5月21日，发布Android 2.2版(Froyo，冻酸奶)。此版本在企业集成、设备管理API、性能、网络共享、浏览器和市场等领域都提供了很多新特性。借助于新的Dalvik JIT编译器，CPU密集型应用的速度要比Android 2.1快2～5倍，并加入对Adobe Flash视频和图片的完美支持。在网络共享方面，通过手机提供的热点，将多个设备节点连接到互联网上。在浏览器方面，由于使用了Chrome V8引擎，JavaScript代码的处理速度要比Android 2.1快2～3倍。Android 2.2的最大改进是可以将应用程序安装在micro SD卡上，应用程序可以在内部存储器和外部存储器上迁移。

2010年12月7日，Android 2.3(Gingerbread，姜饼)正式发布。此版本主要增强了对游戏的支持、多媒体影音和通信功能。在游戏方面，增加了新的垃圾回收和优化处理事件，以提高对游戏的支持能力，原生代码可直接存取输入和感应器事件、EGL/OpenGL ES、OpenSL ES，并增加了新的管理窗口和生命周期的框架。在多媒体影音方面，支持VP8和WebM视频格式，提供AAC和AMR宽频编码，提供了新的音频效果器，例如混响、均衡、虚拟耳机和低频提升。在通信方面，支持前置摄像头、SIP/VoIP 和 NFC（近场通信）功能。

2010 年 12 月 7 日，谷歌公司发布了第二款自主品牌的 Android 手机 Google Nexus S，如图 1-22 所示。Nexus S 用 Cortex A8 处理器，默认频率为 1GHz，512MB 的 RAM 和 16GB 的内置闪存，但不支持存储卡扩展，4.0 英寸 WVGA（480×800）分辨率电容触摸屏幕。Nexus S 支持 AGPS，支持 Bluetooth 2.1＋EDR，支持 Wi-Fi 802.11 n/b/g，支持 NFC 技术，支持三轴陀螺仪、加速计、数字罗盘、光线感应器和距离感应器。Nexus S 搭载最新的 Android 2.3，是第一款具备 NFC 功能的 Android 手机。

2011 年 1 月 6 日，摩托罗拉发布了第一款 Android 3.0（Honeycomb 蜂巢）的平板电脑 Motorola Xoom，如图 1-23 所示。硬件上采用双核 1GHz NVIDIA Tegra 2 处理器，10.1 寸 1280×800 分辨率的触摸屏，内置有 32GB 存储，并配有前置与后置摄像头，支持高清视频录制和播放功能。Motorola Xoom 才是真正意义上的 Android 平板电脑，也是 Android 系统的一个里程碑。

图 1-22　Google Nexus S

图 1-23　Motorola Xoom

2011 年 2 月 3 日，Android 3.0 版本（Honeycomb）正式发布。这是专为平板电脑设计的 Android 系统，因此在界面上更加注重用户体验和良好互动性，重新定义了多任务处理功能，丰富的提醒栏，支持 widgets，并允许用户自定义主界面。Android 3.0 原生支持文件/图片传输协议，允许用户通过 USB 接口连接外部设备同步数据，或通过 USB 或蓝牙连接实体键盘进行更快速的文字输入，改进了 Wi-Fi 连接，搜索信号速度更快，并可通过蓝牙来进行 tether 连接，分享 3G 信号给其他设备。

2011 年 5 月 10 日，Android 3.1 版本正式发布。作为 Android 3.0 的升级版，Android 3.1 界面上做了一些美化与调整，从桌面到程序集菜单的动画转场更为顺畅，界面上的文字颜色与位置也稍微调整，还加入了全系统适用的声音回馈。增加了对 USB 设备的支持，如 USB 鼠标、键盘和游戏控制器等。Widget 加入了可自定改变大小的功能，如果 Widget 本身支持缩放功能，使用者可放大 Widget 以看到更多信息。Android 3.1 除了支持许多新标准与功能外，内建的应用程序也做了一些更新，更适合平板电脑使用。

2011 年 10 月 19 日，Android 4.0 版本（Ice Cream Sandwich，冰激凌三明治）正式发布，如图 1-24 所示。这一版本最显著的特征是同时支持智能手机、平板电脑和电视等设备，而不需要根据设备不同选择不同版本的 Android 系统。该版本取消了底部物理按键的设计，直接使用虚拟按键,增大屏幕面积同时控制手机整体大小,而且这样的操作方式可以使智能手机与平板电脑保持一致。人脸识别功能在 4.0 版本中得到应用,用户可以使用自拍相片设置屏幕锁,Android 系统根据脸部识别结果控制手机的解锁功能。另一个有趣的应用是基于 NFC 的 Android Beam 功能,可以让两部手机在接近到 4 厘米后交换信息,可交换的内容包括网站、联系人、导航、YouTube 视频等,甚至是电子市场的下载链接。

2012 年 6 月 28 日,Android 4.1 Jelly Bean(果冻豆)发布。Android 4.1 更快、更流畅、更灵敏;特效动画的帧速提高至 60fps,增加了三倍缓冲;增强通知栏;全新搜索;搜索将会带来全新的 UI、智能语音搜索和 Google Now 三项新功能;桌面插件自动调整大小;加强无障碍操作;语言和输入法扩展;新的输入类型和功能;新的连接类型。

2012 年 10 月 30 日,Android 4.2 Jelly Bean(果冻豆)发布。Android 4.2 沿用"果冻豆"这一名称,以反映这种最新操作系统与 Android 4.1 的相似性,但 Android 4.2 推出了一些重大的新特性,具体如下:

Photo Sphere 全景拍照功能;键盘手势输入功能;改进锁屏功能,包括锁屏状态下支持桌面挂件和直接打开照相功能等;可扩展通知,允许用户直接打开应用;Gmail 邮件可缩放显示;Daydream 屏幕保护程序;用户连点 3 次可放大整个显示屏,还可用两根手指进行旋转和缩放显示,以及专为盲人用户设计的语音输出和手势模式导航功能等;支持 Miracast 无线显示共享功能;Google Now 可允许用户使用 Gamail 作为新的数据来源,如改进后的航班追踪功能、酒店和餐厅预订功能以及音乐和电影推荐功能等,如图 1-25 所示。

图 1-24　Android 4.0 版本

图 1-25　Android 4.2 Jelly Bean 原生系统用户界面

2013 年 9 月 4 日,凌晨,谷歌对外公布了 Android 新版本 Android 4.4 KitKat(奇巧巧克力),并且于 2013 年 11 月 01 日正式发布,新的 4.4 系统进一步整合了自家服务,力求防止安卓系统继续碎片化、分散化。

2014 年 11 月 15 日,谷歌正式发布 Android 5.0 正式版 Lollipop(棒棒糖)。Android 5.0 最明显的变化是采用了全新的设计语言,被称之为"Material Design"。界面加入了五彩缤纷的颜色、流畅的动画效果,呈现出一种清新的风格。采用这种设计的目的在于统一 Android 设备的外观和使用体验,不论是手机、平板还是多媒体播放器。此外,Android 5.0 还为开发者带来了 5000 个新 API,从而让设备间更具整体感及互联性。

1.3.3 Android 系统特征

Android 广泛支持 GSM、3G 和 4G 的语音与数据业务,支持接收语言呼叫和 SMS 短信,支持数据存储共享和 IPC 消息机制,为地理位置服务(如 GPS)、谷歌地图服务提供易于使用的 API 函数库,提供组件复用和内置程序替换的应用程序框架,提供基于 WebKit 的浏览器,广泛支持各种流行的视频、音频和图像文件格式,支持的格式有 MPEG4、H264、MP3、AAC、AMR、JPG、PNG 和 GIF,为 2D 和 3D 图像处理提供专用的 API 库函数。

Android 系统提供了访问硬件的 API 库函数,用于简化像摄像头、GPS 等硬件的访问过程。只要支持 Android 应用程序框架的手机,对硬件访问的方法是完全一致的,因此即使将应用程序移植到不同硬件配置的手机上,也无须更改应用程序对硬件的访问方法。Android 支持的硬件包括 GPS、摄像头、网络连接、Wi-Fi、蓝牙、NFC、加速度计、触摸屏和电源管理等。

在内存和进程管理方面,Android 具有自己的运行时和虚拟机。与 Java 和.NET 运行时不同,Android 运行时还可以管理进程的生命周期。Android 为了保证高优先级进程运行和正在与用户交互进程的响应速度,允许停止或终止正在运行的低优先级进程,以释放被占用的系统资源。Android 进程的优先级并不是固定的,而是根据进程是否在前台或是否与用户交互而不断变化的。Android 生命周期和调试的相关内容将在第 4 章介绍。

在界面设计上,Android 提供了丰富的界面控件供使用者调用,从而加快了用户界面的开发速度,也保证了 Android 平台上的程序界面的一致性。Android 将界面设计与程序逻辑分离,使用 XML 文件对界面布局进行描述,有利于界面的修改和维护。用户界面的相关内容将在第 5 章介绍。

Android 提供轻量级的进程间通信机制 Intent,使用跨进程组件通信和发送系统级广播成为可能。通过设置组件的 Intent 过滤器,组件通过匹配和筛选机制,可以准确地获取到可以处理的 Intent。组件通信与广播消息的相关内容将在第 6 章介绍。

Android 提供了 Service 作为无用户界面、长时间后台运行的组件。Android 是多任务系统,但受到屏幕尺寸的限制,同一时刻只允许一个应用程序是在前台运行。Service 无需用户干预,可以长时间、稳定的运行,可为应用程序提供特定的后台功能,还可以实现事件处理或数据更新等功能。后台服务相关内容将在第 7 章介绍。

Android 支持高效、快速的数据存储方式,包括快速数据存储方式 SharedPreferences、文件存储和轻量级关系数据库 SQLite,应用程序可以使用适合的方法对数据进程保存和访问。同时,为了便于跨进程共享数据,Android 提供了通用的共享数据接口 ContentProvider,可以无须了解数据源、路径的情况下，对共享数据进行查询、添加、删除和更新等操作。数据存储与访问相关内容将在第 8 章介绍。

Android 支持位置服务和地图应用，可以通过 SDK 提供的 API 直接获取当前的位置，追踪设备的移动路线，或设定敏感区域，并可以将 Google 地图嵌入到 Android 应用程序中，实现地理信息可视化开发。位置服务和地图应用的相关内容将在第 9 章介绍。

Android 支持 Widget 插件，可以方便地在 Android 系统上开发桌面应用，实现比较常见的一些桌面小工具，或在主屏上显示重要的信息。随着 Android 系统可以在平板电脑上使用，Widget 插件的实用性在不断的提高。Android Widget 的相关内容将在第 10 章介绍。

Android 支持使用本地代码(C 或 C++)开发应用程序的部分核心模块，提高了程序的运行效率，并有助于增加 Android 开发的灵活性。Android 本地代码开发的相关内容将在第 11 章介绍。

1.4 Android 平台的技术架构

Android 平台采用了软件堆栈(Software Stack)，又名软件叠层的架构，主要分为 4 部分。底层以 Linux 核心为基础，并包含各种驱动，只提供基本功能。中间层包括程序库(Libraries)和 Android 运行时环境。再往上一层是 Android 提供的应用程序框架。最上层是各种应用软件，包括通话程序、短信程序等，这些应用软件由开发人员自行开发。

Android 平台的架构如图 1-26 所示。

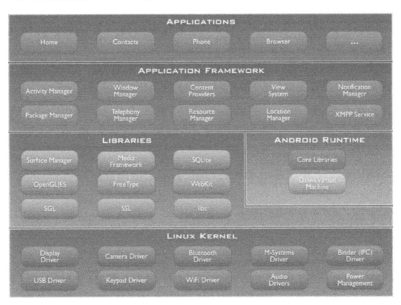

图 1-26 Android 平台的架构

1. 应用程序(Applications)

Android 会附带一系列核心应用程序包，这些应用程序包包括 E-mail 客户端、SMS 短信程序、日历、地图、浏览器和联系人管理程序等。Android 中所有的应用程序都是使用

Java语言编写的。

2．应用程序框架（Application Framework）

开发者也可以访问 Android 应用程序框架中的 API。该应用程序架构简化了组件的重用，任何一个应用程序都可以发布它的功能块，并且任何其他的应用程序都可以使用这些发布的功能块（应该遵循框架的安全性限制）。同样，该应用程序的重用机制也使用户可以方便地替换程序组件。

隐藏在每个应用程序后面的是 Android 提供的一系列的服务和管理器。

- 丰富且可扩展的视图（Views）：有列表（Lists）、网格（Grids）、文本框（Text Boxes）、按钮（Buttons），甚至包括可嵌入的 Web 浏览器，这些视图可用于构建应用程序。
- 内容提供器（Content Providers）：使得应用程序可以访问另一个应用程序的数据（例如联系人数据库），或者可以共享它们自己的数据。
- 资源管理器（Resource Manager）：提供非代码资源的访问，例如本地字符串、图形和布局文件（Layout Files）等。
- 通知管理器（Notification Manager）：使得应用程序可以在状态栏中显示自定义的提示信息。
- 活动管理器（Activity Manager）：用于管理应用程序生命周期，并且提供常用的导航回退功能。

3．程序库（Libraries）

Android 平台包含一些 C/C++ 库，Android 系统中的组件可以使用这些库。它们通过 Android 应用程序框架为开发者提供服务。

- 系统 C 库：一个从 BSD 继承的标准 C 系统函数库，是专门为基于嵌入式 Linux 设备定制的。
- 媒体库：基于 PacketVideo 的 OpenCORE，该库支持多种常用的音频、视频格式文件的回放和录制，同时支持静态图像文件，编码格式包括 MPEG4、H.264、MP3、AAC、AMR、JPG 和 PNG 等。
- Surface Manager：管理显示子系统，并且为多个应用程序提供 2D 和 3D 图层的无缝融合。
- LibWebCore：一个最新的 Web 浏览器引擎，支持 Android 浏览器和一个可嵌入的 Web 视图。
- SGL：底层的 2D 图形引擎。
- 3D 库：基于 OpenGL ES 1.0 API 实现，该库可以使用 3D 硬件加速或者使用高度优化的 3D 软加速。
- FreeType：用于位图和矢量字体显示。
- SQLite 库：一个对于所有应用程序可用的、功能强劲的轻型关系型数据库引擎。

4．Android 运行时环境

Android 运行时环境由一个核心库和 Dalvik 虚拟机组成。核心库提供 Java 编程语言核心库的大多数功能。每一个 Android 应用程序都在自己的进程中运行,都拥有一个独立的 Dalvik 虚拟机实例。Dalvik 被设计成一个设备可以同时高效地运行多个虚拟系统。它依赖于 Linux 内核的一些功能,例如线程机制和底层内存管理机制等。Dalvik 虚拟机执行扩展名为.dex 的 Dalvik 可执行文件,该格式文件针对小内存的使用进行了优化,同时虚拟机是基于寄存器的,所有的类由 Java 编译器编译,然后通过 SDK 中的 dx 工具转化成.dex 格式,最后由虚拟机执行。

5. Linux 内核

Android 核心系统服务依赖于 Linux 2.6 内核,如安全性、内存管理、进程管理、网络协议栈和驱动模型等。Linux 内核也同时作为硬件和软件栈之间的抽象层。

1.5 Android 应用程序的构成

在通常情况下,一个 Android 应用程序是由以下 4 个组件构成的:活动(Activity)、广播(Broadcast)、服务(Service)和内容提供器(Content Provider)。

这 4 个组件是构成 Android 应用程序的基础,但并不是每个 Android 应用程序都必须包含这 4 个组件,除了 Activity 是必要部分之外,其余组件都是可选的,在某些应用程序中,可能只需要其中部分组件构成即可。

1. 活动(Activity)

活动(Activity)是最基本的 Android 应用程序组件。在应用程序中,一个活动通常就是一个单独的屏幕。每个活动都是通过继承活动基类被实现为一个独立的类,活动类将会显示由视图控件组成的用户接口,并对事件做出响应。

大多数的应用程序都是由多个屏幕显示组成的。例如,一个发送信息的应用也许有一个显示发送消息的联系人列表屏幕,第二个屏幕用于写文本消息和选择收件人,第三个屏幕查看历史消息或者消息设置操作等。这里每个屏幕都是一个活动,很容易实现从一个屏幕到一个新屏幕并且完成新的活动。因为 Android 会把每个从主菜单打开的程序保留在堆栈中,所以当打开一个新屏幕时,先前的屏幕会被置为暂停状态并且压入历史堆栈中。用户可以通过回退操作,回到以前打开过的屏幕,也可以有选择性地移去一些没有必要保留的屏幕。

2. 广播(Broadcast)

在 Android 系统中,广播(Broadcast)是在组件之间传播数据(Intent)的一种机制。这些组件甚至可以位于不同的进程中。广播的发送者和接收者事先是不需要知道对方的存在的,这样的优点就是系统的各个组件可以松耦合地组织在一起,使得系统具有高度的可扩展性,容易与其他系统进行集成。

3. 服务(Service)

服务是 Android 应用程序中具有较长的生命周期但是没有用户界面的代码程序。它在后台运行，并且可以进行交互。它与 Activity（活动）的级别差不多，但是不能自己运行，需要通过某一个 Activity 来调用。

Android 应用程序的生命周期是由 Android 系统来决定的，不由具体的应用程序的线程来左右。若应用程序要求在没有界面显示的情况还能正常运行（要求有后台线程，而且直到线程结束，后台线程不会被系统回收），这个时候就需要用到 Service（服务）了。

Service 的典型例子是一个具有播放列表功能的正在播放歌曲的媒体播放器。在媒体播放器应用中，可能会有一个或多个活动，让使用者可以选择并播放歌曲。然而活动本身并不处理音乐播放功能，因为用户期望在切换到其他屏幕后，音乐应该还在后台继续播放。

4．内容提供器（Content Provider）

Android 应用程序可以使用文件或 SQLite 数据库来存储数据。ContentProvider 提供了一种多应用间数据共享的方式。当开发者希望自己的应用数据能与其他应用共享时，内容提供器将会非常有用。一个内容提供器类实现了一组标准的方法，能够让其他应用保存或读取此内容提供器处理的各种类型数据。

也就是说，一个应用程序可以通过实现一个 ContentProvider 抽象接口将自己的数据暴露出去。外界根本看不到，也不用看到这个应用程序暴露的数据在应用程序中是如何存储的，但是外界可以通过这一套标准及统一的接口与应用程序里的数据打交道，可以读取应用程序的数据，也可以删除应用程序的数据。

习题

1．简述各种智能手机操作系统的特点。
2．简述 Android 操作系统的发展史和系统特征。
3．描述 Android 平台的技术架构由哪些部分组成，并说明每一部分的作用。

第 2 章 Android 开发环境与开发工具

搭建 Android 开发环境是开发 Android 应用程序的第一步,也是深入理解 Android 系统的一个良好的途径。掌握安装、配置 Android 开发环境的步骤和注意事项,理解 Android SDK 和 ADT-Bundle 环境的使用,熟悉在应用程序开发过程中可能会使用到的开发工具。

本章主要学习内容:
- 掌握 ADT-Bundle 开发环境的安装配置方法;
- 了解 Android SDK 的目录结构和示例程序;
- 了解各种 Android 开发工具的使用;
- 掌握 Android 程序目录结构;
- 掌握 R.java 文件的用途和生成方法和 AndroidManifest.xml 文件的用途。

2.1 安装 Android 开发环境

Android 5.0 是 Google 于 2014 年 10 月 15 日发布的全新 Android 操作系统。Eclipse 是开发 Android 应用程序的首选集成开发环境。Eclipse 作为开源的 Java 开发环境,功能强大,易于使用。ADT-Bundle for Windows 是由 Google Android 官方提供的集成式 IDE,已经包含了 Eclipse,读者无须再去下载 Eclipse,并且里面已集成了插件,它解决了大部分初学者通过 Eclipse 来配置 Android 开发环境的复杂问题。

安装 Android 开发环境,首先需要安装支持 Java 应用程序运行的 Java 开发工具包 (Java Development Kit,JDK),然后安装集成开发环境 Eclipse,最后安装 Android SDK 和 Eclipse 的 ADT 插件。

2.1.1 JDK 下载及安装

因为 Eclipse 是用 Java 语言编写的应用程序,需要 JRE 才能运行。如果 JRE 没有安装或者没有被检测到,尝试打开 Eclipse 时会有错误提示,如图 2-1 所示。

安装 JRE 的系统可以运行 Java 应用程序,但如果需要进行 Java 应用程序的开发,应该直接安装 JDK。因为 JDK 中包含完整的 JRE,所以只要安装 JDK 后,JRE 也自动安装在操作系统中了。

图 2-1 没有安装 JRE 的错误提示

JDK 的基本组件包括编译器（将源程序转成字节码）、打包工具（将相关的类文件打包成一个文件）、文档生成器（从源码注释中提取文档）和查错工具（用于进行调试和差错）。

JDK 可以到 ORACLE 的官方网站下载，在浏览器中输入下面的网址 http://www.oracle.com/technetwork/java/javase/downloads/index.html，显示的页面如图 2-2 所示，然后单击 Java SE 8 中的 JDK Download，进入 JDK 的下载页面。

图 2-2 ORACLE 的官方网站下载 JDK

在 JDK 的下载页面中，首先需要同意 ORACLE 的 Java SE 二进制代码协议，选择 Accept License Agreement，然后根据用户的系统选择不同版本的 JDK。如果是 32 位的 Windows 系统，选择下载 Windows x86；如果是 64 位的 Windows 系统，选择下载 Windows x64，如图 2-3 所示。

在 JDK 的安装过程中，一般情况下保持 JDK 的默认设置即可，JDK 会安装在 C:\Program File\Java\jdk1.8.0\目录下，如图 2-4 所示。

在 JDK 安装完毕后，安装程序提示 Java(TM) SE Development Kit 8(64-bit)已经成功安装。下一步可以进行添加环境变量的工作了。

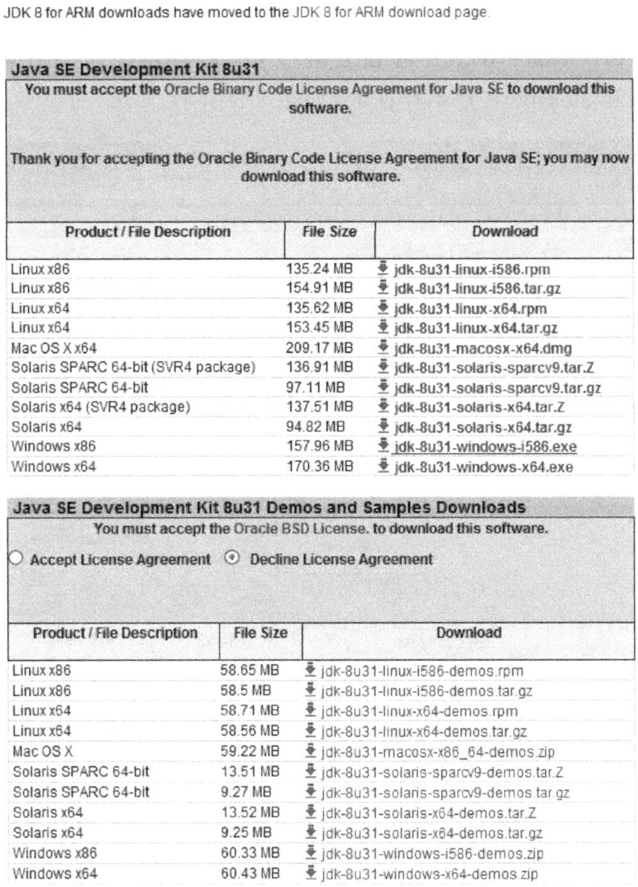

第2章 Android开发环境与开发工具 23

图 2-3 根据用户的系统选择不同版本的 JDK

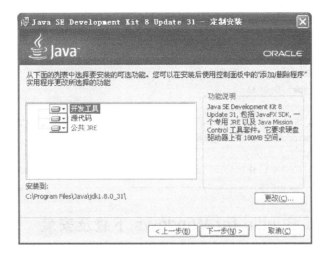

图 2-4 JDK 的默认设置

JDK 安装完成后，需要进行设置使 java 环境能够使用。首先右击"我的电脑"，打开属性，然后选择"高级"里面的"环境变量"，在新的打开界面中的系统变量需要设置 3 个属性 JAVA_HOME、path、classpath，其中在没安装过 jdk 的环境下。path 属性是本来存在的。而 JAVA_HOME 和 classpath 是不存在的，如图 2-5 所示。

图 2-5　环境变量配置窗口

设置 JAVA_HOME 属性主要是为了方便引用，例如，JDK 安装在 C:\Program Files\Java\jdk1.8.0 目录里，则设置 JAVA_HOME 为该目录路径，那么以后在使用这个路径的时候，只需要输入％JAVA_HOME％即可，避免每次引用都输入很长的路径串；设置 CLASSPATH 属性是为了程序能找到相应的.class 文件；设置 path 属性是为了任何路径下都可以仅用 java 来执行命令了，当在命令提示符窗口输入代码时，操作系统会在当前目录和 PATH 变量目录里查找相应的应用程序，并且执行。

设置环境变量如下：

```
变量名：JAVA_HOME
变量值：C:\Java\jdk1.8.0_31
变量名：Path
变量值：.;%JAVA_HOME%\bin;C:\Java\adt-bundle-windows-x86\sdk\tools; C:\Java\adt-
       bundle-windows-x86\sdk\platform-tools
变量名：CLASSPATH
变量值：.;%JAVA_HOME%\lib\tools.jar;%JAVA_HOME%\lib\dt.jar;%JAVA_HOME%\lib;
```

2.1.2　ADT-Bundle for Windows 下载及安装

ADT-Bundle for Windows 是由 Google Android 官方提供的集成式 IDE，已经包含了 Eclipse，配置了 Android SDK，并且里面已集成了 ADT 插件。

Android SDK 是 Android 软件开发工具包（Android Software Development Kit），是 Google 公司为了提高 Android 应用程序开发效率、减少开发周期而提供的辅助开发工具、开发文档和程序范例。ADT 插件是 Eclipse 开发环境的定制插件，为开发 Android 应用程序提供了一个强大的、完整的开发环境，可以快速地建立 Android 工程、用户界面和基于 Android API 的组件，还可以在 Eclipse 中使用 Android SDK 提供的工具进行程序调试，或对 apk 文件进行签名等。

下载 ADT-Bundle 集成开发环境的地址是 http://developer.android.com/sdk/index.html，登录后进入图 2-6 所示的界面。

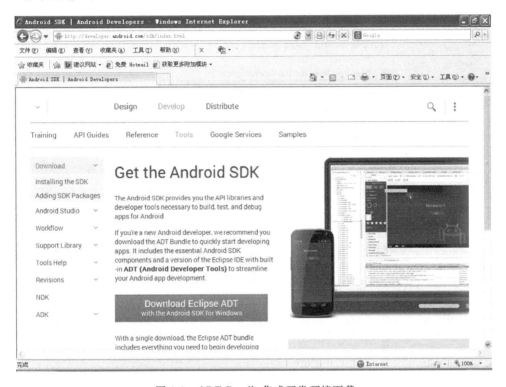

图 2-6　ADT-Bundle 集成开发环境下载

单击 Download Eclipse ADT with the Android SDK Windows 按钮后进入图 2-7 所示的界面。选中 Accepting this License Agreement 复选框，然后在选择 32 位或 64 位进行下载。

解压 adt-bundle-windows-x86-20140702.zip（文件名可能因为版本，略有不一样），内有三部分，如图 2-8 所示。

运行 SDK Manager.exe，SDK Manager 用于下载其他 Android 开发相关的组件，选择要下载 Android 版本的 API。也可以运行 eclipse/eclipse.exe 然后通过 Windows→Android SDK Manager 打开图 2-9 和图 2-10 所示的界面。

至此，Android 应用程序的开发环境就安装完成了。后面会对 Android SDK 的目录结构、示例程序和开发工具进行介绍。

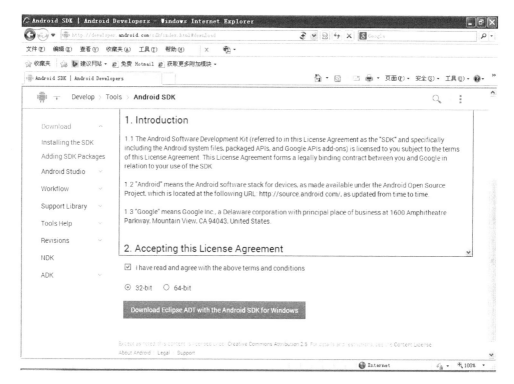

图 2-7　ADT-Bundle 版本选择

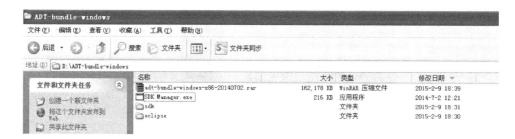

图 2-8　解压 adt-bundle-windows-x86-20140702.zip 文件的内容

图 2-9　运行 Eclipse 界面

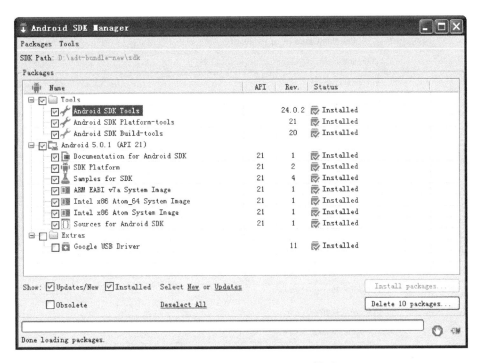

图 2-10　Android SDK Manager 界面

2.2　使用 Android SDK 开发 Android 应用

Android SDK 是程序开发人员学习和开发 Android 程序的宝贵资源,不仅提供了开发所必备的调试、打包和仿真工具,还提供了详尽的说明文档和简单易懂的开发示例。

2.2.1　Android SDK 目录结构

在 Android SDK 安装到本地磁盘后,可以在文件系统中查看到 Android SDK 的目录结构,如图 2-11 所示。

其中,add-one 目录用于存储 Google 提供地图开发包,支持基于 Google Map 的地图开发。docs 目录下的是 Android SDK 的帮助文档,通过目录下的 offline.html 文件启动,帮助文档的首页面如图 2-12 所示。extras\google 目录下保存了 Android 手机的 USB 驱动程序。platforms 目录用于存储 SDK 和 AVD 管理器下载的各种版本的 SDK,笔者的目录中有 4.0 版本的 SDK。platforms-tools 目录中保存了与平台调试相关的工具,如 adb、aapt 和 dx 等。samples 目录示例代码和程序的存储目录。temp 是临时存储文件的目录,在 SDK 和 AVD 管理器下载开发包时,下载文件会临时存储在这个目录中。tools 目录保存了通用的 Android 开发调试工具和 Android 手机模拟器。SDK Manager.exe 和 AVD Manager.exe 分别是 SDK 和 AVD 的管理器,SDK Readme.txt 是 Android SDK 的说明文档。

Android SDK 帮助文档内容非常丰富,详细介绍了 Android 系统中所有 API 函数的使用方法,尤其帮助文档中的开发指南(Dev Guide),系统地介绍了 Android 应用程序的开发

图 2-11　Android SDK 的目录结构

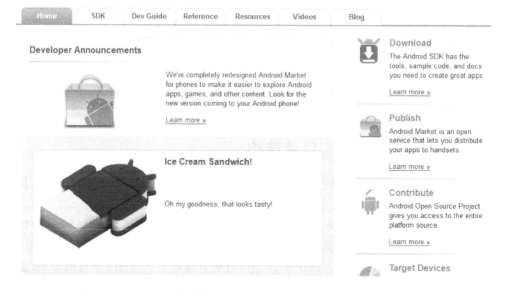

图 2-12　Android SDK 帮助文档

基础、用户界面、资源使用、数据存储、音视频功能、集成开发环境和开发工具等内容,对于学习 Android 程序开发具有指导意义。

2.2.2　Android SDK 中的示例

在＜Android SDK＞\samples\android-14 目录中,有多个基于 Android 4.0 版本的示例程序。这些示例程序多数并不复杂,但可以从不同方面展示 Android SDK 所提供的丰富功能。

1. MultiResolution 示例

MultiResolution 是 Android 程序支持不同尺寸屏幕的示例。根据屏幕解析度不同，Android 程序可以自动加载不同大小的图片，避免图片尺寸对界面布局产生影响。MultiResolution 示例如图 2-13 所示。

2．APIDemos 示例

APIDemos 示例提供了 Android 平台上多数 API 的使用方法，涉及系统、资源、图形、搜索、语音识别和用户界面等方面。程序开发人员可以在 Android 应用程序开发过程中参考 APIDemos 示例，但该示例的代码文件众多，结构上略显混乱，给参考和学习带来不小的阻碍。APIDemos 示例如图 2-14 所示。

3．SkeletonApp 示例

SkeletonApp 示例是一个界面演示程序，说明了如何使用布局和界面控件设计用户界面，以及如何在界面中添加菜单和处理菜单事件。SkeletonApp 示例如图 2-15 所示。

图 2-13　MultiResolution 示例

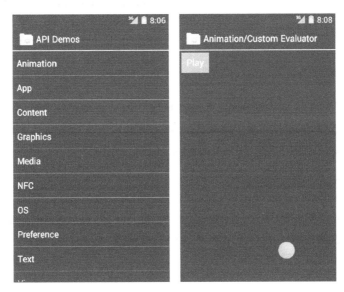

图 2-14　ApiDemos 示例

4．NotesPad 示例

NotesPad 示例是一个记事本程序，可以将文字内容保存在记事本程序中，并支持添加和删除记事本操作。NotesPad 示例说明了如何进行复杂程序设计，以及如何使用 SQLite

数据库保存数据和 ContentProvider 共享数据。NotesPad 示例如图 2-16 所示。

图 2-15　SkeletonApp 示例　　　　　　图 2-16　NotesPad 示例

5．Home 示例

Home 示例是一个桌面主题程序，可以将自定义的桌面主题注册到系统中，用户可以通过单击 Home 键选择不同的桌面主题。此示例说明了如何进行桌面主题程序的开发，以及在开发过程中需要注意的事项。Home 示例如图 2-17 所示。

6．Snake 示例

Snake 示例是贪吃蛇程序，一个经典的小游戏，可以通过导航键控制贪吃蛇的前进方向。该示例演示了如何在 Android 系统中进行游戏开发，对进行游戏开发的程序人员具有一定的参考价值。Snake 示例如图 2-18 所示。

7．LunarLander 示例

LunarLander 示例也是一个小游戏，模拟登陆舱在月球表面着陆。用户通过控制登陆舱的方向和速度，使登陆舱可以平稳地在月球表面着陆。LunarLander 示例实现了简单的碰撞检查功能，值得游戏开发人员学习和参考。LunarLander 示例如图 2-19 所示。

图 2-17　Home 示例

图 2-18　Snake 示例　　　　　　　图 2-19　LunarLander 示例

8. JetBoy 示例

JetBoy 示例是一个支持背景音乐和音效的游戏程序,用户可以控制飞船击碎飞来的陨石。JetBoy 示例如图 2-20 所示。

图 2-20　JetBoy 示例

2.3　Android 常用的开发工具

Android SDK 提供了多个强大的开发工具,便于程序开发人员简化开发和调试过程。这些工具中多数可以在 Eclipse 中直接调用,也有部分是需要在命令行模式下使用的,后面将逐个介绍这些工具。

1. Android 模拟器

Android SDK 中最重要的工具就是 Android 模拟器,如图 2-21 所示,允许程序开发者在没有物理设备的情况,在计算机上对 Android 程序进行开发、调试和仿真。

Android 模拟器可以仿真手机的绝大部分硬件和软件功能,支持加载 SD 卡映像文件、更改模拟网络状态、延迟和速度,模拟电话呼叫和接收短信等,支持将屏幕当成触摸屏使用,可以使用鼠标单击屏幕模拟用户对 Android 设备的触摸操纵。在 Android 模拟器上有普通手机常见的各种按键,如音量键、挂断键、返回键和菜单键等。但目前为止,Android 模拟器仍不支持的功能包括接听真实电话呼叫、USB 链接、摄像头捕获、连接状态检测、电池电量、AC 电源检测、SD 卡插拔检查和蓝牙设备等。AVD 是对 Android 模拟器进行自定义的配置清单,配置 AVD 最简单的方式是通过 Eclipse 的 Window→AVD Manager 启动 AVD 管理器,如图 2-22 所示。

在 AVD 管理器单击 Create,打开 AVD 创建界面,如图 2-22 所示。用户需要在 Name 中输入 AVD 的名称,为了便于区分多个 AVD 的用途,一般在 AVD 命名时会体现 Android

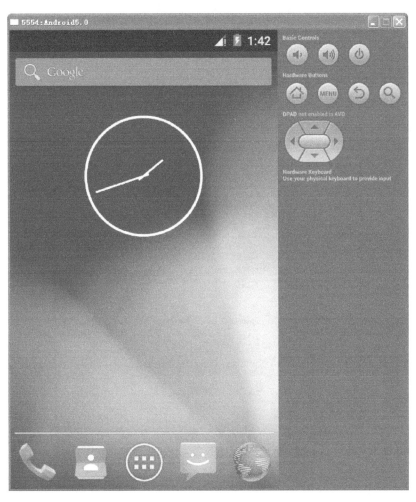

图 2-21 Android 模拟器

的版本信息,建立名为 Android 5.0 的 AVD,Target 是 AVD 支持的 Android 系统,这里选择 Android 5.0.1-API Level 21。SD Card 中输入 128,表示在模拟器中将模拟一个大小为 128M 的 SD 卡。Skin 表示模拟器的外观,可以选择为 Nexus One,支持的分辨率为 480×800。在 Hardware 中可以添加 Android 模拟器的硬件特性,在没有特别要求的情况下,可以使用默认的设置。完成 AVD 的配置后,单击 OK 按钮完成,然后在 AVD 管理器单击 Start 按钮启动 Android 模拟器,如图 2-23 所示。

2. Android 调试桥

Android 调试桥(Android Debug Bridge,ADB)是用于连接 Android 设备或模拟器的工具,负责将应用程序安装到模拟器和设备中,或从模拟器或设备中传输文件。Android 调试桥是一个客户端/服务器程序,包含守护程序、服务器程序和客户端程序。守护程序运行在每个模拟器的后台;服务器程序运行在开发环境中,管理客户端和守护程序的连接;客户端程序通过服务器程序,与模拟器中的守护程序相连接。

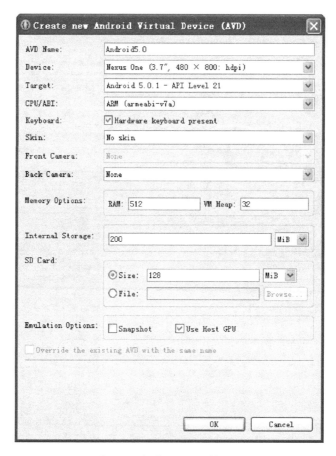

图 2-22　创建 Android 模拟器

图 2-23　配置完成的 Android 模拟器

3. DDMS

DDMS(Dalvik Debug Monitor Service)是 Android 系统中内置了调试工具,可以用于监视 Android 系统中进程、堆栈信息,查看 logcat 日志,实现端口转发服务和屏幕截图功能,模拟器电话呼叫和 SMS 短信,以及浏览 Android 模拟器文件系统等。DDMS 的启动文件是＜Android SDK＞/tools/ddms.bat。

在 Eclipse 中,通过 Window→Open Perspective→Other→DDMS 打开 DDMS 调试界面,然后通过 Window→Show view→Other 打开 Show View 的选择对话框,如图 2-24 所示。这样就可以在 DDMS 调试界面中添加任何希望进行调试和检查的功能。

图 2-24 Show View 选择对话框

DDMS 中的设备管理器(Devices),可以同时监控多个 Android 模拟器,显示每个模拟器中所有正在运行的进程。模拟器使用端口号进行唯一标识,例如监听端口是 5554 的模拟器则标识为 emulator-5554。在选择指定的进程后,可以通过右上角的按钮刷新进程中线程和堆栈的信息,或是单击"STOP"按钮关闭指定进程。另外,这里还提供屏幕截图功能,可以将 Android 模拟器当前的屏幕内容保存成 png 文件。DDMS 中的设备管理器如图 2-25 所示。

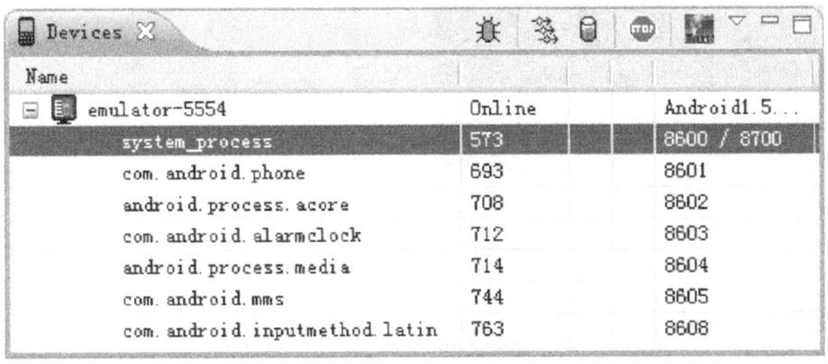

图 2-25 DDMS 中的设备管理器

DDMS 中的模拟器控制器(Emulator Control),可以控制 Android 模拟器的网络速度和延迟,模拟语音和 SMS 短信通信。模拟器控制支持的网络速率包括 GSM、HSCSD、PRS、EDGE、MTS、DPA 和全速率,支持的网络延迟有 GPRS、EDGE、UMTS 和无延迟。DDMS 中的模拟器控制器如图 2-26 所示。

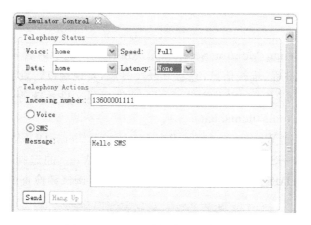

图 2-26　DDMS 中的模拟器控制器

在 Telephony Actions 中的 Incoming number 中输入打入的电话号码,然后选择语言呼叫(Voice)单击 Send 按钮后,模拟器就可以接收到来自输入电话号码的语音电话,如图 2-27 (左)所示。如果选择短信(SMS),在 Message 中填入短信的内容,模拟器就可以接收到来自输入电话号码的 SMS 短信,如图 2-27(右)所示。

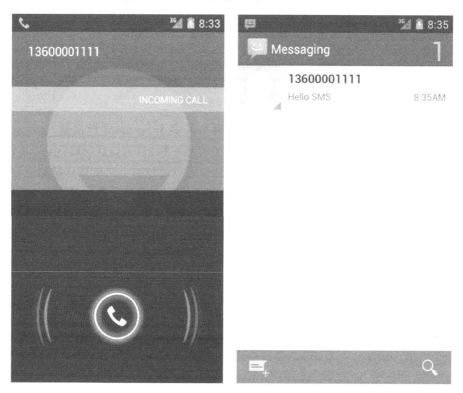

图 2-27　Android 电话呼入和 SMS 短信

DDMS 中的文件浏览器(File Explorer),可以对 Android 内置存储器上的文件进行上传、下载和删除等操作,还可以显示文件和目录的名称、权限、建立时间等信息。DDMS 中的文件浏览器如图 2-28 所示。

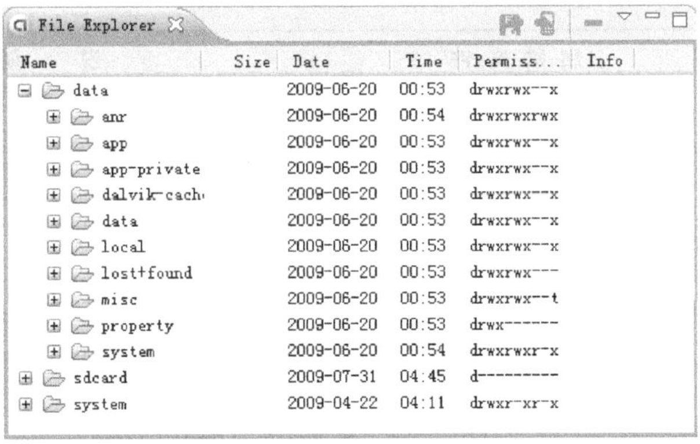

图 2-28 DDMS 中的文件浏览器

DDMS 中的日志浏览器（LogCat），可以浏览 Android 系统、Dalvik 虚拟机或应用程序产生的日志信息，有助于快速定位应用程序产生的错误。DDMS 中的日志浏览器如图 2-29 所示。

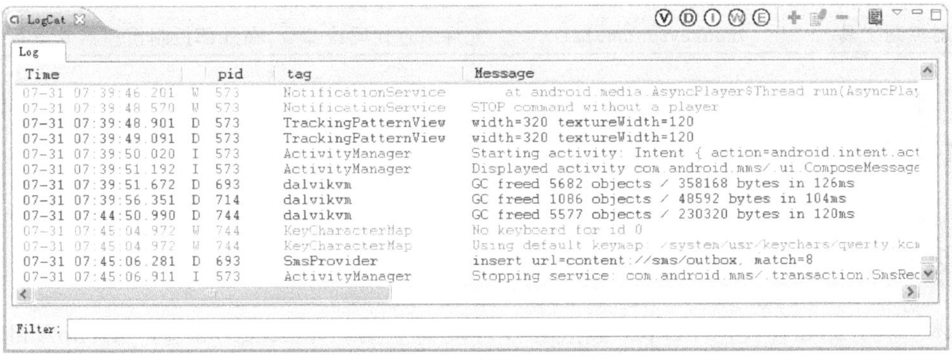

图 2-29 DDMS 中的日志浏览器

除了上面介绍的功能外，DDMS 还能够查看虚拟机的堆栈状态、线程信息和控制台信息。由此可见，DDMS 是进程程序调试和错误定位的强大工具。

4．其他工具

为了便于 Android 程序开发，Android SDK 还提供了一些小工具。这些工具的名称、用途和启动文件如表 2-1 所示。

表 2-1 Android SDK 提供的其他工具

工具名称	启动文件	说　　明
数据库工具	sqlite3.exe	用于创建和管理 SQLite 数据库
打包工具	apkbuilder.bat	将应用程序打包成 .apk 文件
层级观察器	hierarchyviewer.bat	对用户界面进行分析和调试，以图形化的方式展示树形结构的界面布局
跟踪显示工具	traceview.bat	以图形化的方式显示应用程序的执行日志，用于调试应用程序，分析执行效率

续表

工具名称	启动文件	说　明
SD卡映像创建工具	mksdcard.exe	建立SD卡的映像文件
NinePatch文件编辑工具	draw9patch.bat	NinePatch是Android提供的可伸缩的图形文件格式,基于PNG文件
		draw9patch工具可以使用WYSIWYG编辑器建立NinePatch文件
APK程序优化工具	zipalign.exe	经过zipalign优化过的APK程序,Android系统可更高效地根据请求索引APK文件中的资源。使用4字节的边界对齐方式来影射内存,通过空间换时间的方式提高执行效率
代码优化混淆工具	proguard目录	通过删除未使用的代码,并重命名代码中的类、字段和方法名称,使代码较难实施逆向工程
PNG和ETC1转换工具	etc1tool.exe	命令行工具,支持将PNG和ETC1相互转换
界面操作测试工具	Monkey(通过adb运行)	Monkey可在模拟器或设备上产生随机操作事件,包括单击、触摸或手势等,用于对程序的用户界面进行随机操作测试
模拟器控制工具	monkeyrunner.bat	允许通过代码或命令,在外部控制模拟器或设备

2.4　Android程序目录结构

2.4.1　创建第一个Android应用程序

本节将介绍如何使用Eclipse集成开发环境建立第一个Android程序HelloAndroid。首先启动Eclipse,如果在Eclipse中建立过Android工程,工程名称和目录结构将显示在Package Explorer区域内。

有两种方法可以打开Android工程向导,一种是以File→New→Project…|Android→Android Project的顺序,另一种是以File→New→Other…|Android→Android Project的顺序。两种方法只是选择的顺序不同,结果是相同的。在第二种方法中,除了可以建立Android工程外向导,还可以建立Android示例工程,就是保存在＜Android SDK＞/samples目录中的示例。如图2-30所示,选择Android Project建立Android工程。

在Android工程向导中,首先需要输入应用程序名称(Application Name),应用程序名称必须唯一,它是Android程序在手机或模拟器中显示的名称,程序运行时也会显示在屏幕顶部。Eclipse会自动将应用程序名称填写在工程名称这一栏中,用户可以不用更改,使用这个推荐设置。包名称(Package name)是包的命名空间,需要遵循Java包的命名方法。包名称由两个或多个标识符组成,中间用点隔开,例如Hlju.HelloAndroid。使用包主要为了避免命名冲突,为了保证代码的简洁,第一个Android程序的包名称使用Hlju.edu.HelloAndroid。

第二步是选择程序运行的Android系统版本。SDK最低版本(Minimum Required SDK)是指的是Android程序能够运行的最低的API等级,如果手机中的Android系统的API等级低于程序的SDK最低版本,则程序不能够在该Android系统中运行。Target

图 2-30　Eclipse 工程向导

SDK 是程序的目标 SDK 版本。Compile with 中填写程序的编译 SDK 版本,这个一般为默认或者同于 Target SDK。

在图 2-31 中,除了在 Platform 中标识 Android 系统的版本外,还有一个 API Level 的属性。API 等级是 Android 系统中用于标识 API 框架版本的一个整数,用于识别 Android

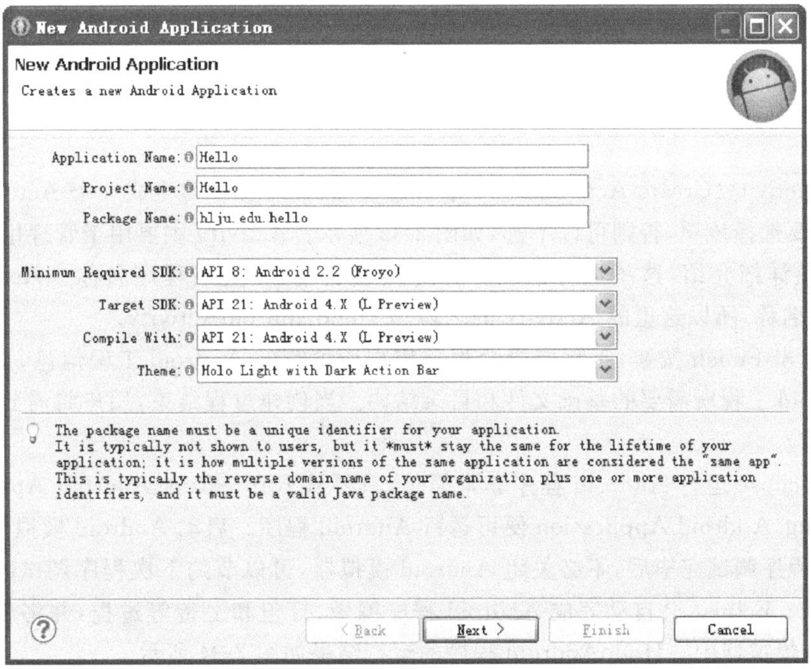

图 2-31　Android 工程建立向导

程序的可运行性。如果 Android 程序标识的 API 等级高于 Android 系统所支持的 API 等级,程序则无法在该 Android 系统中运行。API 等级与系统版本之间的对照关系如表 2-2 所示。

表 2-2　API 等级对照表

系 统 版 本	API 等级	版 本 代 号	支持设备类型
Android 5.0	21	Lollipop	智能手机 平板电脑
Android 4.0	14	ICE CREAM SANDWICH	智能手机 平板电脑
Android 3.2	13	HONEYCOMB_MR2	平板电脑
Android 3.1.x	12	HONEYCOMB_MR1	平板电脑
Android 3.0.x	11	HONEYCOMB	平板电脑
Android 2.3.4 Android 2.3.3	10	GINGERBREAD_MR1	智能手机
Android 2.3.2 Android 2.3.1 Android 2.3	9	GINGERBREAD	智能手机
Android 2.2.x	8	FROYO	智能手机
Android 2.1.x	7	ECLAIR_MR1	智能手机
Android2.0.1	6	ECLAIR_0_1	智能手机
Android 2.0	5	ECLAIR	智能手机
Android 1.6	4	DONUT	智能手机
Android 1.5	3	CUPCAKE	智能手机
Android 1.1	2	BASE_1_1	智能手机
Android 1.0	1	BASE	智能手机

创建 Activity(Create Activity)是一个可选项,如果需要自动生成一个 Activity 的代码文件,则需要选择该项,否则可以不选,如图 2-32 所示。Activity 主要用于管理用户界面,后续章节会做详细介绍,这里选择该项。Eclipse 会自动以"应用程序名称＋Activity"作为 Activity 的名称,所以这里的 Activity 的名称为 HelloAndroidActivity。

最后单击 Finish 按钮,工程向导会根据用户所填写的 Android 工程信息,自动在后台创建 Android 工程所需要的基础文件和目录结构。当创建过程结束,用户将看到图 2-33 所显示的内容。

使用 Eclipse 运行 Android 程序非常简单,只要从 Run→Run|Android Application 或 Run→Debug|Android Application 便可运行 Android 程序。启动 Android 模拟器是一个缓慢的过程,程序调试完毕后,不必关闭 Android 模拟器,可以节约下次程序调试时的启动模拟器的时间。Eclipse 会自动完成 Android 程序编译、打包和上传等过程,并将程序的运行结果显示在模拟器中。HelloAndroid 程序的运行结果如图 2-34 所示。

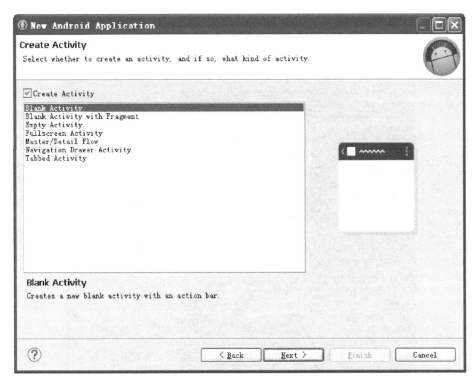

图 2-32　选择创建 Activity 的类型

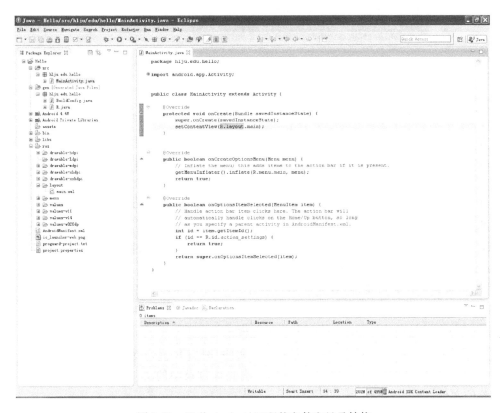

图 2-33　HelloAndroid 工程的文件和目录结构

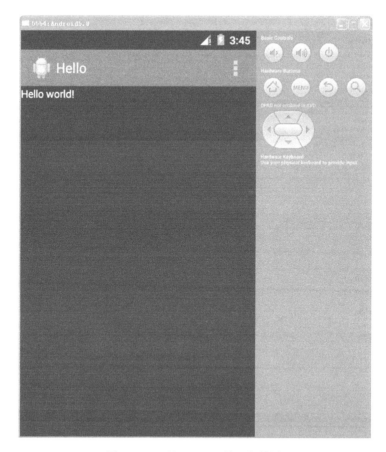

图 2-34　HelloAndroid 的运行结果

2.4.2　Android 程序结构

在建立 HelloAndroid 程序的过程中，ADT 会自动建立一些目录和文件，如图 2-35 所示。这些目录和文件有其固定的作用，有的允许修改，有的则不能修改，了解这些文件和目录对 Android 程序开发非常重要。

在 Package Explore 中，ADT 以工程名称 HelloAndroid 作为根目录，将所有自动生成的和非自动生成的文件都保存在这个根目录下。根目录下主要包含 4 个子目录：src、gen、assets、bin 和 res，一个库文件 android.jar，以及 3 个工程文件 Androidmanifest.xml、project.properties 和 proguard.cfg。

src 目录是源代码目录，所有允许用户修改的 java 文件和用户自己添加的 java 文件都保存在这个目录中。HelloAndroid 工程建立初期，ADT 根据用户在工程向导中的 Create Activity 选项，自动建立 HelloAndroid.java 文件。

gen 目录用于保存 ADT 自动生成的 java 文件，例如 R.java 或 AIDL 文件。这个目录中的文件不建议用户进行任何修改，如果用户删除该目录中的文件，ADT 会自动再次生成被删除的文件。

assets 目录用于存储原始格式的文件，例如音频文件、视频文件等二进制格式文件。此

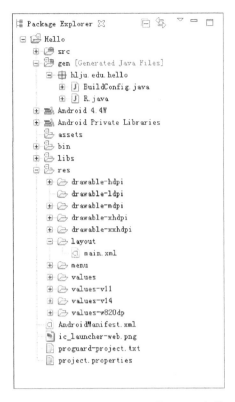

图 2-35　HelloAndroid 工程的目录和文件

目录中的资源不能够被 R.java 文件索引，因此只能以字节流的形式进行读取。默认为空目录。

bin 目录保存了编译过程中的所产生的文件，以及最终生产的 apk 文件。

res 目录是资源目录，Android 程序所有的图像、颜色、风格、主题、界面布局和字符串等资源都保存在其下的几个子目录中。其中，drawable-hdpi、drawable-mdpi 和 drawable-ldpi 目录用于保存同一个程序中针对不同屏幕尺寸需要显示的不同大小的图像文件，layout 目录用于保存与用户界面相关的布局文件，values 目录用于保存颜色、风格、主题和字符串等资源。在 HelloAndroid 工程中，ADT 在每个 drawable 目录中自动引入了一个不同尺寸的 icon.png 文件，Android 系统会根据目标设备的屏幕分辨率，为 HelloAndroid 程序加载不同尺寸的图标文件；在 layout 目录生成了 mail.xml 文件；在 values 目录生成了 strings.xml 文件，将应用程序名称 HelloAndroid 和界面显示的 Hello World, HelloAndroidActivity! 保存在这个文件中。

android.jar 文件是 Android 程序所能引用的函数库文件，Android 系统所支持 API 都包含在这个文件中。

proguard.cfg 文件是供 ProGuard 工具进行代码优化和代码混淆使用的配置文件。

project.properties 文件记录了 Android 工程的相关设置，例如编译目标和 apk 设置等，该文件不能手工修改，如果需要更改其中的设置，必须通过右击工程名称，选择 Properties 进行修改。从 project.properties 文件的代码中可以发现，大部分都是内容注释，仅有第 12 行是有效代码，说明了 Android 程序的编译目标。

project.properties 文件的代码如下：

```
1   # This file is automatically generated by Android Tools.
2   # Do not modify this file -- YOUR CHANGES WILL BE ERASED!
3   #
4   # This file must be checked in Version Control Systems.
5   #
6   # To customize properties used by the Ant build system use,
7   # "build.properties", and override values to adapt the script to your
8   # project structure.
9
10
11  # Project target.
12  target = android-14
```

AndroidManifest.xml 是 XML 格式的 Android 程序声明文件，包含了 Android 系统运行 Android 程序前所必须掌握的重要信息，这些信息包括应用程序名称、图标、包名称、模块组成、授权和 SDK 最低版本等，而且每个 Android 程序必须在根目录下包含一个 AndroidManifest.xml 文件。XML 是一种可扩展标记语言，本身独立于任何编程语言，能够对复杂的数据进行编码，且易于理解。Android 工程中多处使用了 XML 文件，使应用程序开发更加具有弹性，且易于后期的维护和理解。

AndroidManifest.xml 文件的代码如下：

```
1   <?xml version = "1.0" encoding = "utf-8"?>
2   <manifest xmlns:android = "http://schemas.android.com/apk/res/android"
3          package = "edu.hrbeu.HelloAndroid"
4          android:versionCode = "1"
5          android:versionName = "1.0">
6      <application android:icon = "@drawable/icon"
7                 android:label = "@string/app_name">
8          <activity android:name = ".HelloAndroidActivity"
9                 android:label = "@string/app_name">
10             <intent-filter>
11                 <action android:name = "android.intent.action.MAIN" />
12                 <category android:name = "android.intent.category.LAUNCHER" />
13             </intent-filter>
14         </activity>
15     </application>
16     <uses-sdk android:minSdkVersion = "14" />
17  </manifest>
```

在 AndroidManifest.xml 文件中，根元素是 manifest，包含了 xmlns:android、package、android:versionCode 和 android:versionName 共 4 个属性。其中，第 2 行属性 xmlns:android 定义了 Android 的命名空间，值为 http://schemas.android.com/apk/res/android；第 3 行属性 package 定义了应用程序的包名称；第 4 行属性 android:versionCode 定义了应用程序的版本号，是一个整数值，数值越大说明版本越新，但仅在程序内部使用，并不提供给

应用程序的使用者；第 5 行属性 android:versionName 定义了应用程序的版本名称，是一个字符串，仅限于为用户提供一个版本标识。

manifest 元素仅能包含一个 application 元素，application 元素中能够声明 Android 程序中最重要的 4 个组成部分，包括 Activity、Service、BroadcastReceiver 和 ContentProvider，所定义的属性将影响所有组成部分。第 6 行属性 android:icon 定义了 Android 应用程序的图标，其中@drawable/icon 是一种资源引用方式，表示资源类型是图像，资源名称为 icon，对应的资源文件为 icon.png，目录是 res/drawable-hdpi、res/drawable-mdpi 和 res/drawable-ldpi，这 3 个目录中的资源仍通过@drawable 进行调研；第 7 行属性 android:label 则定义了 Android 应用程序的标签名称。

activity 元素是对 Activity 子类的声明，不在 AndroidManifest.xml 文件中声明的 Activity 将不能够在用户界面中显示。第 8 行属性 android:name 定义了实现 Activity 类的名称，可以是完整的类名称，如 edu.hrbeu.HelloAndroidActivity，也可以是简化后的类名称，如.HelloAndroidActivity；第 9 行属性 android:label 则定义了 Activity 的标签名称，标签名称将在用户界面的 Activity 上部显示，@string/app_name 同样属于资源引用，表示资源类型是字符串，资源名称为 app_name，资源保存在 res/values 目录下的 strings.xml 文件中。

intent-filter 中声明了两个子元素 action 和 category，在这里不详细讨论两个字元素的用途，但可以肯定的是，intent-filter 在 HelloAndroid 程序在启动时，将.HelloAndroidActivity 这个 Activity 作为默认启动模块。

ADT 包含了一个可视化的编辑器，如图 2-36 所示，双击 AndroidManifest.xml 文件可直接进入可视化编辑器。用户可以在不接触 XML 的情况下，编辑 Android 工程的应用程序名称、包名称、图标、标签和许可等相关属性。

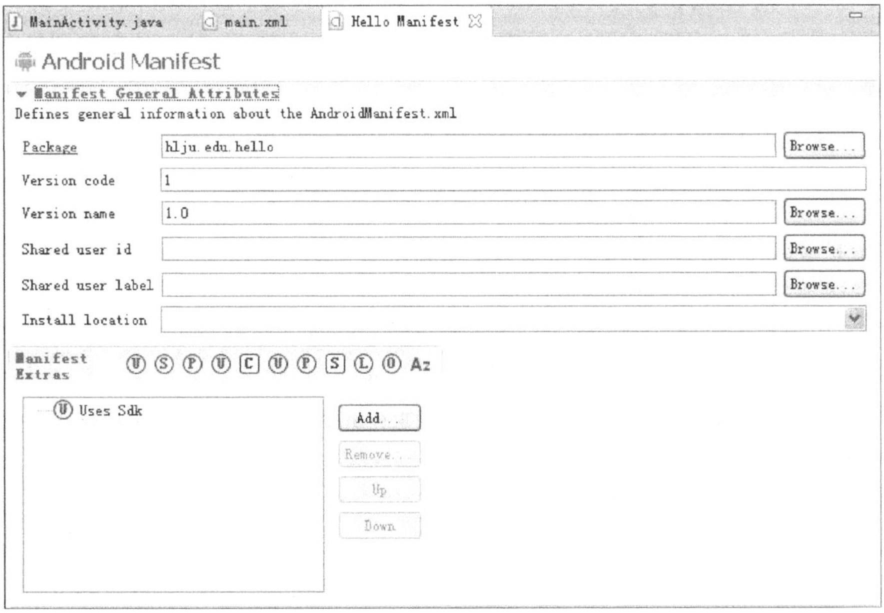

图 2-36　HelloAndroid 工程的目录和文件

R.java 文件是 ADT 自动生成的文件,包含对 drawable、layout 和 values 目录内的资源的引用指针,Android 程序能够直接通过 R 类引用目录中的资源。R.java 文件不能手工修改,所有代码必须由 ADT 自动生成。如果向资源目录中增加或删除了资源文件,则需要在工程名称上右击,选择 Refresh 来更新 R.java 文件中的代码。

HelloAndroid 工程生成的 R.java 文件的代码如下:

```
1   package edu.hrbeu.HelloAndroid;
2
3   public final class R {
4       public static final class attr {
5       }
6       public static final class drawable {
7           public static final int icon = 0x7f020000;
8       }
9       public static final class layout {
10          public static final int main = 0x7f030000;
11      }
12      public static final class string {
13          public static final int app_name = 0x7f040001;
14          public static final int hello = 0x7f040000;
15      }
16  }
```

R 类包含的几个内部类,分别与资源类型相对应,资源 ID 便保存在这些内部类中,例如子类 drawable 表示图像资源,内部的静态变量 icon 表示资源名称,其资源 ID 为 0x7f020000。一般情况下,资源名称与资源文件名相同(不包含扩展名),如 icon 对应 src/drawable 目录下的 icon.png 文件,main 对应 src/layout 目录下的 main.xml 文件。

资源的引用一般有两种方法,一种方法是在代码中引用资源,另一种方法则是在资源中引用资源。

在代码中引用资源时,需要在代码中使用资源 ID,可以通过[R.resource_type.resource_name]或者[andrlid.R.resource_type.resource_name]获取资源 ID。其中 resource_type 代表资源类型,也就是 R 类中的内部类名称;resource_name 代表资源名称,对应资源的文件名(不包含扩展名)或在 XML 文件中定义的资源名称属性。例如在 HelloAndroid.java 中,第 11 行代码便是在代码中对资源的引用。

在资源中引用资源时,一般的引用格式为@[package:]type:name。其中,@表示对资源的引用;package 是包名,如果在相同的包内,package 则可省略;type 是资源的类型,例如 string 或 drawable;name 是资源的名称。例如在 main.xml 文件中,第 10 行代码就是在资源中对资源的引用。

main.xml 文件是界面布局文件,利用 XML 语言描述的用户界面,界面布局的相关内容将在第 5 章用户界面设计中进行详细介绍。main.xml 代码的第 7 行说明在界面中使用 TextView 控件,TextView 控件主要用于显示字符串文本。代码第 10 行说明 TextView 控件需要显示的字符串,非常明显,@string/hello 是对资源的引用。通过 strings.xml 文件的第 3 行代码分析,在 TextView 控件中显示的字符串应是 Hello World,HelloAndroidActivity!。如果读者修改 strings.xml 文件的第 3 行代码的内容,重新编译运行后,模拟器中显示的结果也应该随之更改。

main.xml 文件的代码如下:

```
1   <?xml version = "1.0" encoding = "utf-8"?>
2   <LinearLayout xmlns:android = "http://schemas.android.com/apk/res/android"
3       android:orientation = "vertical"
4       android:layout_width = "fill_parent"
5       android:layout_height = "fill_parent"
6       >
7   <TextView
8       android:layout_width = "fill_parent"
9       android:layout_height = "wrap_content"
10      android:text = "@string/hello"
11      />
12  </LinearLayout>
```

strings.xml 文件的代码如下:

```
13  <?xml version = "1.0" encoding = "utf-8"?>
14  <resources>
15      <string name = "hello">Hello World, HelloAndroidActivity!</string>
16      <string name = "app_name">HelloAndroid</string>
17  </resources>
```

HelloAndroid.java 是 Android 工程向导根据 Activity 名称创建的 java 文件,这个文件完全可以手工修改。为了在 Android 系统上显示图形界面,需要使用代码继承 Activity 类,并在 onCreate()函数中声明需要显示的内容。

HelloAndroid.java 文件的代码如下:

```
1   package edu.hrbeu.HelloAndroid;
2
3   import android.app.Activity;
4   import android.os.Bundle;
5
6   public class HelloAndroid extends Activity {
7       /** Called when the activity is first created. */
8       @Override
9       public void onCreate(Bundle savedInstanceState) {
10          super.onCreate(savedInstanceState);
11          setContentView(R.layout.main);
12      }
13  }
```

代码的第 3 行和第 4 行,通过 android.jar 从 Android SDK 中引入了 Activity 和 Bundle 两个重要的包,用以子类继承和信息传递;第 6 行声明 HelloAndroid 类继承 Activity 类;

第 8 行表明需要重写 onCreate()函数；第 9 行的 onCreate()会在 Activity 首次启动时会被调用，为了便于理解，可以认为 onCreate()是 HelloAndroid 程序的主入口函数；第 10 行调用父类的 onCreate()函数，并将 savedInstanceState 传递给父类，savedInstanceState 是 Activity 的状态信息；第 11 行声明了需要显示的用户界面，此界面是用 XML 语言描述的界面布局，保存在 scr/layout/main.xml 资源文件中。

习题

1. 按照 ADT-Bundle 开发环境的安装配置方法，尝试搭建 Android 开发环境。
2. 创建一个 HelloWorld 应用程序，说明改程序目录结构中包括哪些子目录及文件，各自的作用是什么。
3. 简述 R.java 文件的用途和生成方法和 AndroidManifest.xml 文件的用途。

第3章 Android 界面开发常用控件

Android 程序开发主要分为三个部分：界面设计、代码流程控制和资源建设。代码和资源主要是由开发者进行编写和维护的，而对于用户来讲，最直观的往往是界面设计。Android 系统提供了丰富的界面控件，开发者熟悉这些控件的功能和用法，将可以设计出优秀的图形界面。Android 应用的绝大部分界面(UI)控件都存储在 android.widget 包及其子包、android.view 包及其子包中，Android 应用的所有 UI 控件都继承了 View 类。

本章主要学习内容：
- 了解用户界面基础 View 类的使用；
- 掌握文本显示 TextView 控件的使用；
- 掌握编辑框 EditText 控件的使用；
- 掌握按钮 Button 控件的使用；
- 掌握图片按钮 ImageButton 控件的使用；
- 掌握单选按钮 RadioButton 控件和复选框按钮 CheckBox 控件的使用；
- 掌握信息提示 Toast 的使用；
- 掌握下拉列表框 Spinner 控件的使用；
- 掌握列表显示控件 ListView 控件的使用；
- 掌握进度条 ProgressBar 控件的使用。

3.1 用户界面基础

用户界面(User Interface,UI)是系统和用户之间进行信息交换的媒介，实现信息的内部形式与人类可以接受形式之间的转换。对于 Android 手机应用软件而言，如何从众多的软件中脱颖而出，用户界面的设计是一个不可忽视的因素。Android 系统提供了丰富的界面控件，开发者熟悉这些控件的功能和用法，将可以设计出优秀的图形界面。

3.1.1 手机用户界面应解决的问题

首先，手机的界面设计者和程序开发者是独立且并行工作的，这就需要界面设计与程序逻辑完全分离，不仅有利于并行工作，而且在后期修改界面时也可以避免修改程序的逻辑代码；其次，不同型号手机的屏幕解析度、尺寸和长宽比各不相同，程序界面需要能够根据屏幕信息，自动调整界面控件的位置和尺寸，避免因为屏幕解析度、尺寸或纵横比的变化而出现显示错误;最后,手机屏幕尺寸较小,设计者必须能够合理利用有限的显示空间,构造出符合人机交互规则的用户界面,避免出现凌乱、拥挤的用户界面。

Android 系统已经为使用者解决了界面设计前两个问题。在界面设计与程序逻辑分离方面,Android 程序将用户界面和资源从逻辑代码中分离出来,使用 XML 文件对用户界面进行描述,资源文件独立保存在资源文件夹中。Android 系统的用户界面描述非常灵活,允许模糊定义界面元素的位置和尺寸,通过声明界面元素的相对位置和粗略尺寸,从而使界面元素能够根据屏幕尺寸和屏幕摆放方式动态调整显示方式。

可采用 XML 和代码两种方式控制界面显示,这两种方式各有优缺点。

(1) 完全采用代码控制界面,显得繁杂,不利于解耦、分工;

(2) 完全使用 xml 布局文件虽然方便,但灵活性不好,不利于动态改变属性值。

3.1.2 Android 平台中的 View 类

View 类是所有可视化控件的基类,主要提供控件绘制和事件处理的方法。Android 应用的所有控件都继承了 View 类。Android 应用的绝大部分界面控件都存储在 android.widget 包及其子包 android.view 中。

View 类有一个重要的子类 ViewGroup。ViewGroup 通常作为其他控件的容器使用。Android 的所有 UI 控件都是建立在 View、ViewGroup 基础之上的。对于一个 Android 应用的图形用户界面来说,ViewGroup 作为容器来盛装其他控件,而 ViewGroup 里除了可以包含普通 View 控件之外,还可以再次包含 ViewGroup 控件,如图 3-1 所示。

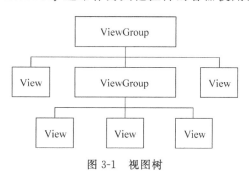

图 3-1 视图树

下面是 View 类常见的属性及对应方法,如表 3-1 所示。

表 3-1 View 类常用属性及对应方法

属性名称	对应方法	描述
android:background	setBackgroundResource(int)	设置背景色/背景图片。可以通过以下两种方法设置背景为透明:@android:color/transparent 和 @null。注意 TextView 默认是透明的,不用写此属性,但是 Buttom、ImageButton 和 ImageView 想透明就要设置该属性了
android:focusable	setFocusable(boolean)	设置是否获得焦点。若有 requestFocus() 被调用时,后者优先处理。注意在表单中想设置某一个控件,如 EditText 获取焦点,仅设置该属性是不行的,需要将 EditText 前面的 focusable 都设置为 false 才行。在 Touch 模式下获取焦点需要设置 focusableInTouchMode 为 true

续表

属性名称	对应方法	描述
android：id	setId(int)	给当前 View 设置一个在当前 layout.xml 中的唯一编号，可以通过调用 View.findViewById()或 Activity.findViewById()，根据该编号查找到对应的 View。不同的 layout.xml 之间定义相同的 id 不会冲突。格式如@+id/btnName
android：minHeight		设置视图最小高度
android：minWidth		设置视图最小宽度
android：padding	setPadding(int,int,int,int)	设置上、下、左、右的边距，以像素为单位填充空白
android：paddingBottom	setPadding(int,int,int,int)	设置底部的边距，以像素为单位填充空白
android：paddingLeft	setPadding(int,int,int,int)	设置左侧边距，以像素为单位填充空白
android：paddingRight	setPadding(int,int,int,int)	设置右侧边距，以像素为单位填充空白
android：paddingTop	setPadding(int,int,int,int)	设置上方的边距，以像素为单位填充空白
android：scrollbarSize		设置滚动条的宽度
android：scrollbarStyle		设置滚动条的风格和位置。设置值：insideOverlay、insideInset、outsideOverlay、outsideInset
android：scrollbars		设置滚动条显示。none(隐藏)、horizontal(水平)、vertical(垂直)。使用该属性使 EditText 内有滚动条。但是其他容器如 LinearLayout 设置了没有效果
android：tag		设置一个文本标签。可以通过 View.getTag()或 View.findViewWithTag()检索含有该标签字符串的 View。但一般最好通过 ID 来查询 View，因为速度更快，并且允许编译时类型检查
android：visibility	setVisibility(int)	设置是否显示 View。设置值：visible(默认值,显示)、invisible(不显示,但是仍然占用空间)、gone(不显示,不占用空间)

在 Android 中每个控件都需要设置 android：layout_height、android：layout_width 这两个属性，用于指定控件的高度和宽度，主要有以下 3 个取值。

- fill_parent：表示控件的高或宽与其父容器的高或宽相同。
- wrap_content：表示控件的高或宽恰好能包裹住内容，随着内容的长宽变化而变化。
- match_parent：该属性值与 fill_parent 完全一样，但 Android 2.2 之后推荐使用该属性代替。

在 Android 中所有控件都可以设置大小，但是在设置时候需要指定其单位，主要单位如下。

- px(像素 pixels)：对应于屏幕上的一个点(1 英寸显示 n 个点)，这个用得比较多。
- dip 或 dp(device independent pixels)：一种基于屏幕密度的抽象单位。在每英寸 160 点的显示器上，1dp＝1px。但随着屏幕密度改变，dp 与 px 的换算也会发生改变。
- sp(scaled pixels—best for text size)：比例像素，主要处理字体的大小，可以根据系统的字体自适应。

- pt：point，是一个标准的长度单位，1pt＝1/72 英寸，用于印刷业，非常简单易用。

为了适应不同分辨率，不同的像素密度，推荐使用 dp，文字使用 sp。例如 xml 界面布局和代码进行界面控制。

Android 系统的界面控件分为定制控件和系统控件。定制控件是用户独立开发的控件，或通过继承并修改系统控件后所产生的新控件，能够提供特殊的功能和显示需求。系统控件是 Android 系统中已经封装的界面控件，是应用程序开发过程中最常见功能控件。系统控件更有利于进行快速开发，同时能够使 Android 应用程序的界面保持一定的一致性。

常见的系统控件包括 TextView、EditText、Button、ImageButton、Checkbox、RadioButton、Spinner、ListView 和 ProgressBar。

3.2 TextView 控件

TextView(文本显示)控件可用于展示文本信息，主要是提供了一个标签的显示操作。可以手动来设置可编辑或不可编辑。在 main.xml 中添加 TextView 配置节来创建，设计基础属性，宽度、高度、颜色和字体大小等。

TextView 类的层次关系如下：

```
java.lang.Object
   ↳ android.view.View
      ↳ android.widget.TextView
```

3.2.1 TextView 控件常见的属性和方法

TextView 控件常见属性及方法如表 3-2 和表 3-3 所示。

表 3-2　TextView 控件常见属性

属 性 名 称	描　　述
android：autoLink	设置是否当文本为 URL 链接/email/电话号码/map 时，文本显示为可单击的链接。可选值(none/web/email/phone/map/all)
android：autoText	如果设置，将自动执行输入值的拼写纠正。此处无效果，在显示输入法并输入的时候起作用
android：digits	设置允许输入哪些字符。如"1234567890.+-*/%\n()"
android：gravity	设置文本位置，如设置成"center"，文本将居中显示
android：ems	设置 TextView 的宽度为 N 个字符的宽度。这里测试为一个汉字字符宽度
android：maxLength	限制显示的文本长度，超出部分不显示
android：password	以密码形式显示小点"."显示文本
android：phoneNumber	设置为电话号码的输入方式
android：text	设置显示文本
android：textColor	设置文本颜色
android：textSize	设置文字大小，推荐度量单位"sp"，如"15sp"
android：textStyle	设置字形[bold(粗体) 0，italic(斜体) 1，bolditalic(又粗又斜) 2]可以设置一个或多个，用"|"隔开

续表

属性名称	描述
android：height	设置文本区域的高度,支持度量单位：px(像素)/dp/sp/in/mm(毫米)
android：maxHeight	设置文本区域的最大高度
android：minHeight	设置文本区域的最小高度
android：width	设置文本区域的宽度,支持度量单位：px(像素)/dp/sp/in/mm(毫米)
android：maxWidth	设置文本区域的最大宽度
android：minWidth	设置文本区域的最小宽度

表 3-3　TextView 控件常见方法

主要方法	功能描述	返回值
TextView	TextView 的构造方法	null
getText	获得 TextView 对象的文本	charSquence
length	获得 TextView 中的文本长度	int
getEditableText	取得文本的可编辑对象,通过这个对象可对 TextView 的文本进行操作,如在光标之后插入字符	void
setTextColor	设置文本显示的颜色	void
setHighlightColor	设置文本选中时显示的颜色	void
setLinkTextColor	设置链接文字的颜色	void
setGravity	设置当 TextView 超出了文本本身时横向以及垂直对齐	void

3.2.2　TextView 控件实例

可以在 XML 布局文件中声明并设置 TextView,也可以在源程序代码中生成 TextView 控件。在 XML 布局文件中主要是完成 XML 文件的属性设置,执行效果如图 3-2 所示。

图 3-2　TextView 控件实例

例 3.1　TextViewDemo 在 XML 文件(/res/layout/activity_main)中的代码如下。

```
1    <TextView
2        android:layout_width = "wrap_content"
3        android:layout_height = "wrap_content"
4        android:text = "@string/text"
5        android:textSize = "15sp"
```

```
6        android:textColor = "#00ff00"
7        android:autoLink = "all"
8   />
```

第 2 行的 android：layout_width 属性用于设置 TextView 的宽度，wrap_content 表示 TextView 的宽度只要能够包含所显示的字符串即可。第 3 行的 android：layout_height 属性用于设置 TextView 的高度。第 4 行表示 TextView 所显示的字符串，在后面将通过代码更改 TextView 的显示内容。第 5 行中 textSize 设置文本的尺寸是 15sp，第 6 行中 textColor 设置文本的颜色，第 7 行中 autoLink 是设置超链接。

3.3 EditText 控件

TextView 控件的功能只是显示一些基础的文字信息，而如果用户要想定义可以输入的文本组件以达到很好的人机交互操作，则只能使用 EditText（编辑框）控件来完成。EditText 控件是用于输入和编辑字符串的控件。

TextView 类的层次关系如下：

```
java.lang.Object
   ↳ android.view.View
      ↳ android.widget.EditText
```

3.3.1 EditText 控件常见的属性和方法

EditText 控件常见的属性及方法如表 3-4 和表 3-5 所示。

表 3-4　EditText 控件常见属性

属 性 名 称	描　　述
android:digits	设置允许输入哪些字符。如"1234567890.+-*/%\n()"
android:editable	设置是否可编辑。仍然可以获取光标，但是无法输入
android:gravity	设置文本位置，如设置成"center"，文本将居中显示
android:ems	设置 TextView 的宽度为 N 个字符的宽度
android:maxLength	限制输入字符数。如设置为 5，那么仅可以输入 5 个汉字/数字/英文字母
android:lines	设置文本的行数，设置两行就显示两行，即使第 2 行没有数据
android:numeric	如果被设置，该 TextView 有一个数字输入法。有如下值设置：integer 正整数、signed 带符号整数、decimal 带小数点浮点数
android:password	以小点"."显示文本
android:text	设置显示文本
android:textColor	设置文本颜色
android:width	设置文本区域的宽度，支持度量单位：px(像素)/dp/sp/in/mm(毫米)
android:maxWidth	设置文本区域的最大宽度
android:minWidth	设置文本区域的最小宽度

表 3-5 EditText 控件常见方法

方　　法	功 能 描 述	返 回 值
setImeOptions	设置软键盘的 Enter 键	void
getDefaultEditable	获取是否默认可编辑	Boolean
setEllipse	设置文件过长时控件的显示方式	void
setFreeezesText	设置保存文本内容及光标位置	void
getFreeezesText	获取保存文本内容及光标位置	Boolean
setGravity	设置文本框在布局中的位置	void
getGravity	获取文本框在布局中的位置	int
setHint	设置文本框为空时,文本框默认显示的字符	void
getHint	获取文本框为空时,文本框默认显示的字符	Charsequence
setIncludeFontPadding	设置文本框是否包含底部和顶端的额外空白	void

3.3.2 EditText 控件实例

EditText 和 TextView 一样,既可以在 xml 中声明实现,也可以在代码中动态的生成,下面以实例来说明 EditText 的使用。

例 3.2 使用 EdietText 控件和 TextView 控件实现一个用户登录界面,包括输入用户名和密码,执行结果如图 3-3 所示。

EditTextDemo01 在 XML 布局文件中 EditText 控件代码如下:

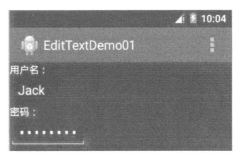

图 3-3 EditTextDemo01 执行结果

```
1   <EditText
2       android:id = "@ + id/UserEdit"
3       android:layout_width = "wrap_content"
4       android:layout_height = "wrap_content"
5       android:inputType = "text"
6       android:labelFor = "@id/UserEdit"
7       android:textColor = " # FFFFFF"
8   />
9   <EditText
10      android:id = "@ + id/PwdEdit"
11      android:layout_width = "wrap_content"
12      android:layout_height = "wrap_content"
13      android:inputType = "textPassword"
14      android:labelFor = "@id/PwdEdit"
15      android:textColor = " # FFFFFF"
16  />
```

在这个 XML 布局用了两个 EditText 控件,第 2 行 android:id 属性声明了 EditText 的 ID,这个 ID 主要用于在 Java 源代码中引用 EditText 对象。"@+id/UserEdit"表示所设置的 ID 值,其中@表示后面的字符串是 ID 资源;加号(+)表示需要建立新资源名称,并添加到 R.java 文件中;斜杠后面的字符串(UserEdit)表示新资源的名称。如果不是新添加的资源,或属于 Android 框架的资源,则不需要使用加号(+),但必须添加 Android 包的命名空间,例如 android:id="@android:id/empty"。

第 5 行 android:inputType="text"表示 EditText 控件中输入类型是文本。第 13 行 android:inputType="textPassword"表示输入的类型是密码'.'形式。

例 3.3 使用 EditText 和 SetOnKeyListener 文本处理方法,实现在 EditText 控件中输入文字,同时获取输入的文字并同步显示于 TextView 控件中,执行结果如图 3-4 所示。

图 3-4 EditTextDemo02 执行结果

EditTextDemo02 在 XML 布局文件中主要代码如下:

```
1   <EditText
2       android:id="@+id/myEditText"
3       android:layout_width="wrap_content"
4       android:layout_height="wrap_content"
5       android:textSize="18sp"
6       android:textColor="#FFFFFF"
7       android:inputType="text"
8       android:labelFor="@id/myEditText"
9   />
10  <TextView
11      android:id="@+id/myTextView"
12      android:layout_width="wrap_content"
13      android:layout_height="wrap_content"
14      android:textColor="#FFFFFF"
15  />
```

在这个 XML 布局用了两个 TextView 控件和 1 个 EditText 控件。

EditTextDemo02 在 Java 源文件中主要代码如下:

```
1   private TextView mTextView01;
2   private EditText mEditText01;
3   @Override
4   protected void onCreate(Bundle savedInstanceState) {
5       super.onCreate(savedInstanceState);
6       setContentView(R.layout.activity_main);
7       mTextView01 = (TextView)findViewById(R.id.myTextView);
8       mEditText01 = (EditText)findViewById(R.id.myEditText);
9       mEditText01.setOnKeyListener(new View.OnKeyListener() {
10
11          @Override
12          public boolean onKey(View v, int keyCode, KeyEvent event) {
13              // TODO Auto-generated method stub
14              mTextView01.setText(mEditText01.getText());
15              return false;
16          }
```

第 1 行和第 2 行是分别声明 TextView 和 EditText 对象变量。第 7 行和第 8 行中 findViewById()函数能够通过 ID 引用界面上的任何控件,只要该控件在 XML 文件中定义过 ID 即可,在这里分别获取 EditText 控件和 TextView 控件。第 14 行 setText()函数用于设置 TextView 所显示的内容,getText()函数用于获取 EditText 控件中的内容。

在 EditText 中,每当任何一个键按下或抬起时,都会引发按键事件。但为了能够使 EditText 处理按键事件,需要使用 setOnKeyListener()函数在代码中设置按键事件监听器,并在 onKey()函数添加按键事件的处理过程。第 9 行中 EditText 控件设置按键事件监听器 OnKeyListener,按键事件先传递到监听器的事件处理函数 onKey()中。事件能否够继续传递给 EditText 控件的其他事件处理函数,完全根据 onKey()函数的返回值来确定。如果 onKey()函数返回 false,事件将继续传递,这样 EditText 控件就可以捕获到该事件,将按键的内容显示在 EditText 控件中。在代码第 12 行的 onKey()函数中,第 1 个参数 view 表示产生按键事件的界面控件;第 2 个参数 keyCode 表示按键代码;第 3 个参数 keyEvent 则包含了事件的详细信息,如按键的重复次数、硬件编码和按键标志等。

EditTextDemo02 执行后的效果如图 3-4 所示,当在编辑框 EditText 中输入字符后,在文本框 TextView 中即时显示输入的字符。

3.4 Button 控件

Button 是按钮控件,用户能够在该控件上单击,引发相应的事件处理函数。Button 按钮在人机交互界面中使用得最广泛,用于用户和应用程序之间的交互行为。

Button 类的层次关系如下:

```
java.lang.Object
   ↳ android.view.View
      ↳ android.widget.TextView
         ↳ android.widget.Button
```

3.4.1 Button 控件常见的属性和方法

Button 控件常见的属性及方法如表 3-6 和表 3-7 所示。

表 3-6 Button 控件常见属性

属　　性	描　　述
android:layout_height	设置控件高度。可选值:fill_parent、warp_content 和 px
android:layout_width	设置控件宽度,可选值:fill_parent、warp_content 和 px
android:text	设置控件名称,可以是任意字符
android:layout_gravity	设置控件在布局中的位置,可选项:top、left、bottom、right、center_vertical、fill_vertica、fill_horizonal、center 和 fill 等
android:layout_weight	设置控件在布局中的比重,可选值:任意的数字
android:textColor	设置文字的颜色
android:bufferType	设置取得的文本类别,normal、spannable、editable
android:hint	设置文本为空是所显示的字符
android:inputType	设置文本的类型,none、text、textWords 等

表 3-7　Button 控件常见方法

方　法	功 能 描 述	返回值
Button	Button 类的构造方法	null
onKeyDown	当用户按键时,该方法调用	Boolean
onKeyUp	当用户按键弹起后,该方法被调用	Boolean
onKeyLongPress	当用户保持按键时,该方法被调用	Boolean
onKeyMultiple	当用户多次调用时,该方法被调用	Boolean
invalidateDrawable	刷新 Drawable 对象	void
scheduleDrawable	定义动画方案的下一帧	void
unscheduleDrawable	取消 scheduleDrawable 定义的动画方案	void
onPreDraw	设置视图显示,例如在视图显示之前调整滚动轴的边界	Boolean
sendAccessibilityEvent	发送事件类型指定的 AccessibilityEvent。发送请求之前,需要检查 Accessibility 是否打开	void
sendAccessibilityEventUnchecked	发送事件类型指定的 AccessibilityEvent。发送请求之前,不需要检查 Accessibility 是否打开	void
setOnKeyListener	设置按键监听	void

3.4.2　Button 控件实例

例 3.4　在例 3.2 的基础上添加两个 Button 按钮,分别是"登录"和"取消",ButtonDemo 执行结果如图 3-5 所示。

ButtonDemo 在 XML 布局文件中主要代码如下:

```
1    <Button
2        android:id = "@ + id/BtnLogin"
3        android:layout_width = "wrap_content"
4        android:layout_height = "wrap_content"
5        android:textColor = "#FFFFFF"
6        android:text = "@string/login"
7    />
8    <Button
9        android:id = "@ + id/BtnCancel"
10       android:layout_width = "wrap_content"
11       android:layout_height = "wrap_content"
12       android:textColor = "#FFFFFF"
13       android:text = "@string/cancel"
14   />
```

图 3-5　ButtonDemo 执行结果

XML 布局文件中定义了两个 Button 按钮,给按钮设置了 ID 分别为 BtnLogin 和 BtnCancel。第 6 行和第 13 行为这两个按钮设置了文本,引用了/res/values/string.xml 中的两个字符串的值。

ButtonDemo 在 Java 源文件中主要代码如下:

```
1    private Button BtnLogin,BtnCacel;
2     private TextView TxtShowResult;
```

```
3    @Override
4    protected void onCreate(Bundle savedInstanceState) {
5        super.onCreate(savedInstanceState);
6        setContentView(R.layout.activity_main);
7        TxtShowResult = (TextView)findViewById(R.id.ShowResult);
8        BtnLogin = (Button)findViewById(R.id.BtnLogin);
9        BtnCacel = (Button)findViewById(R.id.BtnCancel);
10       //(1)按钮注册到各自的监听器上
11       BtnLogin.setOnClickListener(new View.OnClickListener() {
12
13           @Override
14           public void onClick(View v) {
15               // TODO Auto-generated method stub
16               TxtShowResult.setText("登录成功!");
17           }
18       });
19       BtnCacel.setOnClickListener(new View.OnClickListener() {
20
21           @Override
22           public void onClick(View v) {
23               // TODO Auto-generated method stub
24               TxtShowResult.setText("取消登录!");
25           }
26       });
27   }
```

第 1 行和第 2 行是分别声明 Button 和 TextView 对象变量。第 7 行至第 9 行中 findViewById()函数能够通过 ID 分别获取 Button 和 TextView 控件。

为了能够使按钮响应单击事件,在 onCreate()函数中为 Button 控件和 ImageButton 控件添加单击事件的监听器,在第 11 行和第 19 行代码中,Button 对象通过调用 setOnClickListener()函数,注册一个单击(Click)事件的监听器 View.OnClickListener()。第 14 行和第 22 行代码是单击事件的回调函数。第 16 行代码将 TextView 的显示内容更改为选择的结果。

View.OnClickListener()是 View 定义的单击事件的监听器接口,并在接口中仅定义了 onClick()函数。当 Button 从 Android 界面框架中接收到事件后,首先检查这个事件是否单击事件,如果是单击事件,同时 Button 又注册了监听器,则会调用该监听器中的 onClick() 函数。

每个 View 仅可以注册一个单击事件的监听器,如果使用 setOnClickListener()函数注册第二个单击事件的监听器,之前注册的监听器将被自动注销。给每个按钮注册一个单击事件监听器,每个按钮的事件处理程序都在各自的 onClick()函数,这样能够使代码更加清晰、易读,且易于维护。当然,也可以将多个按钮注册到同一个单击事件的监听器上,示例代码如下:

```
1    private Button BtnLogin,BtnCacel;
2        private TextView TxtShowResult;
```

```
3       @Override
4       protected void onCreate(Bundle savedInstanceState) {
5           super.onCreate(savedInstanceState);
6           setContentView(R.layout.activity_main);
7           TxtShowResult = (TextView)findViewById(R.id.ShowResult);
8           BtnLogin = (Button)findViewById(R.id.BtnLogin);
9           BtnCacel = (Button)findViewById(R.id.BtnCancel);
10          //(2)按钮注册到同一个监听器
11          Button.OnClickListener btnListener = new Button.OnClickListener() {
12
13              @Override
14              public void onClick(View v) {
15                  // TODO Auto-generated method stub
16                switch(v.getId()){
17                  case R.id.BtnLogin:
18                    TxtShowResult.setText("登录成功!");
19                    return;
20                  case R.id.BtnCancel:
21                    TxtShowResult.setText("取消登录!");
22                    return;
23                }
24              }
25          };
26          BtnLogin.setOnClickListener(btnListener);
27          BtnCacel.setOnClickListener(btnListener);
28      }
```

第 11 行至第 25 行代码定义了一个名为 btnListener 的单击事件监听器，第 26 行和第 27 行代码分别将该监听器注册到 BtnLogin 和 BtnCacel 上。

3.5 ImageButton 控件

ImageButton（图片按钮）用以实现能够显示图像功能的控件按钮，继承了 Button 用户能够在该类控件上单击，按钮会触发一个 OnClick 事件，与 Button 控件的区别：为 ImageButton 按钮设置 android:text 属性没用。

ImageButton 类的层次关系如下：

```
java.lang.Object
    ↳ android.view.View
        ↳ android.widget.ImageView
            ↳ android.widget.ImageButton
```

3.5.1 ImageButton 控件常见的属性和方法

ImageButton 控件常见的属性及方法如表 3-8 和表 3-9 所示。

表 3-8 ImageButton 控件常见属性

属 性	描 述
android：adjustViewBounds	设置是否保持宽高比，true 或 false
android：cropToPadding	是否截取指定区域用空白代替。单独设置无效果，需要与 scrollY 一起使用，True 或者 false
android：maxHeight	设置图片按钮的最大高度
android：maxWidth	设置图片的最大宽度
android：scaleType	设置图片的填充方式
android：src	设置图片按钮的 drawable
android：tint	设置图片为渲染颜色

表 3-9 ImageButton 控件常见方法

方 法	功 能 描 述	返回值
ImageButton	构造函数	null
setAdjustViewBounds	设置是否保持高宽比，需要与 maxWidth 和 maxHeight 结合起来一起使用	Boolean
getDrawable	获取 Drawable 对象，获取成功返回 Drawable，否则返回 null	Drawable
getScaleType	获取视图的填充方式	ScaleType
setScaleType	设置视图的填充方式，包括矩阵、拉伸等 7 种填充方式	void
setAlpha	设置图片的透明度	void
setMaxHeight	设置按钮的最大高度	void
setMaxWidth	设置按钮的最大宽度	void
setImageURI	设置图片的地址	void
setImageResource	设置图片资源库	void
setOnTouchListener	设置事件的监听	Boolean
setColorFilter	设置颜色过滤	void

3.5.2 ImageButton 控件实例

例 3.5 在屏幕中实现一个背景图片按钮，实现该 ImageButton 按钮获得焦点和失去焦点时能装载不同的图片，ImageButtonDemo 执行结果如图 3-6 和图 3-7 所示。

图 3-6 ImageButtonDemo 初始结果

图 3-7 ImageButtonDemo 单击后的结果

ImageButtonDemo 在 XML 布局文件中主要代码如下：

```
1    <TextView
2        android:id = "@ + id/myTextView1"
3        android:layout_width = "wrap_content"
4        android:layout_height = "wrap_content"
5        android:textColor = "#FFFFFF"
6        android:text = "@string/str_textview1"
7    />
8    <ImageButton
9        android:id = "@ + id/myImageButton1"
10       android:layout_width = "wrap_content"
11       android:layout_height = "wrap_content"
12       android:contentDescription = "@string/desc"
13       android:src = "@drawable/iconempty"
14   />
15   <Button
16       android:id = "@ + id/myButton1"
17       android:layout_width = "wrap_content"
18       android:layout_height = "wrap_content"
19       android:text = "@string/str_button1"
20       android:textColor = "#FFFFFF"
21   />
```

在 XML 布局文件中设置了 1 个 TextView 控件、1 个 ImageButton 控件和 1 个 Button 控件。有许多种方法设置 ImageButton 背景图，在本实例中使用了 ImageButton.setImageResource()方法实现，在此方法中需要传递的是 res/drawable/下面的 Resource ID。除了设置上述方法外，还需要使用 onFocusChange 与 onClick 事件来处理单击按钮后的操作。

ImageButton 在 Java 源文件中主要代码如下：

```
1    private ImageButton mImageButton1;
2    private Button mButton1;
3    private TextView mTextView1;
4    @Override
5    protected void onCreate(Bundle savedInstanceState) {
6        super.onCreate(savedInstanceState);
7        setContentView(R.layout.activity_main);
8        mImageButton1 = (ImageButton) findViewById(R.id.myImageButton1);
9        mButton1 = (Button)findViewById(R.id.myButton1);
10       mTextView1 = (TextView) findViewById(R.id.myTextView1);
11       mImageButton1.setOnFocusChangeListener(new View.OnFocusChangeListener() {
12
13           @Override
14           public void onFocusChange(View v, boolean hasFocus) {
15               // TODO Auto-generated method stub
16               if (hasFocus == true)
17               {
18                mTextView1.setText("图片按钮状态为:Got Focus");
19                mImageButton1.setImageResource(R.drawable.iconfull);
20               }
```

```
21              else
22              {
23               mTextView1.setText("图片按钮状态为:Lost Focus");
24               mImageButton1.setImageResource(R.drawable.iconempty);
25              }
26          }
27      });
28      mImageButton1.setOnClickListener(new View.OnClickListener() {
29
30          @Override
31          public void onClick(View v) {
32              // TODO Auto-generated method stub
33              mTextView1.setText("图片按钮状态为:Got Click");
34              mImageButton1.setImageResource(R.drawable.iconfull);
35          }
36      });
37      mButton1.setOnClickListener(new View.OnClickListener() {
38
39          @Override
40          public void onClick(View v) {
41              // TODO Auto-generated method stub
42              mTextView1.setText("图片按钮状态为:Lost Focus");
43              mImageButton1.setImageResource(R.drawable.iconempty);
44          }
45      });
```

第 1 行至第 3 行声明 3 个对象变量,分别是 ImageButton、Button 和 TextView 对象变量。第 8 行至第 9 行通过 findViewById 获取了以上 3 个对象控件。第 11 行是定义 OnFocusChangeListener 事件,并通过该事件来应答 ImageButton 的 onFous。第 16 行至第 20 行实现了如果 ImageButton 状态为 onFocus 则改变 ImageButton 的图片并改变 textView 的文字,第 22 行至 25 行实现了若 ImageButton 状态为 offFocus 改变 ImageButton 的图片并改变 textView 的文字。第 40 行至 45 行若 Button 状态为 onClick 改变 ImageButton 的图片并改变 textView 的文字。

3.6　RadioButton 控件

RadioButton 指的是一个单选按钮,它有选中和未选中两种状态,RadioGroup 是 RadioButton 的承载体,程序运行时不可见,应用程序中可能包含一个或多个 RadioGroup。RadioGroup 包含多个 RadioButton,在一个 RadioGroup 中,用户仅能够选择其中一个 RadioButton。

RadioButton 的类层次关系如下:

```
java.lang.Object
   ↳ android.view.View
```

```
↳ android.widget.TextView
    ↳ android.widget.Button
        ↳ android.widget.CompoundButton
            ↳ android.widget.RadioButton
```

3.6.1 RadioButton 控件常见的方法

RadioButton 控件常见的方法如表 3-10 所示。

表 3-10　RadioButton 和 RadioGroup 常见的方法

方　　法	功 能 描 述	返回值
toggle()	将单选按钮更改为与当前选中状态相反的状态	void
addView(View child, int index, ViewGroup.LayoutParams params)	使用指定的布局参数添加一个子视图	void
check(int id)	作为指定的选择标识符来清除单选按钮组的选中状态，相当于调用 clearCheck()操作，参数：id 该组中所要选中的单选按钮的唯一标识符(id)	void
getCheckedRadioButtonId()	返回该单选按钮组中所选择的单选按钮的标识 ID，如果没有选中则返回−1	int
setOnCheckedChangeListener(ViewGroup.OnHierarchyChangeListener listener)	注册一个当该单选按钮组中的单选按钮选中状态发生改变时所要调用的回调函数	void

3.6.2 RadioButton 控件实例

例 3.6　用单选按钮 RadioButton 和 RadioGroup 实现在城市列表中选择小吃最多的城市，ImageButton、RadioButtonDemo 执行结果如图 3-8 所示。

ImageButtonDemo 在 XML 布局文件中主要代码如下：

```
 1  <TextView
 2      android:layout_width = "fill_parent"
 3      android:layout_height = "wrap_content"
 4      android:text = "@string/hello"
 5      android:id = "@ + id/textview1"
 6      android:textColor = "#FFFFFF"/>
 7  <RadioGroup
 8      android:id = "@ + id/radiogroup1"
 9      android:layout_width = "wrap_content"
10      android:layout_height = "wrap_content"
11      android:orientation = "vertical"
12      android:textColor = "#FFFFFF"
13  >
14  <RadioButton
15      android:id = "@ + id/radiobutton1"
```

图 3-8　RadioButtonDemo 执行结果

```
16      android:layout_width = "wrap_content"
17      android:layout_height = "wrap_content"
18      android:text = "@string/radiobutton1"
19      android:textColor = "#FFFFFF"/>
20   <RadioButton
21      android:id = "@+id/radiobutton2"
22      android:layout_width = "wrap_content"
23      android:layout_height = "wrap_content"
24      android:text = "@string/radiobutton2"
25      android:textColor = "#FFFFFF"/>
26   <RadioButton
27      android:id = "@+id/radiobutton3"
28      android:layout_width = "wrap_content"
29      android:layout_height = "wrap_content"
30      android:text = "@string/radiobutton3"
31      android:textColor = "#FFFFFF"/>
32   <RadioButton
33      android:id = "@+id/radiobutton4"
34      android:layout_width = "wrap_content"
35      android:layout_height = "wrap_content"
36      android:text = "@string/radiobutton4"
37      android:textColor = "#FFFFFF"/>
38   </RadioGroup>
39   <TextView
40      android:layout_width = "fill_parent"
41      android:layout_height = "wrap_content"
42      android:id = "@+id/textview2"
43      android:textColor = "#FFFFFF"/>
```

在 XML 布局文件中设置了两个 TextView 控件。第 7 行＜RadioGroup＞标签声明了一个 RadioGroup，在第 14 行、第 20 行、第 26 行和第 32 行分别声明了 4 个 RadioButton，这 4 个 RadioButton 是 RadioGroup 的子元素。

RadioButton 和 RadioGroup 的关系如下。

（1）RadioButton 表示单个圆形单选框，而 RadioGroup 是可以容纳多个 RadioButton 的容器。

（2）每个 RadioGroup 中的 RadioButton 同时只能有一个被选中。

（3）不同的 RadioGroup 中的 RadioButton 互不相干，即如果组 A 中有一个选中了，组 B 中依然可以有一个被选中。

（4）一般情况下，一个 RadioGroup 中至少有两个 RadioButton。

（5）一个 RadioGroup 中的 RadioButton 默认会有一个被选中，并建议将它放在 RadioGroup 中的起始位置。

ImageButtonDemo 在 Java 源文件中主要代码如下：

```
1   private TextView textview1,textview2;
2   private RadioGroup radiogroup;
3   private RadioButton radio1,radio2,radio3,radio4;
```

```
4   @Override
5   protected void onCreate(Bundle savedInstanceState) {
6       super.onCreate(savedInstanceState);
7       setContentView(R.layout.activity_main);
8       textview1 = (TextView)findViewById(R.id.textview1);
9       textview2 = (TextView)findViewById(R.id.textview2);
10      radiogroup = (RadioGroup)findViewById(R.id.radiogroup1);
11      radio1 = (RadioButton)findViewById(R.id.radiobutton1);
12      radio2 = (RadioButton)findViewById(R.id.radiobutton2);
13      radio3 = (RadioButton)findViewById(R.id.radiobutton3);
14      radio4 = (RadioButton)findViewById(R.id.radiobutton4);
15      radiogroup.setOnCheckedChangeListener(new RadioGroup.OnCheckedChangeListener() {
16
17          @Override
18          public void onCheckedChanged(RadioGroup group, int checkedId) {
19              // TODO Auto-generated method stub
20              if(checkedId == radio2.getId())
21              {
22                  textview2.setText("正确答案：" + radio2.getText() + ",恭喜你,回答正确!");
23              }else
24              {
25                  textview2.setText("请注意,回答错误!");
26              }
27          }
28      });
29   }
```

第 1 行至第 3 行分别声明了 TextView 控件对象、RadioGroup 控件对象和 RadioButton 控件对象。第 8 行至第 14 行通过 findViewById 获取了以上对象控件。第 18 行是定义 RadioGroup 的 onCheckedChanged 事件响应相对 ID 的 RadioButton 的选项。

3.7 CheckBox 控件

CheckBox(复选框)是同时可以选择多个选项的控件,该组件常用于某选项的打开或者关闭。

CheckBox 类的层次关系如下:

```
java.lang.Object
    ↳ android.view.View
        ↳ android.widget.TextView
            ↳ android.widget.Button
                ↳ android.widget.CompoundButton
                    ↳ android.widget.CheckBox
```

3.7.1 CheckBox 控件常见的方法

CheckBox 控件常见的方法如表 3-11 所示。

表 3-11　CheckBox 控件常见的方法

方　　法	功 能 描 述	返　回　值
dispatchPopulateAccessibilityEvent	在子视图创建时，分派一个辅助事件	boolean（true：完成辅助事件分发 false：没有完成辅助事件分发）
isChecked	判断组件状态是否选中	boolean（true：被选中，false：未被选中）
onRestoreInstanceState	设置视图恢复以前的状态	void
performClick	执行 click 动作，该动作会触发事件监听器	boolean（true：调用事件监听器，false：没有调用事件监听器）
setButtonDrawable	根据 Drawable 对象设置组件的背景	void
setChecked	设置组件的状态	void
setOnCheckedChangeListener	设置事件监听器	void
tooggle	改变按钮当前的状态	void
onCreateDrawableState	获取文本框为空时，文本框里面的内容	CharSequence
onCreateDrawableState	为当前视图生成新的 Drawable 状态	int[]

3.7.2　CheckBox 控件实例

例 3.7　用复选按钮 CheckBox 实现选择所喜欢的运动，并计算运动项目数量，CheckBoxDemo 执行结果如图 3-9 所示。

CheckBoxDemo 在 XML 布局文件中主要代码如下：

```
1    <TextView
2        android:id = "@ + id/TextView1"
3        android:layout_width = "fill_parent"
4        android:layout_height = "wrap_content"
5        android:text = "@string/hello"
6        android:textColor = "#FFFFFF"
7    />
8    <CheckBox
9        android:id = "@ + id/CheckBox1"
10       android:layout_width = "fill_parent"
11       android:layout_height = "wrap_content"
12       android:text = "@string/CheckBox1"
13       android:textColor = "#FFFFFF"
14   >
15   </CheckBox>
16   <CheckBox
17       android:id = "@ + id/CheckBox2"
18       android:layout_width = "fill_parent"
19       android:layout_height = "wrap_content"
20       android:text = "@string/CheckBox2"
21       android:textColor = "#FFFFFF"
22   >
```

图 3-9　CheckBoxDemo 执行结果

```
23      </CheckBox>
24      <CheckBox
25          android:id = "@+id/CheckBox3"
26          android:layout_width = "fill_parent"
27          android:layout_height = "wrap_content"
28          android:text = "@string/CheckBox3"
29          android:textColor = "#FFFFFF"
30      >
31      </CheckBox>
32      <CheckBox
33          android:id = "@+id/CheckBox4"
34          android:layout_width = "fill_parent"
35          android:layout_height = "wrap_content"
36          android:text = "@string/CheckBox4"
37          android:textColor = "#FFFFFF"
38      >
39      </CheckBox>
40      <Button
41          android:id = "@+id/button1"
42          android:layout_width = "wrap_content"
43          android:layout_height = "wrap_content"
44          android:text = "提交"
45          android:textColor = "#FFFFFF"
46      >
47      </Button>
48      <TextView
49          android:id = "@+id/TextView2"
50          android:layout_width = "fill_parent"
51          android:layout_height = "wrap_content"
52          android:textColor = "#FFFFFF"
53      />
```

在 XML 布局文件中设置了两个 TextView 控件。在第 8 行、第 16 行、第 24 行和第 32 行分别声明了 4 个 CheckBox 控件。

CheckBoxDemo 实例在 Java 源文件中主要代码如下:

```
1   private TextView   m_TextView1,m_TextView2;
2   private Button     m_Button1;
3   private CheckBox   m_CheckBox1,m_CheckBox2,m_CheckBox3,m_CheckBox4;
4   @Override
5   protected void onCreate(Bundle savedInstanceState) {
6       super.onCreate(savedInstanceState);
7       setContentView(R.layout.activity_main);
8       m_TextView1 = (TextView) findViewById(R.id.TextView1);
9       m_TextView2 = (TextView) findViewById(R.id.TextView2);
10      m_Button1 = (Button) findViewById(R.id.button1);
11      m_CheckBox1 = (CheckBox) findViewById(R.id.CheckBox1);
12      m_CheckBox2 = (CheckBox) findViewById(R.id.CheckBox2);
13      m_CheckBox3 = (CheckBox) findViewById(R.id.CheckBox3);
14      m_CheckBox4 = (CheckBox) findViewById(R.id.CheckBox4);
```

```java
15      m_CheckBox1.setOnCheckedChangeListener(new CompoundButton.OnCheckedChangeListener() {
16          @Override
17          public void onCheckedChanged(CompoundButton buttonView, boolean isChecked) {
18              // TODO Auto-generated method stub
19              if(m_CheckBox1.isChecked())
20              {
21                  m_TextView2.setText("你选择了: " + m_CheckBox1.getText());
22              }
23          }
24      });
25      m_CheckBox2.setOnCheckedChangeListener(new CompoundButton.OnCheckedChangeListener() {
26
27          @Override
28          public void onCheckedChanged(CompoundButton buttonView, boolean isChecked) {
29              // TODO Auto-generated method stub
30              if(m_CheckBox2.isChecked())
31              {
32                  m_TextView2.setText("你选择了: " + m_CheckBox2.getText());
33              }
34          }
35      });
36      m_CheckBox3.setOnCheckedChangeListener(new CompoundButton.OnCheckedChangeListener() {
37          @Override
38          public void onCheckedChanged(CompoundButton buttonView, boolean isChecked) {
39              // TODO Auto-generated method stub
40              if(m_CheckBox3.isChecked())
41              {
42                  m_TextView2.setText("你选择了: " + m_CheckBox3.getText());
43              }
44          }
45      });
46      m_CheckBox4.setOnCheckedChangeListener(new CompoundButton.OnCheckedChangeListener() {
47          @Override
48          public void onCheckedChanged(CompoundButton buttonView, boolean isChecked) {
49              // TODO Auto-generated method stub
50              if(m_CheckBox4.isChecked())
51              {
52                  m_TextView2.setText("你选择了: " + m_CheckBox4.getText());
53              }
54          }
55      });
56  }
```

第1行至第3行分别声明了 TextView 控件对象、CheckBox 控件对象和 Button 控件对象。第8行至第14行通过 findViewById 获取了以上对象控件。第15行是定义 CheckBox 的 setOnCheckedChangeListener 监听事件。CheckBox 的 setOnClickListener 和 setOnCheckedChangeListener 两种实现都能正常使用,但一般情况下都用 setOnCheckedChangeListener,主要是因为要改变 CheckBox 的状态不一定要通过单击事件,直接调用 setChecked 方法也可以改变,这样 OnClickListener 就监听不到了,而 OnCheckChangedListener 还是能监听到。

3.8 Toast

Toast(信息提示)是 Android 中一种简易的消息提示框。Toast 没有焦点,Toast 提示框不能被用户单击,而且 Toast 显示的时间有限,Toast 会根据用户设置的显示时间后自动消失。

Toast 是直接继承 java.lang.Object 的。因此它的类层次结构如下:

```
java.lang.Object
    ┗ android.widget.Toast
```

3.8.1 Toast 常量和常见的方法

Toast 中有两个关于 Toast 显示时间长短的常量分别如下。

1) int   LENGTH_LONG

持续显示视图或文本提示较长时间。该时间长度可定制。

2) int   LENGTH_SHORT

持续显示视图或文本提示较短时间。该时间长度可定制。

Toast 常见的方法如表 3-12 所示。

表 3-12 Toast 常见的方法

方 法	功 能 描 述	返回值
getDuration()	返回存续期间	int
getGravity()	取得提示信息在屏幕上显示的位置	int
makeText(Context context, int resId, int duration)	生成一个从资源中取得的包含文本视图的标准 Toast 对象。参数:context 使用的上下文。通常是 Application 或 Activity 对象。resId 要使用的字符串资源 ID,可以是已格式化文本。duration 该信息的存续期间。值为 LENGTH_SHORT 或 LENGTH_LONG	Toast
setDuration(int duration)	用于设置消息提示框持续的时间,参数值通常使用 LENGTH_SHORT 或 LENGTH_LONG	void
setGravity(int gravity, int xOffset, int yOffset)	设置提示信息在屏幕上的显示位置,参数 gravity 用于指定对齐方式	Void
setText(int resId)	更新之前通过 makeText()方法生成的 Toast 对象的文本内容。参数:resId 为 Toast 指定的新的字符串资源 ID	void
show()	按照指定的存续期间显示提示信息	void

3.8.2 Toast 实例

Toas 通常用于显示一些快速提示信息,应用范围非常广泛,使用 Toast 来显示消息提示框比较简单。只需要经过以下 3 个步骤即可实现。

(1)创建一个 Toast 对象。通常有两种方法:一种是使用构造方式进行创建;另一种是调用 Toast 类的 makeText 方法创建。

使用构造方法创建一个名称为 toast 的 Toast 对象的基本代码如下：

```
Toast toast = new Toast(this);
```

调用 Toast 类的 makeTextO 方法创建一个名称为 Toast 对象的基本代码如下：

```
Toast toast = Toast.makeText(this,"要显示的内容",Toast.LENGTH_SHORT);
```

（2）调用 Toast 类提供的方法来设置该消息提示框的对齐方式、页边距、显示的内容等。常用的方法如表 3-9 所示。

（3）调用 Toast 类的 show()方法显示消息提示框。需要注意的是，一定要调用该方法，否则设置的消息提示框将不显示。

在例 3.7 中选择结果可以用 Toast 提示框显示，执行结果如图 3-10 所示。

图 3-10　选择结果可以用 Toast 提示框显示

ToastDemo 实例在 Java 源文件中主要代码如下：

```
1    m_Button1.setOnClickListener(new View.OnClickListener() {
2        @Override
3        public void onClick(View v) {
4            … …
5    … …
6            Toast toast = Toast.makeText(MainActivity.this, "谢谢参与!你一共选择了" + num + "项!", Toast.LENGTH_LONG);
7            //设置 toast 显示的位置
8            toast.setGravity(Gravity.TOP,0,220);
9            //显示该 Toast
10           toast.show();
11       }
12   });
```

第 6 行是定义一个 Toast 对象，Toast.makeText(Context context，(CharSequence text)/(int resId)，int duration)参数：context 是指上下文对象，通常是当前的 Activity，text 是指自己写的消息内容，resId 只显示内容引用 Resource 的那条数据，duration 指的是 Toast 的显示时间，Toast 默认有 LENGTH_SHORT 和 LENGTH_LONG 两个常量，分别表示时间长和短，也可以自定义时间，如 5000 表示显示 5 秒。第 8 行是显示 Toast 信息框的位置，正常的 Toast 方法通知窗口水平居中显示在窗口的底部，可以使用 setGravity(int，int，int)方法来调整 Toast 窗口的显示位置，setGravity 方法有 3 个参数，第 1 个参数是一个 Gravity 常量，第 2 个是 x 方向起始的数值，第 3 个是 y 方向起始的数值。第 10 行是得到 Toast 对象之后调用.show()方法即可显示消息。

3.9 Spinner 控件

Spinner(下拉列表框)是从多个选项中选择一个选项的控件，类似于桌面程序的组合框(ComboBox)，但没有组合框的下拉菜单，而是使用浮动菜单为用户提供选择。Spinner 功能类似 RadioGroup，相比 RadioGroup，Spinner 提供了体验性更强的 UI 设计模式。一个 Spinner 对象包含多个子项，每个子项只有两种状态，选中或未被选中。

Spinner 类的层次关系如下：

```
java.lang.Object
    ↳ android.view.View
        ↳ android.view.ViewGroup
            ↳ android.widget.AdapterView
                ↳ android.widget.AbsSpinner
                    ↳ android.widget.Spinner
```

3.9.1 Spinner 控件常见的属性和方法

Spinner 常见 XML 属性 prompt 用法如表 3-13 和表 3-14 所示。

表 3-13 Spinner 控件常见的属性

属性名称	描述
android：prompt	该提示在下拉列表对话框显示时显示（也就是对话框的标题）
	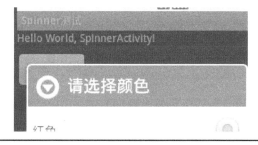

表 3-14　Spinner 控件常见的方法

方　　法	功　能　描　述	返回值
setPromptId(CharSequence prompt)	设置对话框弹出的时候显示的提示参数：prompt 设置的提示	void
onClick（DialogInterface dialog，int which）	当单击弹出框中的项时这个方法将被调用。参数：dialog 单击弹出的对话框，which 单击按钮（如 Button）或者单击位置	void
getBaseline()	返回这个控件文本基线的偏移量	int
CharSequence getPrompt()	当对话框弹出的时候显示的提示	void
onDetachedFromWindow()	当 Spinner 脱离窗口时被调用	void

3.9.2　Spinner 控件实例

例 3.8　用 Spinner 控件实现从城市列表中选择你所居住的城市，SpinnerDemo 示例如图 3-11 和图 3-12 所示。

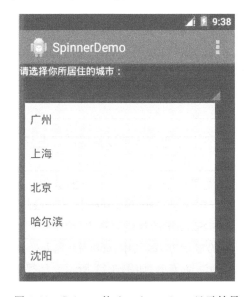

图 3-11　Spinner 的 dropdown_item 显示结果　　　　图 3-12　Spinner 的 item 显示结果

SpinnerDemo 在 XML 布局文件中主要代码如下：

```
1   <TextView
2       android:id = "@ + id/TextView01"
3       android:layout_width = "wrap_content"
4       android:layout_height = "wrap_content"
5       android:text = "@string/hello_world"
6       android:textColor = "#FFFFFF"
7   />
8   <Spinner
9       android:id = "@ + id/Spinner01"
```

```
10              android:layout_width = "300dip"
11              android:layout_height = "wrap_content"
12              android:textColor = "#FFFFFF"
13  />
```

从上至下分别是 TextView01 和 Spinner01，第 8 行使用＜Spinner＞标签声明了一个 Spinner 控件，并在第 10 行代码中指定该控件的宽度为 300dip。

在 SpinnerDemo.java 源文件文件中，定义一个 ArrayAdapter 适配器，在 ArrayAdapter 中添加在 Spinner 中可以选择的内容。为了使程序能够正常运行，需要在代码中引入 android.widget.ArrayAdapter 和 android.widget.Spinner。

SpinnerDemo 实例在 Java 源文件中主要代码如下：

```
1   Spinner spinner = (Spinner) findViewById(R.id.Spinner01);
2   List<String> list   = new ArrayList<String>();
3   list.add("广州");
4   list.add("上海");
5   list.add("北京");
6   list.add("哈尔滨");
7   list.add("沈阳");
8   ArrayAdapter<String> adapter = new ArrayAdapter<String>(this,
9       android.R.layout.simple_spinner_item, list);
10      adapter.setDropDownViewResource(android.R.layout.simple_spinner_item);
11      spinner.setAdapter(adapter);
```

第 2 行代码建立了一个字符串数组列表（ArrayList），这种数组列表可以根据需要进行增减，＜String＞表示数组列表中保存的是字符串类型的数据。在代码的第 3、4、5 行中，使用 add() 函数分别向数组列表中添加 3 个字符串。第 8 行代码建立了一个 ArrayAdapter 的数组适配器，数组适配器能够将界面控件和底层数据绑定在一起。在这里 ArrayAdapter 将 Spinner 和 ArrayList 绑定在一起，所有 ArrayList 中的数据，将显示在 Spinner 的浮动菜单中，绑定过程由第 8 行代码实现。第 10 行代码设定了 Spinner 浮动菜单的显示方式，其中，android.R.layout.simple_spinner_dropdown_item 是 Android 系统内置的一种浮动菜单，如图 3-11 所示。如果将其改为 android.R.layout.simple_spinner_item，则显示效果如图 3-12 所示。

为了保证用户界面显示的内容与底层数据一致，应用程序需要监视底层数据的变化，如果底层数据更改了，则用户界面也需要修改显示内容。在使用适配器绑定界面控件和底层数据后，应用程序就不需要再监视底层数据的变化，从而极大地简化了代码的复杂性。

3.10 ListView 控件

ListView（列表显示）是用于垂直显示的列表控件，如果显示内容过多，则会出现垂直滚动条。ListView 是在界面设计中经常使用的界面控件，其原因是 ListView 能够通过适配器将数据和显示控件绑定，在有限的屏幕上提供大量内容供用户选择；而且支持单击事件，可以用少量的代码实现复杂的选择功能。

ListView 类的层次关系：

```
java.lang.Object
   ↳ android.view.View
      ↳ android.view.ViewGroup
         ↳ android.widget.AdapterView
            ↳ android.widget.AbsListView
               ↳ android.widget.ListView
```

3.10.1 ListView 控件常见的属性和方法

ListView 控件常见的属性及方法如表 3-15 和表 3-16 所示。

表 3-15 ListView 控件常见的 XML 属性

属 性 名 称	描 述
android:choiceMode	规定此 ListView 所使用的选择模式。默认状态下，list 没有选择模式。属性值必须设置为下列常量之一：none，值为 0，表示无选择模式；singleChoice，值为 1，表示最多可以有一项被选中；multipleChoice，值为 2，表示可以多项被选中。可参看全局属性资源符号 choiceMode
android:dividerHeight	分隔符的高度。若没有指明高度，则用此分隔符固有的高度。必须为带单位的浮点数，如 "14.5sp"。可用的单位如 px（pixel，像素）、dp（density-independent pixels 与密度无关的像素）、sp（scaled pixels based on preferred font size 基于字体大小的固定比例的像素）、in（inches，英寸）和 mm（millimeters，毫米）
android:entries	引用一个将使用在此 ListView 里的数组。若数组是固定的，使用此属性将比在程序中写入更为简单。必须以 "@[+][package:]type:name"或者"?[package:][type:]name"的形式来指向某个资源
android:footerDividersEnabled	设成 flase 时，此 ListView 将不会在页脚视图前画分隔符。此属性默认值为 true。属性值必须设置为 true 或 false。可以用 "@[package:]type:name"或者"?[package:][type:]name"（主题属性）的格式来指向某个包含此类型值的资源
android:headerDividersEnabled	设成 flase 时，此 ListView 将不会在页眉视图后画分隔符。此属性默认值为 true。属性值必须设置为 true 或 false。可以用 "@[package:]type:name"或者"?[package:][type:]name"（主题属性）的格式来指向某个包含此类型值的资源

表 3-16 ListView 控件常见的方法

方　　法	功 能 描 述	返回值
addHeaderView（View v）	加一个固定显示于 list 顶部的视图，在调用 setAdapter 之前调用此方法	void
clearChoices()	取消之前设置的任何选择	void
getAdapter()	返回当前用于显示 ListView 中数据的适配器	ListAdapter
getCheckItemIds()	返回被选中项目的索引集合，只有当选择模式没有被设置为 CHOICE_MODE_NONE 时才有效	long[]
getCheckedItemPosition()	返回当前被选中的项目。只有当选择模式已被设置为 CHOICE_MODE_SINGLE 时，结果才有效	int

续表

方　　法	功 能 描 述	返回值
isItemChecked（int position）	对于由 position 指定的项目，返回其是否被选中，只有当选择模式已被设置为 CHOICE_MODE_SINGLE 或 CHOICE_MODE_MULTIPLE 时，结果才有效	boolean
setAdapter（ListAdapter adapter）	设置 ListView 背后的数据。根据 ListView 目前使用的特性	void
setItemChecked（int position，boolean value）	设置 position 所指定项目的选择状态，参数 position 需要改变选择状态的项目的索引，value 是新的选择状态	void
setSelection（int position）	选中 position 指定的项目，参数 position 需要选中的项目的索引（从 0 开始）	void

3.10.2　ListView 控件实例

例 3.9　单击选中 ListView 列表中的一项，把选中的结果显示在文本框中，SpinnerDemo 示例如图 3-13 所示。

图 3-13　ListViewDemo 执行结果

ListViewDemo 在 XML 布局文件中主要代码如下：

```
1   <TextView
2       android:id = "@+id/TextView01"
3       android:layout_width = "wrap_content"
4       android:layout_height = "wrap_content"
5       android:text = "@string/hello_world"
6       android:textColor = "#FFFFFF"
7   />
8   <ListView
9       android:id = "@+id/ListView01"
10          android:listSelector = "#FFFFFF"
11          android:layout_width = "wrap_content"
12          android:layout_height = "wrap_content"
13   />
```

在 ListViewDemo.java 文件中，首先需要为 ListView 创建适配器，并添加 ListView 中所显示的内容。

ListViewDemo 实例在 Java 源文件中主要代码如下：

```
1   final TextView textView = (TextView)findViewById(R.id.TextView01);
2   ListView listView = (ListView)findViewById(R.id.ListView01);
3   
4   List<String> list = new ArrayList<String>();
5   list.add("ListView 子项 1");
6   list.add("ListView 子项 2");
7   list.add("ListView 子项 3");
8   ArrayAdapter<String> adapter = new ArrayAdapter<String>(this,
9       android.R.layout.simple_list_item_1, list);
10  listView.setAdapter(adapter);
11  AdapterView.OnItemClickListener listViewListener = new AdapterView.OnItemClickListener(){
12      @Override
13      public void onItemClick(AdapterView<?> arg0, View arg1, int arg2, long arg3) {
14          String msg = "父 View: " + arg0.toString() + "\n" +
15              "子 View: " + arg1.toString() + "\n" +
16              "位置: " + String.valueOf(arg2) + ",ID:
17              " + String.valueOf(arg3);
18          textView.setText(msg);
19      }};
20  listView.setOnItemClickListener(listViewListener);
```

第 2 行代码通过 ID 引用了 XML 文件中声明的 ListView。第 4 行～第 7 行声明了数组列表。第 8 行声明了适配器 ArrayAdapter，第 3 个参数 list 说明适配器的数据源为数组列表。第 10 行将 ListView 和适配器绑定。第 11 行的 AdapterView.OnItemClickListener 是 ListView 子项的单击事件监听器，同样是一个接口，需要实现 onItemClick() 函数。在 ListView 子项被选择后，onItemClick() 函数将被调用。第 13 行的 onItemClick() 函数中一共有 4 个参数，参数 1 表示适配器控件，这里就是 ListView；参数 2 表示适配器内部的控件，这里是 ListView 中的子项；参数 3 表示适配器内部的控件，也就是子项的位置；参数 4 表示子项的行号。第 14 行和第 18 行代码用于显示信息，选择子项确定后，在 TextView 中显示子项父控件的信息、子控件信息、位置信息和 ID 信息。第 20 行代码表示 ListView 指定刚声明的监听器。

3.11　ProgressBar 控件

ProgressBar(进度条)是一个显示进度的控件，Android 提供了两大类进度条样式，长形进度条样式 progress-BarStyleHorizontal 和圆形进度条 progressBarStyleLarge。

ProgressBar 类的层次关系如下：

```
java.lang.Object
   ↳ android.view.View
       ↳ android.widget.ProgressBar
```

3.11.1 ProgressBar 常见方法

ProgressBar 控件常见的方法如表 3-17 所示。

表 3-17 ProgressBar 控件常见的方法

方　　法	功　能　描　述	返回值
ProgressBar	3 个构造函数： ProgressBar(Context contex) ProgressBar(Context contex,AttributeSet attrs) ProgressBar(Context contex,AttributeSet attrs,intdefStyle)	null
onAttachedToWindow()	视图附加到窗体时调用	void
onDraw(Canvas canvas)	绘制视图时,该方法被调用	void
onMeasure(int widthMeasureSpec, int heightMeasureSpec)	该方法被重写时,必须调用 setMeasedDimension(int,int)来存储已测量视图的高度和宽度,否则将通过 measure(int,int)抛出一个 IllegalStateException 异常	void
onDetachedFromWindow()	从窗体分离事件的响应方法。当视图 Progress 从窗体上分离或移除时调用	void
addFocusables(ArrayList＜View＞views,int direction)	继承自 android.view.View 的方法,用于为当前 ViewGroup 中的所有子视图添加焦点获取能力	void
addTouchables (ArrayList＜View views＞)	继承 android.view.View 的方法,用于为子视图添加触摸能力	void
getBaseline	继承 android.view.View 的方法。返回窗口空间的文本基准线到其顶边界的偏移量	int
dispatchDisplayHint(int hint)	继承 android.view.View 的方法。分发视图是否显示的提示	void
dispatchDraw(Canvas canvas)	继承 android.view.View 的方法。调用此方法来绘制子视图	void

3.11.2 ProgressBar 控件实例

例 3.10　编程实现显示进度条的长条形和圆形两种不同状态,SpinnerDemo 示例如图 3-14 所示。

图 3-14 ProgressBarDemo 执行结果

ProgressBarDemo 在 XML 布局文件中主要代码如下：

```
1   <TextView
2       android:layout_width = "wrap_content"
3       android:layout_height = "wrap_content"
4       android:text = "@string/hello_world"
5       android:textColor = "#FFFFFF"/>
6   <ProgressBar
7       android:id = "@+id/FirstBar"
8       style = "?android:attr/progressBarStyleHorizontal"
9       android:layout_width = "200dp"
10      android:layout_height = "wrap_content"
11      android:visibility = "gone"
12  />
13  <ProgressBar
14      android:id = "@+id/secondBar"
15      style = "?android:attr/progressBarStyle"
16      android:layout_width = "wrap_content"
17      android:layout_height = "wrap_content"
18      android:visibility = "gone"
19  />
20  <Button
21      android:id = "@+id/MyButton"
22      android:layout_width = "wrap_content"
23      android:layout_height = "wrap_content"
24      android:text = "@string/begin"
25      android:textColor = "#FFFFFF"
26  />
```

从上至下分别是 1 个 TextView 控件、两个 ProgressBar 控件和 1 个 Button 控件，第 9 行代码中指定 FirstBar 控件的宽度为 200dp，第 11 行和第 18 行分别设置了 FirstBar 和 SecondBar 初始状态是不可见的。

ProgressBarDemo 实例在 Java 源文件中主要代码如下：

```
1   private ProgressBar FirstBar;
2   private ProgressBar SecondBar;
3   private Button Mybutton;
4   int i = 0;
5   @Override
6   protected void onCreate(Bundle savedInstanceState) {
7       super.onCreate(savedInstanceState);
8       setContentView(R.layout.activity_main);
9       FirstBar = (ProgressBar)findViewById(R.id.FirstBar);
10          SecondBar = (ProgressBar)findViewById(R.id.secondBar);
11          Mybutton = (Button)findViewById(R.id.MyButton);
12          Mybutton.setOnClickListener(new View.OnClickListener() {
13          @Override
14          public void onClick(View v) {
15              // TODO Auto-generated method stub
16              if(i == 0)
```

```
17              {
18                      FirstBar.setVisibility(View.VISIBLE);
19                      FirstBar.setMax(150);
20                      SecondBar.setVisibility(View.VISIBLE);
21              }
22              else if(i<FirstBar.getMax())
23              {
24                      FirstBar.setProgress(i);
25                      FirstBar.setSecondaryProgress(i+10);
26              }
27              else
28              {
29                      FirstBar.setVisibility(View.GONE);
30                      SecondBar.setVisibility(View.GONE);
31              }
32              i=i+10;
33      }
34 });
```

第 1 行至第 3 行分别声明了 ProgressBar 控件对象 Button 控件对象。第 9 行至第 11 行通过 findViewById 获取了以上对象控件。第 14 行是定义按钮的事件监听器,在单击事件代码中,如果变量 i 的值为 0 时,开始显示两个进度条,同时设定 FirstBar 的最大值为 150,每单击一次按钮,FirstBar 增加值为 10,SecondBar 增加值为 $i+10$,最后显示进度条的状态。

习题

1. 说明在 Android 常用控件中 TextView 和 EditView 控件用法的不同之处。
2. Android 中 Button 按钮控件响应单击事件方式有哪两种。
3. Android 中定义控件的方式大都类似,首先要声明它的类型,然后通过哪种方法获取控件的 Id 索引。
4. 在 Android 中每个控件都要指定控件的高度和宽度,主要有哪 3 种取值。
5. 在 Android 中所有控件都可以设置大小,设置时可以指定的单位有哪些。
6. 编写代码练习本章中学习过的控件,熟悉每种控件的主要属性和方法。

第 4 章 Android 界面布局与菜单处理

Android 应用程序开发中界面布局和菜单处理是不可忽视的因素。在 Android 中，View 有六大布局方式，分别是线性布局、表格布局、相对布局、帧布局、绝对布局和网格布局，布局方式使用 XML 语言进行描述。常见的菜单处理包括选项菜单、子菜单和快捷菜单。

本章主要学习内容：
- 了解界面布局概述；
- 掌握线性布局 LinearLayout 类和线性布局实例；
- 掌握表格布局 TableLayout 类和表格布局实例；
- 掌握相对布局 RelativeLayout 类和相对布局实例；
- 掌握绝对布局 AbsoluteLayout 类和绝对布局实例；
- 了解帧布局；
- 了解网格布局。

4.1 界面布局概述

界面布局（Layout）是用户界面结构的描述，定义了界面中所有的元素、结构和相互关系。一般声明 Android 程序的界面布局有两种方法，第一种是使用 XML 文件描述界面布局，另一种是在程序运行时动态添加或修改界面布局。

Android 系统在声明界面布局上提供了很好的灵活性，用户既可以独立使用任何一种声明界面布局的方式，也可以同时使用两种方式。一般情况下，使用 XML 文件来描述用户界面中的基本元素，而在代码中动态修改需要更新状态的界面元素。当然，用户也可以将所有的界面元素，无论在程序运行后是否需要修改其内容，都放在代码中进行定义和声明。很明显这不是一种良好的界面设计模式，会给后期界面修改带来不必要的麻烦，而且界面元素较多时，程序的代码也会显得凌乱不堪。

使用 XML 文件声明界面布局，能够更好地将程序的表现层和控制层分离，在修改界面时将不再需要更改程序的源代码。例如，在程序开发完成后，为了让程序能够支持不同屏幕尺寸、规格和语言的手机，则可以声明多个 XML 布局，而无须修改程序代码。不仅如此，使用 XML 文件声明的界面布局，用户还能够通过可视化工具直接看到所设计的用户界面，有利于加快界面设计的过程，并且为界面设计与开发带来极大的便利性。

4.2 线性布局

线性布局(LinearLayout)是较简单的一个布局,它提供了控件水平或者垂直排列的模型。本节将会对线性布局进行简单介绍,首先介绍 LinearLayout 类的相关知识,然后通过一个实例说明 LinearLayout 的使用方法。

4.2.1 LinearLayout 类简介

LinearLayout 通过设置的垂直或水平的属性值,来排列所有的子元素。所有的子元素都被堆放在其他元素之后,因此一个垂直列表的每一行只会有一个元素,而不管它们有多宽,而一个水平列表将会只有一个行高(高度为最高子元素的高度加上边框高度)。LinearLayout 保持子元素之间的间隔以及互相对齐(相对一个元素的右对齐、中间对齐或者左对齐)。

LinearLayout 的常用属性及对应设置方法如表 4-1 所示。

表 4-1 LinearLayout 的常用属性及对应设置方法

属性名称	设置方法	描述
android:orientation	setOrientation(int)	设置线性布局的朝向,可设置为 horizontal、vertical 两种排列方式
android:gravity	setGravity(int)	设置线性布局的内部元素的对齐方式

1. orientation 属性

在线性布局中可以使用 orientation 属性来设置布局的朝向,可取值及说明如下。
- horizontal:定义横向布局。
- vertical:定义纵向布局。

对于纵向布局与横向布局而言,控件的排列方式分别如图 4-1 和图 4-2 所示。

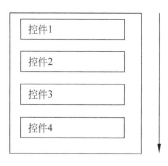

图 4-1 纵向布局

图 4-2 横向布局

2. gravity 属性

在线性布局中可以使用 gravity 属性设置控件的对齐方式,可取的值及说明如表 4-2 所示。

表 4-2　gravity 属性

常　量	描　述
top	不改变控件大小,对齐到容器顶部
bottom	不改变控件大小,对齐到容器底部
left	不改变控件大小,对齐到容器左侧
right	不改变控件大小,对齐到容器右侧
center_vertical	不改变控件大小,对齐到容器纵向中央位置
fill_vertical	纵向拉伸以填充满容器
center_horizontal	不改变控件大小,对齐到容器横向中央位置
fill_horizontal	横向拉伸以填充满容器
center	不改变控件大小,放置在容器的正中间
fill	横向和纵向同时拉伸以填充满容器

4.2.2　线性布局实例

本节将通过一个实例来说明 LinearLayout 的使用方法。在本实例中,最上层的纵向线性布局中嵌套了一个纵向线性布局和一个横向线性布局。在嵌套的纵向线性布局中,摆放了一个 TextView 和一个 Button 控件;在嵌套的横向线性布局中摆放了两个 TextView 控件。本实例开发步骤如下。

(1) 创建项目 LinearLayoutDemo。

(2) 修改主 Activity 的布局文件 activity_main.xml,编写代码如下:

```
1  < LinearLayout xmlns:android = "http://schemas.android.com/apk/res/android"
2  android:orientation = "vertical"
3  android:layout_width = "fill_parent"
4  android:layout_height = "fill_parent"
5  android:background = "♯000000"
6  >
7  < TextView
8  android:layout_width = "fill_parent"
9  android:layout_height = "wrap_content"
10 android:text = "本案例演示 LinearLayout 线性布局"
11 android:textSize = "20px"
12 android:background = "♯000000"
13 android:textColor = "♯FFFFFF"
14 />
15 < LinearLayout
16 android:orientation = "vertical"
17 android:layout_width = "fill_parent"
18 android:layout_height = "wrap_content"
19 android:background = "♯000000"
20 >
21 < TextView
22 android:layout_width = "fill_parent"
23 android:layout_height = "wrap_content"
24 android:text = "这是纵向布局的第一个 TextView."
```

```
25    android:textSize = "20px"
26    android:background = " # 000000"
27    android:textColor = " # FFFFFF"
28    />
29    < Button
30    android:layout_width = "wrap_content"
31    android:layout_height = "wrap_content"
32    android:layout_gravity = "right"
33    android:text = "纵向一个按钮"
34    />
35    </LinearLayout >
36    < LinearLayout
37    android:orientation = "horizontal"
38    android:layout_width = "fill_parent"
39    android:layout_height = "fill_parent"
40    android:background = " # 000000"
41    >
42    < Button
43    android:layout_width = "200px"
44    android:layout_height = "wrap_content"
45    android:text = "横向第一个按钮"
46    android:textSize = "20px"
47    android:padding = "2px"
48    />
49    < Button
50    android:layout_width = "200px"
51    android:layout_height = "wrap_content"
52    android:text = "横向第二个按钮"
53    android:textSize = "20px"
54    android:padding = "2px"
55    />
56    </LinearLayout >
57    </LinearLayout >
```

其中第 2 行代码声明该布局为一个纵向布局,第 3～4 行代码设置该布局高度和宽度填充满整个容器。对于最顶层的布局来说,它的容器就是手机屏幕,所以该布局会填充手机屏幕进行显示。第 7～12 行中,在最顶层的布局中声明第一个控件 TextView。第 8 行代码定义 TextView 控件的高度,wrap_content 的含义是根据视图内部内容自动扩展以适应其大小;第 11 行代码定义 TextView 控件的字体大小为 20px。第 15～35 中,在最顶层的布局中嵌套一个纵向线性布局。第 16 行代码定义该布局的朝向为纵向。在该布局中包含一个 TextView 控件与一个 Button 控件;第 32 行代码定义 Button 控件的对齐方式为右对齐(即 Button 放在该布局的最右侧)。第 36～57 行中,在最顶层的布局中嵌套一个横向线性布局。第 37 行代码定义该布局的朝向为横向;第 38 行代码定义该布局填充满顶层布局的剩余空间。在该布局中包含两个 Button 控件;第 47 行代码定义 Button 的内容与父容器边界的距离为 2px(2 个像素)。android:padding 规定父 view 里面的内容与父 view 边界的距离。本实例运行结果如图 4-3 所示。

图 4-3　LinearLayoutDemo 运行结果

4.3　帧布局

帧布局(FrameLayout)是最简单的界面布局,用于存储一个元素的空白空间,且子元素的位置是不能够指定的,只能够放置在空白空间的左上角。如果有多个子元素,后放置的子元素将遮挡先放置的子元素。

为了更好地理解帧布局,这里使用 Android SDK 中提供的层级观察器(Hierarchy Viewer)进一步分析界面布局。层级观察器能够对用户界面进行分析和调试,并以图形化的方式展示树形结构的界面布局,另外,还提供了一个精确的像素观察器(Pixel Perfect View),以栅格的方式详细观察放大后的界面。

在模拟器上运行上垂直排列的线性布局示例,控件属性如表 4-3 所示,在层级观察器中获得示例界面布局的树形结构图,如图 4-4 所示。

表 4-3　线性布局界面控件的属性设置

编　号	类　型	属　性	值
1	TextView	Id	@+id/label
		Text	用户名：
2	EditText	Id	@+id/entry
		Layout width	fill_parent
		Text	[null]
3	Button	Id	@+id/ok
		Text	确认
4	Button	Id	@+id/cancel
		Text	取消

结合界面布局的树形结构图(如图 4-4 所示)和示意图(如图 4-5 所示),分析不同界面布局和界面控件的区域边界。用户界面的根结点(♯0@43599ee0)是线性布局,其边界是整个界面,也就是示意图的最外层的实心线。根结点右侧的子结点(♯0@43599a730)是帧布局,仅有一个结点元素(♯0@4359ad18),这个子元素是 TextView 控件,用于显示 Android 应用程序名称,其边界是示意图中的区域 1。因此框架布局元素♯0@43599a730 的边界是同区域 1 的高度相同,宽度充满整个根结点的区域。这两个界面元素是系统自动生成的,一般情况下用户不能够修改和编辑。

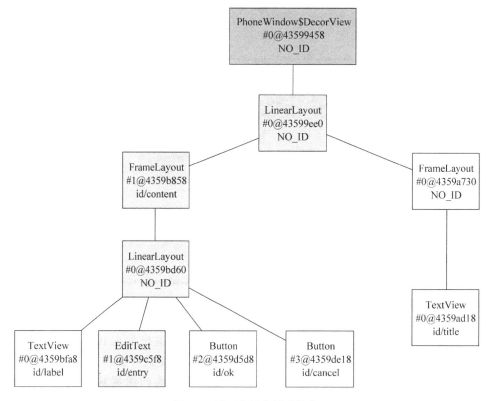

图 4-4 界面布局的树形结构

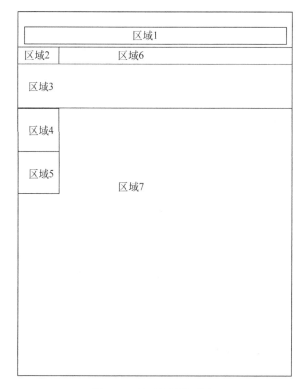

图 4-5 界面布局的示意

根结点左侧的子结点(♯1@4359b858)也是框架布局,边界是区域 2 到区域 7 的全部空间。其下仅有一个子结点(♯0@4359bd60)元素是线性布局,因为线性布局的 Layout width 属性设置为 fill_parent,Layout height 属性设置为 wrap_content,因此该线性布局的宽度就是其父结点♯1@4359b858 的宽度,高度等于所有子结点元素的高度之和。线性布局♯0@4359bd60 的 4 个子结点元素♯0@4359bfa8、♯1@4359c5f8、♯2@4359d5d8 和♯3@4359de18 的边界,分别是界面布局示意图中的区域 2、区域 3、区域 4 和区域 5。

4.4 表格布局

表格布局(TableLayout)是按照行列来组织子视图的布局,包含一系列的 TableRow 对象,用于定义行。本节将对表格布局进行介绍,首先介绍 TableLayout 类的相关知识,然后通过一个实例说明 TableLayout 的使用方法。

4.4.1 TableLayout 类简介

表格布局也是一种常用的界面布局,它将屏幕划分为表格,通过指定行和列可以将界面元素添加到表格中。对比网格布局的示意图(如图 4-6 所示),可以发现网格的边界对用户是不可见的。表格布局还支持嵌套,可以将另一个表格布局放置在前一个表格布局的网格中,也可以在表格布局中添加其他界面布局,例如线性布局、相对布局等。

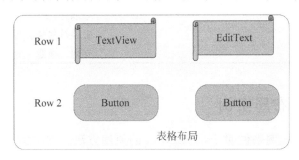

图 4-6 表格布局的示意

表格布局包含一系列的 TableRow 对象,用于定义行。表格布局不为它的行、列和单元格显示表格线。每个行可以包含 0 个以上(包括 0)的单元格;每个单元格可以设置一个 View 对象。与行包含很多单元格一样,表格包含很多列。表格的单元格既可以为空,也可以像 HTML 那样跨列。

无论是在代码还是在 XML 布局文件中,单元格必须按照索引顺序加入表格行。列号从 0 开始,如果不为子单元格指定列号,其将自动增值,使用下一个可用列号。虽然表格布局典型的子对象是表格行,但实际上可以使用任何视图类的子类,作为表格视图的直接子对象,视图会作为一行并合并了所有列的单元格显示。

列的宽度由该列所有行中最宽的一个单元格决定,而表格的总宽度由其父容器决定。不过表格布局可以通过 setColumnShrinkable() 或者 setColumnStretchable() 方法来标记哪些列可以收缩或拉伸。如果标记为可以收缩,列宽可以收缩以使表格适合容器的大小;如果标记为可以拉伸,列宽可以拉伸以占用多余的空间。可以通过调用 setColumnCollapsed()方法来隐藏列。

在表格布局中,可以为列设置以下 3 种属性。

(1) Shrinkable:表示列的宽度可以进行收缩,以使表格能够适应其父容器的大小。

(2) Stretchable:表示列的宽度可以进行拉伸,以填满表格中空闲的空间。

(3) Collapsed:表示列将会被隐藏。

TableLayout 继承 LinearLayout 类,除了继承 LinearLayout 类的属性和方法外,TableLayout 类中还包含表格布局自身的属性和方法。TableLayout 的常用属性及对应设置方法如表 4-4 所示。

表 4-4 TableLayout 的常用属性及对应设置方法

属性名称	相关方法	描述
android:collapseColumns	setColumnCollapsed(int,boolean)	隐藏从 0 开始的索引列。列号必须用逗号隔开,如 1,2,…。非法或重复的设置将被忽略
android:shrinkColumns	setShrinkAllColumns(boolean)	收缩从 0 开始的索引列。列号必须用逗号隔开,如 1,2,…。非法或重复的设置将被忽略。可以通过"*"代替收缩所有列。注意一列能同时表示收缩和拉伸
android:stretchColumns	setStretchAllColumns(boolean)	拉伸从 0 开始的索引列。列号必须用逗号隔开,如 1,2,…。非法或重复的设置将被忽略。可以通过"*"代替拉伸所有列。注意一列能同时表示收缩和拉伸

4.4.2 表格布局实例

本节将通过一个实例来说明 TableLayout 的使用方法。在本实例中,实现一个登录的界面。本实例开发步骤如下。

创建项目 TableLayoutDemo,在/res/layout/activity_main.xml 文件中设计基于表格布局的用户界面。在表格布局中设计一个 2×2 的网格,每个网格中置放一个界面控件,实现效果如图 4-7 所示。

建立表格布局的示例并不困难,要注意以下几点。

(1) 向界面中添加一个表格布局,无须修改布局的属性值。其中,Id 属性为 TableLayout01,Layout width 和 Layout height 属性都为 wrap_content。

(2) 在 Outline 视图中(如图 4-8 所示),在 TableLayout01 上右击,选择 Add Row 向 TableLayout01 中添加两个 TableRow。TableRow 代表一个单独的行,每行被划分为几个小的单元,单元中可以添加一个界面控件。其中,Id 属性分别为 TableRow01 和 TableRow02,Layout width 和 Layout height 属性都为 wrap_content。

(3) 在界面可视化编辑器上,向 TableRow01 中拖曳 TextView 和 EditText。

图 4-7 TableLayoutDemo 的效果

图 4-8 Outline 视图中的表格布局

(4) 在界面可视化编辑器上,再向 TableRow02 中拖曳两个 Button。

(5) 参考表 4-5 所示设置 TableRow 中 4 个界面控件的属性值。

表 4-5 表格布局界面控件的属性设置

编 号	类 型	属 性	值
1	TextView	Id	@+id/label
		Text	用户名:
		Gravity	right
		Padding	3dip
		Layout width	160dip
2	EditText	Id	@+id/entry
		Text	[null]
		Padding	3dip
		Layout width	160dip
3	Button	Id	@+id/ok
		Text	确认
		Padding	3dip
4	Button	Id	@+id/cancel
		Text	取消
		Padding	3dip

activity_main.xml 文件的完整代码如下:

```
1   <?xml version = "1.0" encoding = "utf - 8"?>
2
3   < TableLayout android:id = "@ + id/TableLayout01"
4       android:layout_width = "fill_parent"
5       android:layout_height = "fill_parent"
6       xmlns:android = "http://schemas.android.com/apk/res/android">
7       < TableRow android:id = "@ + id/TableRow01"
8           android:layout_width = "wrap_content"
9           android:layout_height = "wrap_content">
10          < TextView android:id = "@ + id/label"
11              android:layout_height = "wrap_content"
12              android:layout_width = "160dip"
13              android:gravity = "right"
```

```
14              android:text = "用户名："
15              android:padding = "3dip" >
16          </TextView>
17          <EditText android:id = "@ + id/entry"
18              android:layout_height = "wrap_content"
19              android:layout_width = "160dip"
20              android:padding = "3dip" >
21          </EditText>
22      </TableRow>
23      <TableRow android:id = "@ + id/TableRow02"
24          android:layout_width = "wrap_content"
25          android:layout_height = "wrap_content">
26          <Button android:id = "@ + id/ok"
27              android:layout_height = "wrap_content"
28              android:padding = "3dip"
29              android:text = "确认">
30          </Button>
31          <Button android:id = "@ + id/Button02"
32              android:layout_width = "wrap_content"
33              android:layout_height = "wrap_content"
34              android:padding = "3dip"
35              android:text = "取消">
36          </Button>
37      </TableRow>
38 </TableLayout>
```

第 3 行代码使用了＜TableLayout＞标签声明表格布局；第 7 行和第 23 行代码声明了两个 TableRow 元素，用于表示布局中的两行；第 12 行代码利用宽度属性 android：layout_width，将 TextView 元素的宽度指定为 160dip；第 13 行代码使用属性 android：gravity，将 TextView 中的文字对齐方式指定为右对齐；第 15 行代码使用属性 android：padding，声明 TextView 元素与其他元素的间隔距离为 3dip。

4.5 相对布局

相对布局（RelativeLayout）是指在容器内部的子元素可以使用彼此之间的相对位置或者和容器间的相对位置来进行定位。相对布局和线性布局有着共同的优点，能够最大程度保证在各种屏幕类型的手机上正确显示界面布局。本节将对相对布局进行介绍，首先介绍 RelativeLayout 类的相关知识，然后通过一个实例说明 RelativeLayout 的使用方法。

4.5.1 RelativeLayout 类简介

在相对布局中，控件的位置是相对其他控件或者父容器而言的。在进行设计时，需要按照控件之间的依赖关系排列，例如，控件 B 的位置相对于控件 A 决定，则在布局文件中控件 A 需要在控件 B 的前面进行定义。

在设计相对布局时，会用到很多属性，下面对属性分别进行说明，如表 4-6 所示。

表 4-6 RalativeLayout 属性

属性	值	描述
android:layout_alignParentTop	true 或 false	如果为 true,该控件的顶部与其父控件的顶部对齐
android:layout_alignParentBottom	true 或 false	如果为 true,该控件的底部与其父控件的底部对齐
android:layout_alignParentLeft	true 或 false	如果为 true,该控件的左部与其父控件的左部对齐
android:layout_alignParentRight	true 或 false	如果为 true,该控件的右部与其父控件的右部对齐
android:layout_alignWithParentIfMissing	true 或 false	参考控件不存在或不可见时参照父控件
android:layout_centerHorizontal	true 或 false	如果为 true,该控件置于父控件的水平居中位置
android:layout_centerVertical	true 或 false	如果为 true,该控件置于父控件的垂直居中位置
android:layout_centerInParent	true 或 false	如果为 true,该控件置于父控件的中央位置
android:layout_above	某控件的 id 属性	将该控件的底部置于给定 ID 控件的上方
android:layout_below	某控件的 id 属性	将该控件的底部置于给定 ID 控件的下方
android:layout_toLeftOf	某控件的 id 属性	将该控件的右边缘与给定 ID 的控件左边缘对齐
android:layout_toRightOf	某控件的 id 属性	将该控件的左边缘与给定 ID 的控件右边缘对齐
android:layout_alignBaseline	某控件的 id 属性	将该控件的 baseline 与给定 ID 的 baseline 对齐
android:layout_alignTop	某控件的 id 属性	将该控件的顶部边缘与给定 ID 的顶部边缘对齐
android:layout_alignBottom	某控件的 id 属性	将该控件的底部边缘与给定 ID 的底部边缘对齐
android:layout_alignLeft	某控件的 id 属性	将该控件的左边缘与给定 ID 的左边缘对齐
android:layout_alignRight	某控件的 id 属性	将该控件的右边缘与给定 ID 的右边缘对齐
android:layout_marginTop	int 类型数值	上偏移的值
android:layout_marginBottom	int 类型数值	下偏移的值
android:layout_marginLeft	int 类型数值	左偏移的值
android:layout_marginRight	int 类型数值	右偏移的值

需要注意的是能在 RelativeLayout 容器本身及其子元素之间产生循环依赖。例如,不能将 RelativeLayout 的高设置为 WRAP_CONTENT 时,将子元素的高设置为 ALIGN_PARENT_BOTTOM。

4.5.2 相对布局实例

本节将用一个实例来说明相对布局的使用方法。在本实例中,采用相对布局来实现登录的界面。创建项目 RalativeLayoutDemo,在/res/layout/activity_main.xml 文件中设计基于相对布局的用户界面。执行效果如图 4-9 所示。

图 4-9 是相对布局的一个示例,下面先用文字对界面元素的添加顺序和相互关系进行描述。首先添加 TextView 控件("用户名"),相对布局会将 TextView 控件放置在屏幕的最上方;然后添加 EditText 控件(输入框),并声明该控件的位

图 4-9 相对布局

置在 TextView 控件的下方,相对布局会根据 TextView 的位置确定 EditText 控件的位置;之后添加第 1 个 Button 控件("取消"按钮),声明在 EditText 控件的下方,且在父控件的最右边;最后,添加第 2 个 Button 控件("确认"按钮),声明该控件在第 1 个 Button 控件的左方,且与第 1 个 Button 控件处于相同的水平位置。

main.xml 文件的完整代码如下:

```xml
1  <?xml version = "1.0" encoding = "utf - 8"?>
2
3  < RelativeLayout android:id = "@ + id/RelativeLayout01"
4      android:layout_width = "fill_parent"
5      android:layout_height = "fill_parent"
6      xmlns:android = "http://schemas.android.com/apk/res/android">
7      < TextView android:id = "@ + id/label"
8          android:layout_height = "wrap_content"
9          android:layout_width = "fill_parent"
10         android:text = "用户名: ">
11     </TextView>
12     < EditText android:id = "@ + id/entry"
13         android:layout_height = "wrap_content"
14         android:layout_width = "fill_parent"
15         android:layout_below = "@id/label">
16     </EditText>
17     < Button android:id = "@ + id/cancel"
18         android:layout_height = "wrap_content"
19         android:layout_width = "wrap_content"
20         android:layout_alignParentRight = "true"
21         android:layout_marginLeft = "10dip"
22         android:layout_below = "@id/entry"
23         android:text = "取消" >
24     </Button>
25         < Button android:id = "@ + id/ok"
26         android:layout_height = "wrap_content"
27         android:layout_width = "wrap_content"
28         android:layout_toLeftOf = "@id/cancel"
29         android:layout_alignTop = "@id/cancel"
30         android:text = "确认">
31     </Button>
32  </RelativeLayout>
```

在上面的代码中,首先在第 3 行使用了 <RelativeLayout> 标签声明一个相对布局;第 15 行使用位置属性 android:layout_below,确定 EditText 控件在 ID 为 label 的元素下方;第 20 行使用属性 android:layout_alignParentRight,声明该元素与其父元素的右边界对齐;第 21 行使用属性 android:layout_marginLeft,将该元素向左移动 10dip;第 22 行声明该元素在 ID 为 entry 的元素下方;第 28 行声明使用属性 android:layout_toLeftOf,声明该元素在 ID 为 cancel 元素的左边;第 29 行使用属性 android:layout_alignTop,声明该元素与 ID 为 cancel 的元素在相同的水平位置。

4.6 绝对布局

绝对布局(AbsoluteLayout)是指所有控件的排列由开发人员通过控件的坐标来指定，容器不再负责管理其子控件的位置。本节将对绝对布局进行介绍，首先介绍 AbsoluteLayout 类的相关知识，然后通过一个实例说明 AbsoluteLayout 的使用方法。

4.6.1 AbsoluteLayout 类简介

绝对布局(AbsoluteLayout)能通过指定界面元素的坐标位置，来确定用户界面的整体布局。绝对布局是一种不推荐使用的界面布局，因为通过绝对位置确定的界面元素，Android 系统不能够根据不同屏幕对界面元素的位置进行调整，降低了界面布局对不同类型和尺寸屏幕的适应能力。使用绝对布局往往在目标手机上非常完美，但在其他不同类型的手机上，界面布局却变得混乱不堪。

绝对布局缺乏灵活性，在没有绝对定位的情况下相比其他类型的布局更难维护，并且采用绝对布局设计的界面有可能在不同的手机设备上显示完全不同的结果。因此，在选择设计布局时，不推荐使用绝对布局。

AbsoluteLayout 类的常用属性及对应设置方法如表 4-7 所示。

表 4-7 AbsoluteLayout 类的常用属性及对应设置方法

属　　性	描　　述
android:layout_x	指定控件的 x 坐标
android:layout_y	指定控件的 y 坐标

需要注意的是对于手机屏幕而言，坐标原点为屏幕左上角。当向右或者向下移动时，坐标值将变大。

4.6.2 绝对布局实例

本节用一个实例来说明绝对布局的使用方法。在本实例中，采用绝对布局来实现登录的界面。创建项目 AbsoluteLayoutDemo，在/res/layout/activity_main.xml 文件中设计基于相对布局的用户界面。执行效果如图 4-10 所示。

图 4-10 是绝对布局的一个示例，每一个界面控件都必须指定坐标(x,y)，例如"确认"按钮的坐标是(40,120)，"取消"按钮的坐标是(120,120)。坐标原点(0,0)在屏幕的左上角。

下面给出 activity_main.xml 文件的完整代码：

图 4-10 AbsoluteLayoutDemo 执行效果

```xml
1  <?xml version = "1.0" encoding = "utf-8"?>
2
3  <AbsoluteLayout android:id = "@ + id/AbsoluteLayout01"
4      android:layout_width = "fill_parent"
5      android:layout_height = "fill_parent"
6      xmlns:android = "http://schemas.android.com/apk/res/android">
7      <TextView android:id = "@ + id/label"
8          android:layout_x = "40dip"
9          android:layout_y = "40dip"
10         android:layout_height = "wrap_content"
11         android:layout_width = "wrap_content"
12         android:text = "用户名: ">
13     </TextView>
14     <EditText android:id = "@ + id/entry"
15         android:layout_x = "40dip"
16         android:layout_y = "60dip"
17         android:layout_height = "wrap_content"
18         android:layout_width = "150dip">
19     </EditText>
20     <Button android:id = "@ + id/ok"
21         android:layout_width = "70dip"
22         android:layout_height = "wrap_content"
23         android:layout_x = "40dip"
24         android:layout_y = "120dip"
25         android:text = "确认">
26     </Button>
27     <Button android:id = "@ + id/cancel"
28         android:layout_width = "70dip"
29         android:layout_height = "wrap_content"
30         android:layout_x = "120dip"
31         android:layout_y = "120dip"
32         android:text = "取消">
33     </Button>
34 </AbsoluteLayout>
```

4.7 网格布局

网格布局(GridLayout)是 Android SDK 4.0(API Level 14)新支持的布局方式,将用户界面划分为网格,界面元素可随意摆放在这些网格中。网格布局比表格布局(TableLayout)在界面设计上更加灵活,在网格布局中界面元素可以占用多个网格的,而在表格布局却无法实现,只能将界面元素指定在一个表格行(TableRow)中,不能跨越多个表格行。

下面用 GroidLayoutDemo 示例说明网格布局的使用方法。图 4-11(a)和图 4-11(b)所示分别是在 Eclipse 界面设计器中的界面图示和在 Android 模拟器运行后的用户界面。

在界面设计器中可以看到虚线网格,但在模拟器的运行结果中是看不到的。网格布局将界面划分成多个块,这些块是根据界面元素动态划分的。具体地讲,GroidLayoutDemo

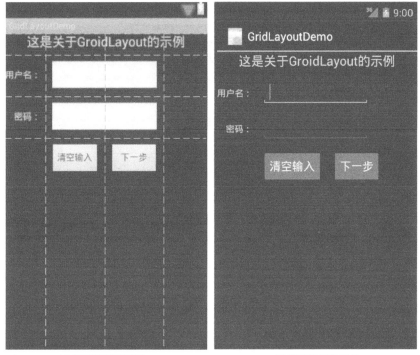

(a) 界面设计器图示　　　　(b) 模拟器运行结果

图 4-11　GroidLayoutDemo 示例

示例的左边第 1 列的宽度,是综合分析在第 1 列中的两个界面元素"用户名"和"密码"TextView 的宽度而进行设定的,选择两个元素中最宽元素的宽度,作为第 1 列的宽度。同意,最上方第 1 行的高度,也是分析"这是关于 GridLayout 的示例"这个 TextView 元素的高度进行设定的。因此,网格布局中行的高度和列的宽度,完全取决于本行或本列中,高度最高或宽对最宽的界面元素。

但在网格布局中,界面元素是可以跨越多个块的,例如"这是关于 GridLayout 的示例"这个 TextView 元素就占据的纵向 4 个块,"用户名:"这个 TextView 在纵向仅占用了 1 个块,而用户名输入框控件 EditText 在纵向上占用了两个块。这个示例中没有横向占用多个块的界面元素,但这样设计在网格布局中是允许的。

下面给出 main.xml 文件的全部代码:

```
1    <?xml version = "1.0" encoding = "utf - 8"?>
2    <GridLayout
3        xmlns:android = "http://schemas.android.com/apk/res/android"
4
5        android:layout_width = "match_parent"
6        android:layout_height = "match_parent"
7        android:useDefaultMargins = "true"
8        android:columnCount = "4"   >
9
10       <TextView
```

```
11          android:layout_columnSpan = "4"
12          android:layout_gravity = "center_horizontal"
13          android:text = "这是关于 GroidLayout 的示例"
14          android:textSize = "20dip" />
15
16      <TextView
17          android:text = "用户名："
18          android:layout_gravity = "right" />
19
20      <EditText
21          android:ems = "8"
22          android:layout_columnSpan = "2"/>
23
24      <TextView
25          android:text = "密码："
26          android:layout_column = "0"
27          android:layout_gravity = "right"/>
28
29      <EditText
30          android:ems = "8"
31          android:layout_columnSpan = "2" />
32
33      <Button
34          android:text = "清空输入"
35          android:layout_column = "1"
36          android:layout_gravity = "fill_horizontal"/>
37
38      <Button
39          android:text = "下一步"
40          android:layout_column = "2"
41          android:layout_gravity = "fill_horizontal"/>
42
43  </GridLayout>
```

代码第 7 行的 useDefaultMargins 表示网格布局中的所有元素都遵循默认的边缘规则，就是说所有元素之间都会留有一定的边界空间。代码第 8 行的 columnCount 表示纵向分为 4 列，从第 0 列到第 3 列，程序开发人员也可以在这里定义横向的行数，使用 rowCount 属性。

代码第 11 行的 layout_columnSpan 属性表示 TeixtView 控件所占据的列的数量。代码第 12 行的 layout_gravity = center_horizontal 表示文字内容在所占据的块中居中显示。

代码第 10 行到第 14 行定义了第 1 个界面控件，虽然定义了纵向所占据的块的数量，但却没有定义元素起始位置所在的块，原因是网格布局中第 1 个元素默认在第 0 行第 0 列。

代码第 16 行到第 18 行定义了第 2 个界面控件，仍然没有定义元素起始位置所在的块。根据网格布局界面元素的排布规则，如果没有明确说明元素所在的块，那么当前元素会放置在前一个元素的同一行右侧的块上；如果前一个元素已经是这一行的末尾块，则当前元素放置在下一行的第 1 个块上；如果当前元素在纵向上占据多个块，而前一个元素右侧没有足够数量的块,则当前元素的起始位置也会放置在下一行的第 1 个块上。

代码第 26 行的 layout_column 属性表示当前元素列的起始位置。如果 layout_column 所指定的列的位置在当前行已经被占用,则当前元素也会放置在下一行的这一列中。

在网格布局中没有定义的属性是具有默认值的,具体默认值如表 4-8 所示。

表 4-8　网格布局中属性的默认值

属　　性	默　认　值	备　　注
width	WRAP_CONTENT	
height	WRAP_CONTENT	
topMargin	0	当用户将 useDefaultMargins 设置为 false
leftMargin	0	当用户将 useDefaultMargins 设置为 false
bottomMargin	0	当用户将 useDefaultMargins 设置为 false
rightMargin	0	当用户将 useDefaultMargins 设置为 false
rowSpec.row	UNDEFINED	
rowSpec.rowSpan	1	
rowSpec.alignment	BASELINE	
columnSpec.column	UNDEFINED	
columnSpec.columnSpan	1	
columnSpec.alignment	LEFT	

4.8　菜单

菜单是应用程序中非常重要的组成部分,能够在不占用界面空间的前提下,为应用程序提供统一的选择功能和设置界面,并为程序开发人员提供易于使用的编程接口。Android 系统支持 3 种菜单模式,分别是选项菜单(Option Menu)、子菜单(Submenu)和快捷菜单(Context Menu)。

4.8.1　菜单资源

Android 程序的菜单可以在代码中动态生成,也可以使用 XML 文件制作菜单资源,然后通过 inflate()函数映射到程序代码中。使用 XML 文件描述菜单是较好的选择,可以将菜单的内容与代码分离,且有利于分析和调整菜单结构。

下面是 MenuResource 示例 main_menu.xml 文件的全部代码:

```
1   <?xml version = "1.0" encoding = "utf - 8"?>
2   < menu xmlns:android = "http://schemas.android.com/apk/res/android">
3       < item android:id = "@ + id/main_menu_0"
4           android:icon = "@drawable/pic0"
5           android:title = "打印" />
6       < item android:id = "@ + id/main_menu_1"
7           android:icon = "@drawable/pic1"
8           android:title = "新建" />
9       < item android:id = "@ + id/main_menu_2"
```

```
10            android:icon = "@drawable/pic2"
11            android:title = "邮件" />
12    < item android:id = "@ + id/main_menu_3"
13            android:icon = "@drawable/pic3"
14            android:title = "设置" />
15    < item android:id = "@ + id/main_menu_4"
16            android:icon = "@drawable/pic4"
17            android:title = "订阅" />
18  </menu >
```

上面的代码生成的菜单如图 4-12 所示，生成具有 5 个子项的菜单。代码第 2 行的 menu 是菜单的容器，菜单资源必须以 menu 作为根元素。代码第 3 行 item 是菜单项，其属性值 id、icon 和 title 分别是菜单项的 ID 值、图标和标题。代码第 4 行以及后续的代码中，虽然定义了菜单项的图标，但菜单资源选择的菜单模式不同，菜单项图标可能会不显示。MenuResource 示例使用的是选项菜单，因此菜单项的图标没有显示，而仅显示了菜单项的标题。

图 4-12 MenuResource 示例界面

4.8.2 选项菜单

选项菜单是一种经常被使用的 Android 系统菜单，用户可以通过"菜单键"（MENU key）打开选项菜单。

在 Android 2.3 之前的系统中，选项菜单分为图标菜单（Icon Menu）和浮动菜单（Overflow Menu），通过"菜单键"直接打开的是图标菜单，如图 4-13 所示。顾名思义，图标菜单就是能够同时显示文字和图标的菜单，最多支持 6 个子项，如果子项多于 6 个，则需要扩展菜单显示其他的子项。

浮动菜单是垂直的列表型菜单，如图 4-14 所示，仅在图标菜单子项多于 6 个时才出现，通过单击图标菜单最后的子项 More 才能打开。浮动菜单不能够显示图标，但支持单选框和复选框；相反，图标菜单支持显示图标，但不支持单选框和复选框。

图 4-13　图标菜单

图 4-14　浮动菜单

在 Android 4.0 系统中，选项菜单只出现浮动菜单，而不再出现图 4-14 所示的选项菜单，图标菜单的部分功能由操作栏代替实现。

在 Android 4.0 系统中，Activity 在创建时便会调用 onCreateOptionsMenu()函数初始化自身的菜单系统。在 Activity 的整个生命周期中，选项菜单是一直被重复利用的，直到 Activity 被销毁。在 Android 2.3 之前的系统中，onCreateOptionsMenu()函数只有在用户单击"菜单键"后才被调用，就是说选项菜单是在需要的时候才被创建的。但 Android 4.0 系统需要在程序的顶部显示操作栏，操作栏的初始化代码也在 onCreateOptionsMenu()函数中，因此该函数在 Activity 创建时就会被调用。

重载 onCreateOptionsMenu()函数主要目的是初始化菜单，可以使用 XML 文件的菜单资源，也可以使用代码动态加载菜单。下面的代码是使用 main_menu.xml 文件作为菜单资源初始化 Activity 的菜单。

```
1  @Override
2  public boolean onCreateOptionsMenu(Menu menu) {
3      MenuInflater inflater = getMenuInflater();
4      inflater.inflate(R.menu.main_menu, menu);
5      return true;
6  }
```

在用户选择菜单项后，Android 系统会调用 onOptionsItemSelected()函数，一般将菜单选择事件的响应代码放置在 onOptionsItemSelected()函数中。onOptionsItemSelected()函数会返回用户选择的 MenuItem，可以通过 getItemId()函数获取到 MenuItem 的 ID，这个 ID 就是用户在 XML 文件中为每个菜单项所设定的 android:id 属性值。

onOptionsItemSelected()函数在每次用户单击菜单子项时都会被调用。下面的代码说明如何通过菜单子项的子项 ID 执行不同的操作：

```
1   @Override
2   public boolean onOptionsItemSelected(MenuItem item) {
3       TextView label = (TextView)findViewById(R.id.label);
4
5       switch (item.getItemId()) {
6           case R.id.main_menu_0:
7               label.setText("打印,菜单 ID: " + item.getItemId());
8               return true;
9           case R.id.main_menu_1:
10              label.setText("新建,菜单 ID: " + item.getItemId());
11              return true;
12          case R.id.main_menu_2:
13              label.setText("邮件,菜单 ID: " + item.getItemId());
14              return true;
15          case R.id.main_menu_3:
16              label.setText("设置,菜单 ID: " + item.getItemId());
17              return true;
18          case R.id.main_menu_4:
19              label.setText("订阅,菜单 ID: " + item.getItemId());
20              return true;
21          default:
22              return false;
23      }
24  }
```

函数 onOptionsItemSelected() 的返回值表示是否需求其他事件处理函数菜单选择事件进行处理，如果不需要其他函数处理该事件则返回 true，否则返回 false。

代码第 5 行的 getItemId() 函数获取到被选择菜单子项的 ID。代码第 7 行通过在 XML 文件中定义的菜单 ID 与 getItemId() 函数的返回值进行匹配。

选项菜单的代码可参考 OptionsMenu 示例，程序运行后通过单击"菜单键"可以调出选项菜单，如图 4-15(a)所示。

开发人员除了可以使用 XML 文件的菜单资源以外，还可以在代码中动态生成菜单。OptionsMenu2 示例说明如何使用代码生成的菜单，所生产的菜单内容与 OptionsMenu 示例的菜单完全一样，如图 4-15(b)所示。

开发人员首先要在代码中定义菜单 ID，然后在 onCreateOptionsMenu() 函数中添加选项菜单，并设置菜单的标题和图标等信息。

```
1   final static int MENU_00 = Menu.FIRST;
2   final static int MENU_01 = Menu.FIRST + 1;
3   final static int MENU_02 = Menu.FIRST + 2;
4   final static int MENU_03 = Menu.FIRST + 3;
5   final static int MENU_04 = Menu.FIRST + 4;
6
```

```
7   @Override
8   public boolean onCreateOptionsMenu(Menu menu) {
9       menu.add(0,MENU_00,0,"打印").setIcon(R.drawable.pic0);
10      menu.add(0,MENU_01,1,"新建").setIcon(R.drawable.pic1);
11      menu.add(0,MENU_02,2,"邮件").setIcon(R.drawable.pic2);
12      menu.add(0,MENU_03,3,"设置").setIcon(R.drawable.pic3);
13      menu.add(0,MENU_04,4,"订阅").setIcon(R.drawable.pic4);
14      return true;
15  }
```

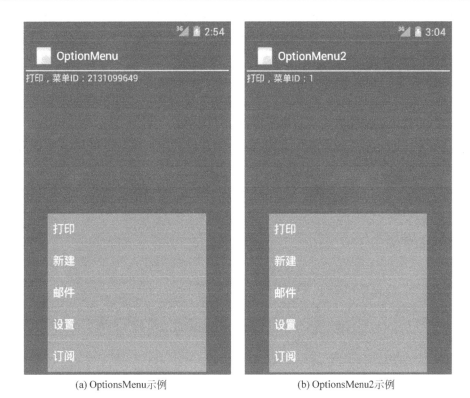

(a) OptionsMenu示例   (b) OptionsMenu2示例

图 4-15　选项菜单示例界面

一般将菜单项的 ID 定义成静态常量(代码第 1 行至第 5 行),并使用静态常量 Menu.FIRST(整数类型,值为 1)定义第 1 个菜单子项,以后的菜单项仅需要在 Menu.FIRST 增加相应的数值即可。

在 onCreateOptionsMenu()函数中,函数的返回类型为布尔值(代码第 14 行),返回 true 则可显示在函数中设置的菜单,否则将不能够显示菜单。

Menu 对象作为一个参数被传递到函数内部,因此在 onCreateOptionsMenu()函数中,用户可以使用 Menu 对象的 add()函数添加菜单项。

add()函数的语法:

```
MenuItem android.view.Menu.add(int groupId, int itemId, int order, CharSequence title)
```

add()函数的第 1 个参数 groupId 是一组 ID,用以批量的对菜单子项进行处理和排序;第 2 个参数 itemId 是子项 ID,是每一个菜单子项的唯一标识,通过子项 ID 使应用程序能够定位到用户所选择的菜单子项;第 3 个参数 order 是定义菜单子项在选项菜单中的排列顺序;第 4 个参数 title 是菜单子项所显示的标题。

另外,通过 setIcon()函数可以为菜单子项添加图标,需要将图像资源文件复制到/res/drawable 目录下。

4.8.3 子菜单

子菜单就是二级菜单,用户单击选项菜单或快捷菜单中的菜单项,就可以打开子菜单。当程序具有大量的功能时,可以将相似的功能划分成组,选项菜单可用于表示功能组,而具体功能则可由子菜单进行选择。

传统的子菜单一般采用树型的层次化结构,但 Android 系统却使用浮动窗体的形式显示菜单子项。采用与众不同的显示方式,主要是为了更好适应小屏幕的显示方式。子菜单不支持嵌套,也就是说不能够在子菜单中再使用子菜单。

下面以 SubMenu 示例说明如何使用 XML 文件设计子菜单。SubMenu 示例的用户界面如图 4-16 所示。

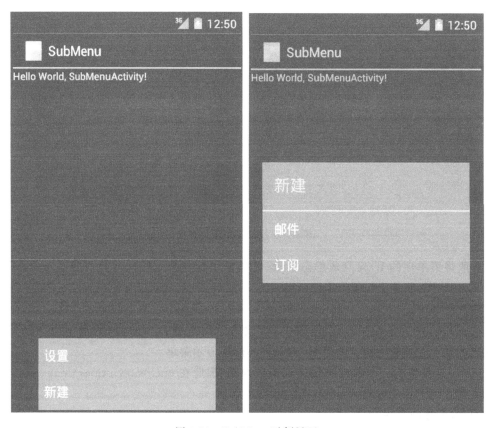

图 4-16　SubMenu 示例界面

SubMenu 示例选项菜单包含两个菜单项："设置"和"新建"。菜单项"设置"包含子菜单,子菜单中只有 1 个菜单项"打印"。菜单项"新建"包含子菜单,子菜单中有两个菜单项,分别是"邮件"和"订阅"。SubMenu 示例的菜单结构如下:

```
(+)设置
    (-)打印
(+)新建
    (-)邮件
    (-)订阅
```

SubMenu 示例使用 XML 文件描述菜单结构,sub_menu.xml 文件代码如下:

```xml
1   <?xml version = "1.0" encoding = "utf-8"?>
2   <menu xmlns:android = "http://schemas.android.com/apk/res/android">
3       <item android:id = "@+id/main_menu_0"
4           android:icon = "@drawable/pic0"
5           android:title = "设置" >
6           <menu>
7               <item android:id = "@+id/sub_menu_0_0"
8                   android:icon = "@drawable/pic4"
9                   android:title = "打印" />
10          </menu>
11      </item>
12      <item android:id = "@+id/main_menu_1"
13          android:icon = "@drawable/pic1"
14          android:title = "新建" >
15          <menu>
16              <item android:id = "@+id/sub_menu_1_0"
17                  android:icon = "@drawable/pic2"
18                  android:title = "邮件" />
19              <item android:id = "@+id/sub_menu_1_1"
20                  android:icon = "@drawable/pic3"
21                  android:title = "订阅" />
22          </menu>
23      </item>
24  </menu>
```

代码第 6 行至代码第 10 行是一个子菜单的描述。子菜单也使用<menu>标签进行声明,内部使用<item>标签描述菜单项。

Android 系统的子菜单使用起来非常灵活,除了可以 XML 文件描述菜单结构,也可以通过代码在选项菜单或快捷菜单中使用子菜单。SubMenu2 是使用代码实现子菜单的示例,界面如图 4-17 所示。

SubMenu2 子菜单结构与 SubMenu 示例是完全相同的,不同之处在于子菜单上多了标题图标。下面首先给出 SubMenu2 示例的核心代码:

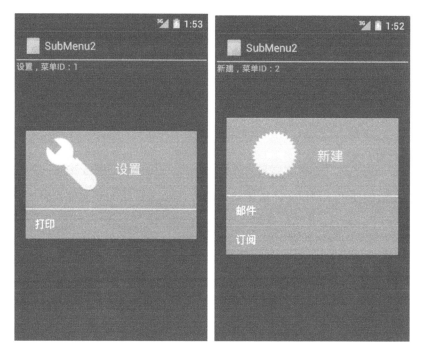

图 4-17　SubMenu2 示例界面

```
1   final static int MENU_00 = Menu.FIRST;
2   final static int MENU_01 = Menu.FIRST + 1;
3   final static int SUB_MENU_00_01 = Menu.FIRST + 2;
4   final static int SUB_MENU_01_00 = Menu.FIRST + 3;
5   final static int SUB_MENU_01_01 = Menu.FIRST + 4;
6
7   SubMenu sub1 = (SubMenu) menu.addSubMenu(0,MENU_00,0,"设置")
8               .setHeaderIcon(R.drawable.pic3);
9   sub1.add(0,SUB_MENU_00_01 ,0,"打印").setIcon(R.drawable.pic0);
10
11  SubMenu sub2 = (SubMenu) menu.addSubMenu(0,MENU_01,1,"新建")
12              .setHeaderIcon(R.drawable.pic1);
13  sub2.add(0,SUB_MENU_01_00 ,0,"邮件").setIcon(R.drawable.pic2);
14  sub2.add(0,SUB_MENU_01_01 ,0,"订阅").setIcon(R.drawable.pic4);
```

代码第 1 行至第 5 行是定义选项菜单和子菜单所有菜单项的 ID。代码第 7 行使用 addSubMenu()函数在选项菜单中增加了 1 个菜单项 MENU_00，当用户单击这个菜单项后会打开子菜单。addSubMenu()函数共有 4 个参数，参数 1 是组 ID，如果不分组则可以使用 0；参数 2 是菜单项的 ID；参数 3 是显示排序，数字越小越靠近列表上方；参数 4 是菜单项显示的标题。代码第 8 行设置了子菜单的图标。代码第 9 行在子菜单中添加了菜单项。

4.8.4　快捷菜单

快捷菜单类似于计算机程序中的"右键菜单"，当用户单击界面上某个元素超过两秒后，将启动注册到该界面元素的快捷菜单。快捷菜单同样采用浮动的显示方式,虽然快捷菜单的现实方式与子菜单相同,但两种菜单的启动方式却截然不同。

下面用 ContextMenu 示例说明如何使用快捷菜单,如何将快捷菜单注册到某个界面元素上。ContextMenu 示例的用户界面如图 4-18 所示。

图 4-18　ContextMen 用户界面

快捷菜单的使用方法与选项菜单极为相似,只是重载的函数不同而已。快捷菜单需要重载 onCreateContextMenu() 函数初始化菜单项,包括添加快捷菜单所显示的标题、图标和菜单子项等内容。

下面的代码说明如何使用 onCreateContextMenu() 函数初始化菜单项:

```
1   final static int CONTEXT_MENU_1 = Menu.FIRST;
2   final static int CONTEXT_MENU_2 = Menu.FIRST + 1;
3   final static int CONTEXT_MENU_3 = Menu.FIRST + 2;
4   @Override
5   public void onCreateContextMenu(ContextMenu menu, View v,
            ContextMenuInfo menuInfo){
6       menu.setHeaderTitle("快捷菜单标题");
7       menu.add(0, CONTEXT_MENU_1, 0,"菜单子项 1");
8       menu.add(0, CONTEXT_MENU_2, 1,"菜单子项 2");
9       menu.add(0, CONTEXT_MENU_3, 2,"菜单子项 3");
10  }
```

上面的代码实现了一个具有 3 个菜单项的子菜单,子菜单的标题是"快捷菜单标题"。ContextMenu 类支持 add() 函数(代码第 7 行)和 addSubMenu() 函数,可以在快捷菜单中添加菜单子项和子菜单。onCreateContextMenu() 函数(代码第 5 行)的第 1 个参数 menu 是需要显示的快捷菜单;第 2 个参数 v 是用户单击的界面元素;第 3 个参数 menuInfo 是所选择界面元素的额外信息。

重载 onContextItemSelected()函数响应菜单选择事件。该函数在用户选择快捷菜单中的菜单项后被调用,与 onOptionsItemSelected()函数的使用方法基本相同。

下面代码说明如何重载 onContextItemSelected()函数响应子菜单事件:

```
1   @Override
2   public boolean onContextItemSelected(MenuItem item){
3       switch(item.getItemId()){
4           case CONTEXT_MENU_1:
5               LabelView.setText("菜单子项 1");
6               return true;
7           case CONTEXT_MENU_2:
8               LabelView.setText("菜单子项 2");
9               return true;
10          case CONTEXT_MENU_3:
11              LabelView.setText("菜单子项 3");
12      return true;
13      }
14      return false;
15  }
```

最后,还需要使用 registerForContextMenu()函数,将快捷菜单注册到界面中的某个控件上(下方代码第 7 行)。在用户长时间单击该界面控件时,便会启动快捷菜单。同时,为了能够在界面上直接显示用户所选择快捷菜单的菜单项,在代码中引用了界面元素 TextView(下方代码第 6 行),通过更改 TextView 的显示内容(上方代码第 5、8 和 11 行),显示用户所选择的菜单子项。

```
1   TextView LabelView = null;
2   @Override
3   public void onCreate(Bundle savedInstanceState) {
4       super.onCreate(savedInstanceState);
5       setContentView(R.layout.main);
6       LabelView = (TextView)findViewById(R.id.label);
7       registerForContextMenu(LabelView);
8   }
```

下方代码是/src/layout/main.xml 文件的部分内容,第 1 行声明了 TextView 的 ID 为 label,在上方代码的第 6 行中,通过 R.id.label 将 ID 传递给 findViewById()函数,这样用户便能够引用该界面元素,并能够修改该界面元素的显示内容。

```
1   <TextView   android:id = "@+id/label"
2       android:layout_width = "fill_parent"
3       android:layout_height = "fill_parent"
4       android:text = "@string/hello"
5   />
```

还有一点需要注意,上方代码的第 2 行,将 android:layout_width 设置为 fill_parent,这样 TextView 将填满父结点所有剩余屏幕空间,用户单击屏幕 TextView 下方任何位置都可以启动快捷菜单。如果将 android:layout_width 设置为 wrap_content,则用户必须准确单击 TextView 才能启动快捷菜单。

习题

1. Android 开发中,界面布局包括哪几种,各种的特点是什么。
2. 简述 Android 定义用户界面的两种方式,不同之处是什么。
3. Android 程序设计中用 XML 文档定义布局有何优点。
4. 简述 Android 中常见的菜单处理有哪几种方式,各自的特点是什么。

第 5 章

Android 生命周期

Android 应用程序组件中有一个生命周期，贯穿于创建到结束的整个周期。周期里面含有各种状态，这些状态对组件的生命周期起着至关重要的影响。深入理解 Android 系统管理生命周期，以 Activity 为例说明 Android 系统如何管理程序组件的生命周期。

本章主要学习内容：
- 了解 Android 系统的进程优先级的变化方式；
- 了解 Android 系统的四大基本组件；
- 了解 Activity 的生命周期中各状态的变化关系；
- 掌握 Activity 事件回调函数的作用和调用顺序；
- 掌握 Android 应用程序的调试方法和工具。

5.1 Android 应用程序组件

应用程序组件是 Android 应用的主要组成。每个组件是应用不同的入口点。不是所有的组件都是给用户的入口点，有些组件会依赖于其他组件，但是，每个组件有自己的实体，并扮演一个特定的角色。Android 系统有 4 个重要的组件，分别是 Activity、Service、Broadcast receiver 和 Content provider。

Activity 是 Android 程序的呈现层，显示可视化的用户界面，并接收与用户交互所产生的界面事件，一个 Activity 表示一个单一的用户界面，一个应用程序是由一个或者多个 Activity 组成。例如，一个邮件应用使用一个 Activity 来显示新邮件列表，使用另一个 Activity 来撰写邮件，其他 Activity 来阅读邮件。虽然这些 Activity 一起才构成这个邮件应用，但是每一个 Activity 与其他的是独立的。同样，其他应用也能够启动这些 Activity。例如，一个照相机的应用可以启动一个撰写邮件的 Activity 来分享照片。一个 Activity 的实现必须是 Activity 类的子类。

Services 是一个后台运行的并且执行长时间操作（或者执行远程访问）的组件。服务不需要提供用户界面。例如，当用户在使用其他应用的时候，一个服务可以在后台播放音乐；一个服务能够在后台获取网络数据而不打断用户操作。另外的组件，如 Activity，能够启动一个服务或者绑定一个服务（和服务交互）。一个服务的实现必须是 Service 类的子类。

Broadcast receiver 是广播接收器，用于响应系统范围内广播消息的组件。许多广播消息来源于系统，例如，屏幕关闭的通知、电量低的通知、图片截取的通知等。应用程序也能够发出广播,例如,通知其他应用数据已经下载完成可以使用。虽然,广播接收器不用显示界面,它们一般会在系统状态栏中创建一个通知图标。一般地,一个广播接收器仅是一个到其他组件的 gateway,且只执行很少量的工作,例如,它可以创建一个服务来执行更多的工作。一个广播接收器的实现必须是 BroadcastReceiver 类的子类。

Content providers 是内容提供者,一个内容提供者管理了一个共享的数据集合。这些数据可以保存在文件系统、SQLite 数据库、网站或者其他存储介质中。通过内容提供者,其他应用程序能够查询甚至修改(如果允许修改)这些数据。例如,Android 系统实现了一个内容提供者来管理联系人信息。这样,经过授权的应用可以通过这个内容提供者来读取或者修改特定联系人的信息。内容提供者对于读取和写入非共享数据也很有用。例如,NotePad 的实例程序使用一个内容提供者来保存笔记。一个内容提供者的实现必须是 ContentProvider 的子类,并且必须实现一个标准的 API 集合(实现事务)。

Android 系统设计的一个独特的方面是:任何应用都能启动其他应用的组件。例如,如果需要使用照相机拍摄一张照片,应用程序不需要实现自己实现一个拍照的 Activity,而可以使用现有的照相机的组件。在编写程序的时候,甚至不需要和照相机程序进行链接,只需要简单地启动其他照相机应用对应的 Activity 就可以了。对于应用的用户来说,照相机就像当前应用的一部分一样。

当系统启动了一个组件,系统会为此应用启动一个进程并且初始化此组件所需要的类。例如,如果一个应用启动了照相机应用中拍照的 Activity,这个 Activity 是运行在照相机应用的进程中,而不是当前进程中。因此,和其他的系统的应用不一样,Android 应用没有单一的入口点,即没有 min 函数。

因为系统在一个独立的进程中执行应用程序,一个应用不能直接激活另一个应用中的某个组件,而必须通过 Android 系统。要启动其他应用中的组件,必须发送一个信息给系统,告诉系统要启动特定组件的打算,然后系统会完成启动工作。

5.2 Android 程序生命周期

Android 是一个构建在 Linux 之上的开源移动开发平台。Android 程序也如同自然界的生命体一样,有自己的生命周期,在 Android 中,多数情况下每个程序都是在各自独立的 Linux 进程中运行的。应用程序的生命周期即程序的存活时间,即在指定时间内有效,超过这个期限就没有作用了。当一个程序或其某些部分被请求时,它的进程就开始了;当这个程序没有必要再运行下去且系统需要回收这个进程的内存用于其他程序时,这个进程就结束了。可以看出,Android 程序的生命周期是由系统控制而非程序自身直接控制。

Android 系统一般是运行在资源受限的硬件平台上,因此资源管理对 Android 系统至关重要。Android 系统主动管理资源,为了保证高优先级程序正常运行,可以在无任何警告的情况下终止低优先级程序,并回收其使用的系统资源。因此 Android 程序并不能控制自身的生命周期,而完全是由 Android 系统进行调度和控制的。

Android 系统尽可能地不主动终止应用程序,即使生命周期结束的程序也会保存在内存中,以便再次快速启动。但在内存紧张时,系统会根据进程的优先级清除进程,回收系统资源。Android 系统中的进程优先级如图 5-1 所示，优先级由高到低分别为前台进程、可见进程、服务进程、后台进程和空进程。

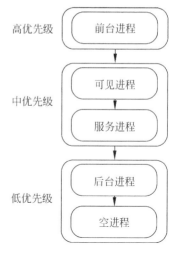

图 5-1　进程优先级

1. 前台进程

前台进程与用户正在做的事情密切相关，是 Android 系统中最重要的进程，前台进程拥有一个在屏幕上显示并和用户交互的 Activity 正在运行。这样的程序重要性最高，只有在系统内存非常低，万不得已时才会被结束。不同的应用程序组件能够通过不同的方法将它的主进程移到前台，包含以下 4 种情况，当满足：

- 进程正在屏幕的最前端运行一个与用户交互的活动 Activity；
- 进程服务被 Activity 调用，而且这个 Activity 正在与用户进行交互；
- 进程有一个 Service，并且在服务的某个回调函数，如 onCreate()、onStart() 或者 onDestroy()，内有正在执行的代码中的任一条件时，系统将把进程移到前台；
- 进程的 BroadcastReceiver 正在执行 onReceive() 函数。

Android 系统在多个前台进程同时运行时，可能会出现资源不足的情况，此时会清除部分前台进程，保证主要的用户界面能够及时响应。

2. 可见进程

可见进程是在屏幕上显示，但不在前台的程序。例如一个前台进程以对话框的形式显示在该进程前面。这样的进程也很重要，它们只有在系统没有足够内存运行所有前台进程时，才会被结束。例如，如果前台的活动是一个对话框，当前的活动隐藏在对话框之后时，就会出现这种进程。可见进程非常重要，一般不允许被终止，除非是为了保证前台进程的运行而不得不终止它。

3. 服务进程

服务进程在后台持续运行，虽然用户无法直接看到这些进程，但它们做的事情却是用户关心的。因此，系统将直接运行这些进程，除非内存不足以维持所有的前台进程和可见进程。例如后台音乐播放、后台数据上传下载等。这样的进程对用户来说一般很有用，所以只有当系统没有足够内存来维持所有的前台和可见进程时，才会被结束。

4. 后台进程

当前用户看不到的活动 Acitivity。这些进程对用户体验没有直接的影响。如果它们正确执行了活动生命周期，系统可以在任意时刻终止该进程以回收内存，并提供给前面 3 种类型的进程使用。例如一个仅有 Activity 组件的进程，当用户启动了其他应用程序时这个进程的 Activity 完全被遮挡，则这个进程便成为了后台进程。一般情况下，Android 系统中存在数量较多的后台进程,在系统资源紧张时,系统将优先清除用户较长时间没有见到的后台进程。

5. 空进程

不拥有任何活动的应用程序组件的进程叫作空进程。保留这种进程的唯一原因是在下次应用程序的某个组件需要运行时,不需要重新创建进程,这样可以提高启动速度。

在 Android 中,进程的优先级取决于所有组件中的优先级最高的部分。例如,在进程中同时包含部分可见的 Activity 和已经启动的服务,则该进程是可见进程,而不是服务进程。另外,进程的优先级会根据与其他进程的依赖关系而变化。例如,进程 A 的服务被进程 B 调用,如果调用前进程 A 是服务进程,进程 B 是前台进程,则调用后进程 A 也具有前台进程的优先级。

5.3 Activity 生命周期

了解 Activity 的生命周期的根本目的就是为了设计用户体验更加良好的应用。因为 Activity 就相当于 MVC 中的 View 层,是为了更好地向用户展现数据,并与之交互。一个 Activity 的生命周期直接影响与它结合的其他 Activitys 和它的任务返回堆栈。对于开发一个强大和灵活的应用程序,实现 Activity 的回调方法来管理 Activity 的生命周期至关重要。Activity 生命周期指 Activity 从启动到销毁的过程。Activity 表现为 4 种状态,分别是活动状态、暂停状态、停止状态和非活动状态。

1)活动状态

此时 Activity 程序显示在屏幕前台,并且具有焦点,可以与用户的操作进行交互,如向用户提供信息、捕获用户单击按钮的事件并做处理。Activity 在用户界面中处于最上层,完全能被用户看到,能够与用户进行交互,则这个 Activity 处于活动状态。

2)暂停状态

此时 Activity 程序失去了焦点,并被其他处于运行态的 otherActivity 取代在屏幕显示,但 otherActivity 程序并没有覆盖整个屏幕或者具有半透明的效果,此状态即为暂停态。处于暂停态的 Activity 仍然对用户可见,并且是完全存活的。如果系统处于内存不足的情况下,会杀死这个 Activity。当 Activity 在界面上被部分遮挡,该 Activity 不再处于用户界面的最上层,且不能够与用户进行交互,则这个 Activity 处于暂停状态。

3)停止状态

当 Activity 完全被另一个 otherActivity 覆盖时(此时 otherActivity 显示在屏幕前台),则处于停止态。处于停滞态的 Activity 依然是存活的(此时 Activity 对象依然存留在内存里,保留着所有的状态和与成员信息,但没有与窗口管理器保持连接),而且它对用户是不可见的,如果其他地方需要内存,系统会销毁这个 Activity。当 Activity 在界面上完全不能被用户看到,也就是说这个 Activity 被其他 Activity 全部遮挡,则这个 Activity 处于停止状态。

4)非活动状态

活动状态、暂停状态和停止状态是 Activity 的主要状态,不在以上 3 种状态下 Activity

的则处于非活动状态。

Activity 的 4 种状态的变换关系如图 5-2 所示。Activity 启动后处于活动状态,此时的 Activity 位于界面的最上层,是与用户正在进行交互的组件,因此 Android 系统会努力保证处于活动状态 Activity 的资源需求,资源紧张时可终止其他状态的 Activity;如果用户启动了新的 Activity,部分遮挡了当前的 Activity,或新的 Activity 是半透明的,则当前的 Activity 转换为暂停状态,Android 系统仅在为处于活动状态的 Activity 释放资源时才终止处于暂停状态的 Activity;如果用户启动新的 Activity 完全遮挡了当前的 Activity,则当前的 Activity 转变为停止状态,停止状态的 Activity 将优先被终止;活动状态的 Activity 被用户关闭后,以及暂停状态或停止状态的 Activity 被系统终止后,Activity 便进入了非活动状态。

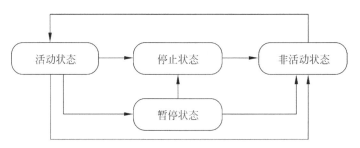

图 5-2　Activity 状态变换图

为能够更好地理解 Activity 的生命周期,还需要对 Activity 栈做一下简要介绍。Activity 栈保存了已经启动且没有终止的所有 Activity,并遵循"后进先出"的规则。如图 5-3 所示,栈顶的 Activity 处于活动状态,除栈顶以外的其他 Activity 处于暂停状态或停止状态,而被终止的 Activity 或已经出栈的 Activity 则不在栈内。

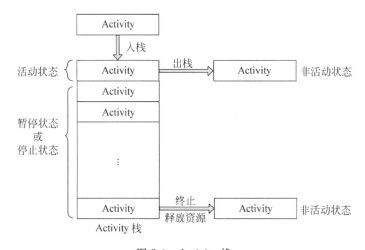

图 5-3　Activity 栈

Activity 的状态与其在 Activity 栈的位置有着密切的关系,不仅如此,Android 系统在资源不足时,也是通过 Activity 栈来选择哪些 Activity 是可以被终止的。一般来讲,Android 系统会优先选终止处于停止状态,且位置靠近栈底的 Activity,因为这些 Activity 被用户再次调用的机会最小,且在界面上用户是看不到的。

随着用户在界面进行的操作,以及 Android 系统对资源的动态管理,Activity 不断变化其在 Activity 栈的位置,状态也不断在 4 种状态中转变。随着 Activity 自身状态的变化,Android 系统会调用不同的事件回调函数,开发人员在事件回调函数中添加代码,就可以在 Activity 状态变化时完成适当的工作。

下面的代码给出了 Activity 的主要事件回调函数:

```
1   public class ExampleActivity extends Activity {
2       @Override
3       public void onCreate(Bundle savedInstanceState) {
4           super.onCreate(savedInstanceState);
5           // The activity is being created. }
6       @Override
7       protected void onStart() {
8           super.onStart();
9           // The activity is about to become visible. }
10      @Override
11      protected void onResume() {
12          super.onResume();
13          // The activity has become visible (it is now "resumed"). }
14      @Override
15      protected void onPause() {
16          super.onPause();
17          // Another activity is taking focus (this activity is about to be "paused"). }
18      @Override
19      protected void onStop() {
20          super.onStop();
21          // The activity is no longer visible (it is now "stopped")}
22      @Override
23      protected void onDestroy() {
24      super.onDestroy();
25      // The activity is about to be destroyed. }}
```

这些事件回调函数何时被调用,具体用途是什么以及是否在可以被 Android 系统终止,下面详细说明。

1) onCreate()函数

在这个回调函数中创建界面,做一些数据的初始化工作。通过调用 setContentView 函数创建界面,还可以在里面初始化各控件、设置监听和并初始化一些全局的变量。因为在 Activity 的一次生命周期中,onCreate 方法只会执行一次。在 Paused 和 Stopped 状态下恢复或重启的情况下,这些控件、监听和全局变量也不会丢失。即便是内存不足,被回收了,再次 Recreate,又是一次新的生命周期的开始,又会执行 onCreate 函数。

另外还可以在 onCreate 执行数据操作,例如从 Cursor 中检索数据等,但是如果每次进入这个 Activity 都可能需要更新数据,那么最好放在 onStart()里面。这个需要根据实际情况来确定。

2) onStart()函数

当执行 onStart()函数时变成用户可见不可交互的,即当 Activity 显示在屏幕上时,该函数被调用。

3) onRestart()函数

当Activity从停止状态进入活动状态前,调用该函数。

4) onResume()函数

当执行onResume()函数时变成和用户可交互的,在Activity栈系统通过栈的方式管理这些个Activity的最上面,运行完弹出栈,则回到上一个Activity。当Activity可以接受用户输入时,该函数被调用。此时的Activity位于Activity栈的栈顶。

5) onPause()函数

onPause和onResume中做的操作差不多,只不过是要更轻量级的,因为onPause不能阻塞转变到下一个Activity。例如停止动画、取消broadcast receivers。当Activity进入暂停状态时,该函数被调用。主要用于保存持久数据、关闭动画、释放CPU资源等。该函数中的代码必须简短,因为另一个Activity必须等待该函数执行完毕后才能显示在界面上。

6) onStop()函数

当Activity不对用户可见后,该函数被调用,Activity进入停止状态。Activity进入Stopped状态之后,它极有可能被系统所回收,在某些极端情况下,系统可能是直接杀死应用程序的进程,而不是调用onDestory函数,所以需要在onStop函数中尽可能的释放那些用户暂时不需要使用的资源,防止内存泄露。尽管onPause在onStop之前执行,但是onPause只适合做一些轻量级的操作,更多的耗时、耗资源的操作还是要放在onStop里面,例如说对数据保存,需要用到的数据库操作。

7) onDestory()函数

在Activity被终止前,即进入非活动状态前,该函数被调用。这是Activity结束前最后一个被调用函数了,可能是外面类调用finish函数,或者是系统为了节省空间将它暂时性地干掉,可以用isFinishing()来判断它。有两种情况该函数会被调用:

(1) 当程序主动调用finish()函数;

(2) 程序被Android系统终结。

除了Activity生命周期的事件回调函数以外,还有onRestoreInstanceState()和onSaveInstanceState()两个函数经常会被使用,用于保存和恢复Activity的界面临时信息,如用户在界面中输入的数据或选择的内容等,而onPause()一般被用于保存界面的持久信息。

这两个函数不属于生命周期的事件回调函数,onSaveInstanceState()在Activity被暂时停止时(被其他程序中断或锁屏)被调用,而Activity在完全关闭时(调用finishi()函数)则不会被调用。当暂停的Activity被恢复时,系统会调用onRestoreInstanceState()函数。

举个例子说明这两个函数是如何被调用的。如用户启动Activity A,然后直接又启动Activity B,这时系统需要停止Activity A,则会调用Activity A的onSaveInstanceState()来保存Activity A的界面临时信息。当用户主动关闭Activity B时,Activity B的onSaveInstanceState()不会被调用,因为是用户主动关闭Activity B而不是系统暂停的,所以当Activity A重新显示在屏幕上后,Activity A可以选择调用onRestoreInstanceState()用以恢复之前保存的Activity A的状态信息。

Activity状态保存和恢复函数onRestoreInstanceState()和onSaveInstanceState()。其中onSaveInstanceState()函数是暂停或停止Activity前调用该函数,用以保存Activity的

临时状态信息,onRestoreInstanceState()函数是恢复 onSaveInstanceState()保存的 Activity 状态信息。

onSaveInstanceState()函数会将界面临时信息保存在 Bundle 中,onCreate()函数和 onRestoreInstanceState()函数都可以恢复这些保存的信息。一般简化的作法是在 onCreate()函数中恢复保存的信息,但有些特殊的情况下还是需要使用 onRestoreInstanceState()函数恢复保存信息,如必须在界面完全初始化完毕后才能进行的操作,或需要由子类来确定是否采用默认设置等。

在 Activity 的生命周期中,并不是所有的事件回调函数都会被执行,但如果被调用则会遵循图 5-4 所示描述的调用顺序。

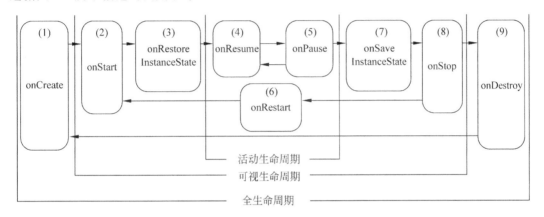

图 5-4　Activity 事件回调函数的调用顺序

从图 5-4 中可知,Activity 的生命周期可分为完全生命周期、可视生命周期和活动生命周期。每种生命周期中包含不同的事件回调函数。

完全生命周期开是从 Activity 建立到销毁的全部过程,始于 onCreate(),结束于 onDestroy()。一般情况下,使用者在 onCreate()中初始化 Activity 所能使用的全局资源和状态,并在 onDestroy()中释放这些资源。例如,Activity 中使用后台线程,则需要在 onCreate()中创建线程,在 onDestroy()中停止并销毁线程。在一些极端的情况下,Android 系统会不调用 onDestroy()函数,而直接终止进程。

可视生命周期是 Activity 在界面上从可见到不可见的过程,开始于 onStart(),结束于 onStop()。onStart()一般用于初始化或启动与更新界面相关的资源。onStop()一般用于暂停或停止一切与更新用户界面相关的线程、计时器或 Service 等,因为在调用 onStop()后,Activity 对用户不再可见,更新用户界面也就没有任何实际意义。onRestart()函数在 onSart()前被调用,用于在 Activity 从不可见变为可见的过程中,进行一些特定的处理过程。因为 Activity 不断从可见变为不可见,再从不可见变为可见,所以 onStart()和 onStop()会被多次调用。另外,onStart()和 onStop()也经常被用于注册和注销 BroadcastReceiver,例如使用者可以在 onStart()中注册一个 BroadcastReceiver,用于监视某些重要的广播消息,并使用这些消息更新用户界面中的相关内容,并可以在 onStop()中注销 BroadcastReceiver。

活动生命周期是 Activity 在屏幕的最上层,并能够与用户进行交互的阶段,开始于 onResume(),结束于 onPause()。因为在 Activity 的状态变换过程中 onResume()和

onPause()经常被调用,因此这两个函数中应使用简单、高效的代码。

从图 5-4 的 Activity 事件回调函数的调用顺序上分析,onStop()是第一个被标识为"可终止"的函数,因此在 onStop()和 onDestroy()函数的执行过程中随时可能被 Android 系统终止。因此,onPause()常用于保存持久数据,如界面上的用户的输入信息等。很多时候使用者不清楚何时该使用 onPause(),何时该使用 onSaveInstanceState(),因为两个函数都可以用于保存界面的用户输入数据。其主要区别在于两个函数保存数据的性质和方法不同,onPause()一般用于保存持久性数据,并将数据保存在存储设备上的文件系统或数据库系统中的;而 onSaveInstanceState()主要用于保存动态的状态信息,信息一般保存在 Bundle 中。Bundle 是能够保存多种格式数据的对象,在 onSaveInstanceState()保存在 Bundle 中的数据,系统在调用 onRestoreInstanceState()和 onCreate()时,会同样利用 Bundle 将数据传递给函数。

为了能够更好地理解 Activity 事件回调函数的调用顺序,下面以 Activity 生命周期示例来进行说明,Activity 生命周期示例的运行界面如图 5-5 所示。

下面给出 ActivityLifeperiod.java 文件的全部代码:

图 5-5　ActivityPeriod 示例用户界面

```
1   public class ActivityPeriodActivity extends Activity {
2   /** Called when the activity is first created. */
3   @Override
4   public void onCreate(Bundle savedInstanceState) {
5   super.onCreate(savedInstanceState);
6   setContentView(R.layout.main);
7   System.out.println("Acity--->OnCreat()");
8   Button button = (Button)findViewById(R.id.btn_finish);
9   button.setOnClickListener(new View.OnClickListener() {
10  public void onClick(View v) {
11  // TODO Auto-generated method stub
12  finish();
13  }
14  });
15  }
16  @Override
17  protected void onDestroy() {
18  // TODO Auto-generated method stub
19  System.out.println("Acity--->OnDestroy()");
20  super.onDestroy();
21  }
22  @Override
23  protected void onPause() {
24  // TODO Auto-generated method stub
25  System.out.println("Acity--->OnPause()");
26  super.onPause();
27  }
28  @Override
```

```
29  protected void onRestart() {
30  // TODO Auto-generated method stub
31  System.out.println("Acitity--->OnRestart()");
32  super.onRestart();
33  }
34  @Override
35  protected void onResume() {
36  // TODO Auto-generated method stub
37  System.out.println("Acitity--->OnResume()");
38  super.onResume();
39  }
40  @Override
41  protected void onStart() {
42  // TODO Auto-generated method stub
43  System.out.println("Acitity--->OnStart()");
44  super.onStart();
45  }
46  @Override
47  protected void onStop() {
48  // TODO Auto-generated method stub
49  System.out.println("Acitity--->OnStop()");
50  super.onStop();
51  }
52  }
```

上面的程序主要通过在生命周期函数中添加"日志点"的方法进行调试，程序的运行结果将会显示在 LogCat 中。为了显示结果易于观察和分析，在 LogCat 设置过滤器 Life，过滤器的条件为"标签=LIFTCYCLE"。

1. 完全生命周期

为了观察 Activity 从启动到关闭所调用的全部生命周期函数的顺序，首先正常启动 ActivityLifeCycle，然后单击用户界面的"结束程序"按钮关闭程序。LogCat 的输出结果如图 5-6 所示。

```
09-23 04:41:09.820    552   hlju.edu.ActivityPeriod    System.out    Acitity--->OnCreat()
09-23 04:41:09.880    552   hlju.edu.ActivityPeriod    System.out    Acitity--->OnStart()
09-23 04:41:09.880    552   hlju.edu.ActivityPeriod    System.out    Acitity--->OnResume()
09-23 04:41:16.990    552   hlju.edu.ActivityPeriod    System.out    Acitity--->OnPause()
09-23 04:41:17.740    552   hlju.edu.ActivityPeriod    System.out    Acitity--->OnStop()
09-23 04:41:17.740    552   hlju.edu.ActivityPeriod    System.out    Acitity--->OnDestroy()
```

图 5-6 完全生命周期的 LogCat 输出

从图 5-6 可以得知，函数调用顺序如下：(1)onCreate→(2)onStart→(4)onResume→(7)onPause→(8)onStop→(9)onDestroy。

在 Activity 启动时，系统首先调用 onCreate()函数分配资源，然后调用 onStart()将 Activity 显示在屏幕上，之后调用 onResume()获取屏幕焦点，使 Activity 能够接受用户的输入，这时用户就能够正常使用这个 Android 程序了。

用户单击"结束程序"按钮，会导致 Activity 关闭，系统会相继调用 onPause()、onStop() 和 onDestroy()，释放资源并销毁进程。因为 Activity 关闭后，除非用户重新启动应用程序，否则这个 Activity 不会在出现在屏幕上，因此系统直接调用 onDestroy() 销毁了进程，且没有调用 onSaveInstanceState() 函数来保存 Acitivity 状态。

2．可视生命周期

在 Activity 启动后，如果启动其他的程序，原有的 Activity 会被新启动程序的 Activity 完全遮挡，因此原有 Activity 会进入停止状态。如果将新启动的程序关闭，则原有 Activity 从停止状态恢复到活动状态。

为了能够分析上述状态转换过程中的函数调用顺序，首先正常启动 ActivityLifeCycle，然后通过"拨号键"启动内置的拨号程序，再通过"回退键"退出拨号程序，使 ActivityLifeCycle 重新显示在屏幕中。LogCat 的输出结果如图 5-7 所示。

```
9-23 04:43:41.241    552    hlju.edu.ActivityPeriod    System.out    Acitity--->OnStart()
9-23 04:43:41.241    552    hlju.edu.ActivityPeriod    System.out    Acitity--->OnResume()
9-23 04:43:52.715    552    hlju.edu.ActivityPeriod    System.out    Acitity--->OnPause()
9-23 04:43:54.470    552    hlju.edu.ActivityPeriod    System.out    Acitity--->OnStop()
9-23 04:43:57.741    552    hlju.edu.ActivityPeriod    System.out    Acitity--->OnRestart()
9-23 04:43:57.741    552    hlju.edu.ActivityPeriod    System.out    Acitity--->OnStart()
9-23 04:43:57.750    552    hlju.edu.ActivityPeriod    System.out    Acitity--->OnResume()
```

图 5-7 可视生命周期的 LogCat 输出

从图 5-7 可以得知，函数调用顺序如下：(1) onCreate→(2) onStart→(4) onResume→(7) onPause→(5) onSaveInstanceState→(8) onStop→(6) onRestart→(2) onStart→(4) onResume。

Activity 启动时的函数调用顺序仍为 (1)→(2)→(4)，当内置的拨号程序被启动时，原有的 Activity 被完全覆盖，系统首先调用 onPause() 函数，然后调用 onSaveInstanceState() 函数保存 Activity 状态；最后调用 onStop()，停止对不可见 Activity 的更新。

在用户关闭拨号程序后，系统调用 onRestart() 恢复界面上需要更新的信息，然后调用 onStart() 和 onResume() 重新显示 Activity，并接受用户交互。

Android 系统虽然调用了 onSaveInstanceState() 保存 Activity 的状态，但因为 Activity 并没有被销毁，所以没有必要调用 onRestoreInstanceState() 恢复保存的 Activity 状态。

如果用户在 Dev Tools→Development Settings→Immediately destroy activities 下开启 IDA，如图 5-8 所示，被其他程序遮挡的 Activity 会被立即终止，这样被遮挡的 Activity 重新显示在屏幕上时，系统会调用 onRestoreInstanceState() 恢复 Activity 销毁前的状态。

IDA 未开启前，用户单击"回退"按钮后的函数调用顺序是 (6)→(2)→(4)，开启 IDA 后的函数调用顺序是 (1)→(2)→(3)→(4)。由此可见开启 IDA 导致 Android 系统在用户打开其他程序时销毁了原来已经打开的

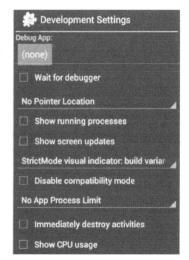

图 5-8 开启 IDA 选项

Activity，这个被销毁的 Activity 重新出现在用户的屏幕上前，系统额外调用了 onCreate() 和 onRestoreInstanceState() 函数，用以恢复 Activity 在销毁前所保存的数据。

5.4 程序调试

在 Android 程序开发过程中，出现错误（Bug）是不可避免的事情。一般情况下，语法错误会被集成开发环境检测到，并提示使用者错误的位置以及修改方法。但逻辑错误就很难发现了，通常只有将程序在模拟器或硬件设备上运行才能够发现。逻辑错误的定位和分析是件困难的事情，尤其程序是代码量较大且结构复杂的应用程序，仅凭直觉很难快速找到并解决问题。因此，Android 系统提供了几种调试工具，用于定位、分析及修复程序中出现的错误，这些工具包括 LogCat 和 DevTools。

5.4.1 LogCat

LogCat 是用于获取系统日志信息的工具，并可以显示在 Eclipse 集成开发环境中。LogCat 能够捕获的信息包括 Dalvik 虚拟机产生的信息、进程信息、ActivityManager 信息、PackagerManager 信息、Homeloader 信息、WindowsManager 信息、Android 运行时信息和应用程序信息等。在 Eclipse 的默认开发模式下没有 LogCat 的显示页，用户可以使用 Window→Show View→Other 打开 Show View 的选择菜单，然后在 Andoird→LogCat 中选择 LogCat，如图 5-9 所示。

这样，LogCat 便显示在 Eclipse 的下方区域，如图 5-10 所示。LogCat 的右上方的 5 个字母 [V]、[D]、[I]、[W] 和 [E]，表示 5 种不同类型的日志信息，分别是详细（Verbose）信息、调试（Debug）信息、通告（Info）信息、警告（Warn）和错误（Error）信息调试（Debug）信息。不同类型日志信息的级别是不相同的，级别最高的是错误信息，其次是警告信息，然后是通知信息和调试信息，级别最低的是详细信息。在 LogCat 中，用户可以通过 5 个字

图 5-9 在 Show View 中选择 LogCat

母图标选择显示的信息类型，同时级别比选择类型高的信息也可以在 LogCat 中显示，但级别低于选定的信息则会被忽略掉。

即使用户指定了所显示日志信息的级别，仍然会产生很多日志信息，很容易让用户不知所措。LogCat 还提供了"过滤"功能，在右上角的"＋"号和"－"号，分别是添加和删除过滤器。用户可以根据日志信息的标签（Tag）、产生日志的进程编号（Pid）或信息等级（Level），对显示的日志内容进行过滤。

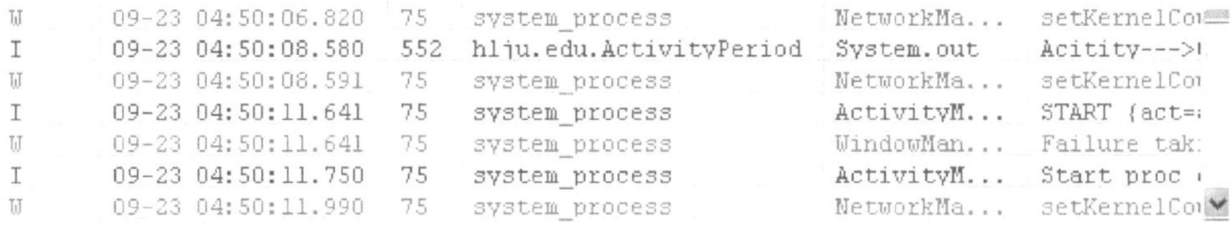

```
W    09-23 04:50:06.820    75     system_process              NetworkMa...  setKernelCo
I    09-23 04:50:08.580    552    hlju.edu.ActivityPeriod     System.out    Acitity--->
W    09-23 04:50:08.591    75     system_process              NetworkMa...  setKernelCo
I    09-23 04:50:11.641    75     system_process              ActivityM...  START {act=
W    09-23 04:50:11.641    75     system_process              WindowMan...  Failure tak
I    09-23 04:50:11.750    75     system_process              ActivityM...  Start proc
W    09-23 04:50:11.990    75     system_process              NetworkMa...  setKernelCo
```

图 5-10　clipse 中的 LogCat

在 Android 程序调试过程中，首先需要引入 android.util.Log 包，然后使用 Log.v()、Log.d()、Log.i()、Log.w()和 Log.e() 5 个函数在程序中设置"日志点"。每当程序运行到"日志点"时，应用程序的日志信息便被发送到 LogCat 中，使用者可以根据"日志点"信息是否与预期的内容一致，判断程序是否存在错误。之所以使用 5 个不同的函数产生日志，主要是为了区分日志信息的类型，其中，Log.v()用于记录详细信息，Log.d()用于记录调试信息，Log.i()用于记录通告信息，Log.w()用于记录警告信息，Log.e()用于记录错误信息。

在下面的程序中，演示了 Log 类的具体使用方法。

```
1   package edu.hrbeu.LogCat;
2
3   import android.app.Activity;
4   import android.os.Bundle;
5   import android.util.Log;
6
7   public class LogCatActivity extends Activity {
8       final static String TAG = "LOGCAT";
9       @Override
10      public void onCreate(Bundle savedInstanceState) {
11          super.onCreate(savedInstanceState);
12          setContentView(R.layout.main);
13
14          Log.v(TAG,"Verbose");
15          Log.d(TAG,"Debug");
16          Log.i(TAG,"Info");
17          Log.w(TAG,"Warn");
18          Log.e(TAG,"Error");
19      }
20  }
```

为了使用 Log 类中的函数，首先在程序第 5 行引入 android.util.Log 包；然后在第 8 行定义标签，标签帮助用户在 LogCat 中找到目标程序生成的日志信息，同时也能够利用标签对日志进行过滤；第 14 行记录一个详细信息，Log.v()函数的第 1 个参数是日志的标签，第 2 个参数是实际的信息内容；第 15 行到第 18 行分别产生了调试信息、通告信息、警告信息和错误信息。

程序运行后，LogCat 捕获得到应用程序发送的日志信息，显示结果如图 5-11 所示。在 LogCat 中显示了标签为"LOGCAT"的日志信息共 5 条，并以不同颜色加以显示。可见，LogCat 对不同类型的信息使用了不同的颜色加以区别。

V	09-23 05:07:24.020	1176	hlju.edu.LogcatTest	LOGCAT	Verbose
D	09-23 05:07:24.020	1176	hlju.edu.LogcatTest	LOGCAT	Debug
I	09-23 05:07:24.020	1176	hlju.edu.LogcatTest	LOGCAT	Info
W	09-23 05:07:24.020	1176	hlju.edu.LogcatTest	LOGCAT	Warn
E	09-23 05:07:24.020	1176	hlju.edu.LogcatTest	LOGCAT	Error
D	09-23 05:07:24.130	1176	hlju.edu.LogcatTest	gralloc_q...	Emulator without GPU

图 5-11　LogCat 工程的运行结果

如果能够使用 LogCat 的过滤器,则可以使显示的结果更加清晰。下面使用在 LogCat 右侧的"＋"号,添加一个名为 LogcatFilter 的过滤器,并设置过滤条件为"标签＝LOGCAT",具体设置方法如图 5-12 所示。

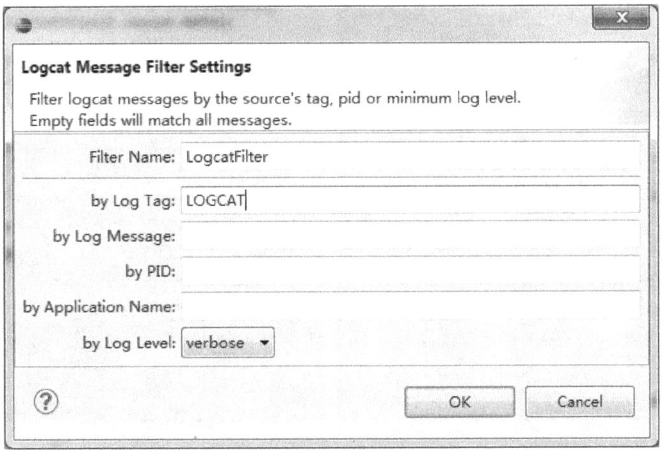

图 5-12　LogCat 过滤器

过滤器设置好后,LogcatFilter 过滤后的日志信息如图 5-13 所示。在这之后,无论什么类型的日志信息,属于哪一个进程,只要标签为 LOGCAT,都将显示在 LogcatFilter 区域内。

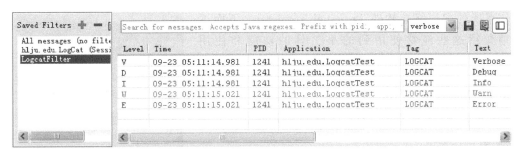

图 5-13　LogCat 过滤后的输入结果

5.4.2　DevTools

在 Android 模拟器中,内置了一个用于调试和测试的工具 DevTools。DevTools 包括了一系列用户各种用途的小工具,包括 AccountsTester、Bad Behavior、Configuration、Connectivity、Development Settings、Google Login Service、Instrumentation、Media Scanner、

Package Browser、Pointer Location、Running processes、Sync Tester 和 Terminal Emulator。从模拟器的应用程序列表中可以找到启动 DevTools 的图标，启动 DevTools 后的显示界面如图 5-14 所示。

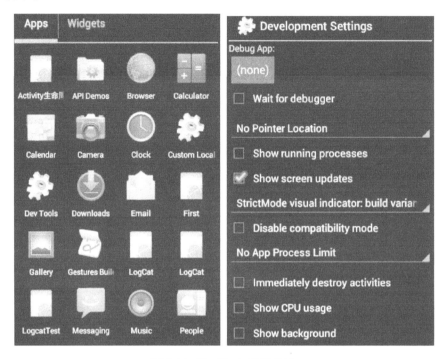

图 5-14　DevTools 使用界面

在这些工具里，经常用到的有设置调试选项的 Development Settings，查看已经安装程序包的 Package Browser，确定触摸点位置的 Pointer Location，查看当前运行进程的 Running processes，还有连接底层 Linux 操作系统的虚拟终端软件 Terminal Emulator。下面逐一地介绍这些经常使用到的小工具的功能和使用方法。

1. Development Settings

Development Settings 中包含了程序调试的相关选项，如图 5-15 所示。

如果希望启动 Development Settings 中某项功能，只需要单击功能前面选择框，出现绿色的"对号"表示功能启用。功能启用后，模拟器会自动保存设置，即使再次启动模拟器用户的选择内容仍会存在。Development Settings 每一项的具体说明如表 5-1 所示。

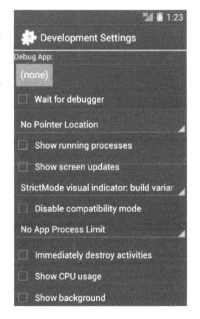

图 5-15　Development Settings

表 5-1　Development Settings 选项

选项	说明
Debug App	为 Wait for debugger 选项指定应用程序,如果不指定(选择 none), Wait for debugger 选项将适用于所有应用程序。Debug App 可以有效地防止 Android 程序长时间停留在断点而产生异常
Wait for debugger	阻塞加载应用程序,直到关联到调试器(Debugger)。用于在 Activity 的 onCreate()函数进行断点调试
Show running processes	在屏幕右上角显示运行中的进程
Show screen updates	选中该选项时,界面上任何被重绘的矩形区域都会闪现粉红色,有利于发现界面中不必要的重绘区域
No App Process limit	允许同时运行进程的数量上限
Immediately destroy activities	Activity 进入停止状态后立即销毁,用于测试在函数 onSaveInstanceState()、onRestoreInstanceState()和 onCreate()中的代码
Show CPU usage	在屏幕顶端显示 CPU 使用率,上层红线显示总的 CPU 使用率,下层绿线显示当前进程的 CPU 使用率
Show background	应用程序没有 Activity 显示时,直接显示背景面板,一般这种情况仅在调试时出现
Show sleep state on LED	在休眠状态下开启 LED
Windows Animation Scale	窗口动画模式
Transition Animation Scale	渐变动画模式
Light Hinting	提示模式
Show GTalk service connection status	显示 GTalk 服务连接状态

2. Package Browser

Package Browser 是 Android 系统中的程序包查看工具,能够详细显示已经安装到 Android 系统中的程序信息,包括包名称、应用程序名称、图标、进程、用户 ID、版本、.apk 文件保存位置和数据文件保存位置等,而且能够进一步查看应用程序所包含 Activity、Service、BroadcastReceiver 和 Provider 的详细信息。图 5-16 所示是在 Package Browser 中查看 Android keyboard 程序的相关信息。

3. Pointer Location

Pointer Location 是屏幕点位置查看工具,能够显示触摸点的 x 轴坐标和 y 轴坐标。图 5-17 所示是 Pointer Location 的使用界面。

4. Running processes

Running processes 能够查看在 Android 系统中正在运行的进程,并能查看进程的详细信息,包括进程名称和进程所调用的程序包。图 5-18 所示是 Android 模拟器所运行进程的列表和 com.android.phone 进程的详细信息。

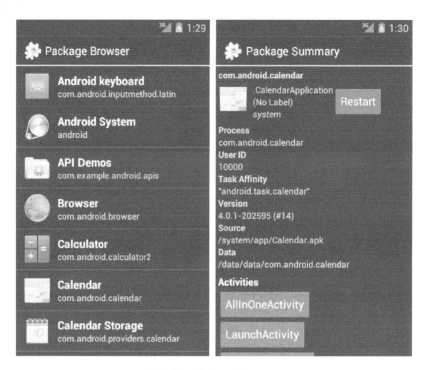

图 5-16　Package Browser

图 5-17　Pointer Location

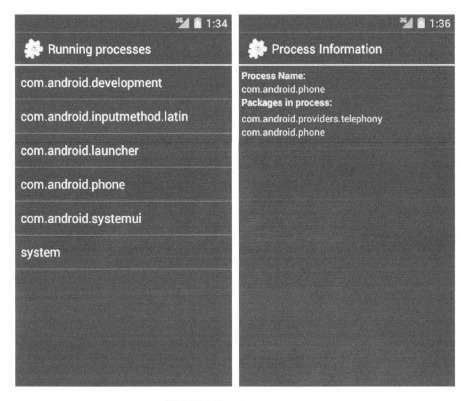

图 5-18 Running processes

5. Connectivity

Connectivity 允许用户控制 Wi-Fi、屏幕锁定界面、MMS 和导航的开启与关闭，并可以设置 Wi-Fi 和屏幕锁定界面的开启与关闭的周期。其中 Enable Wifi 和 Disable Wifi 分别是控制了 Wi-Fi 的开启和关闭，Start Wifi Toggle 和 Stop Wifi Toggle 是 Wi-Fi 周期性开启和关闭的开关，Cycles done 后面的数值记录了 Wi-Fi 开启和关闭的次数。图 5-19 所示是 Connectivity 的运行界面。

6. Configuration

Configuration 中详细列出了 Android 系统的配置信息，包括屏幕解析度、字体缩放比例、屏幕初始方向、触屏类型、导航、本地语言和键盘等信息，如图 5-20 所示。

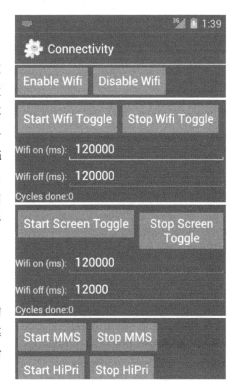

图 5-19 Connectivity

7. Bad Behavior

Bad Behavior 中可以模拟各种程序崩溃和失去响应的情况，如主程序崩溃、系统服务崩溃、启动 Service 时失去响应和启动 Activity 时失去响应等。Bad Behavior 界面如图 5-21 所示。

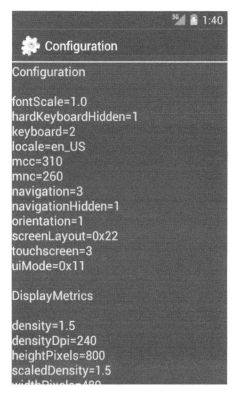

图 5-20　Configuration

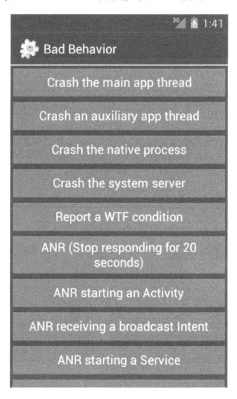

图 5-21　Bad Behavior

表 5-2 列出了 Bad Behavior 中所有的可以模拟的事件，并对每个事件给出了简要的说明。

表 5-2　Bad Behavior 选项

选　　项	说　　明
Crash the main app thread	应用程序主线程崩溃
Crash an auxiliary app thead	应用程序工作线程崩溃
Crash the native process	本地进程崩溃
Crash the system server	系统服务器崩溃
Report a WTF condition	报告 WTF
ANR(Stop responding for 20 seconds)	应用程序无响应（Application Not Responding，ANR）20 秒
ANR starting an Activity	启动 Activity 时应用程序无响应
ANR starting a broadcast Intent	启动 Intent 时应用程序无响应
ANR starting a Service	启动 Service 时应用程序无响应
System ANR (in ActivityManager)	Activity 管理器级别 ANR
Wedge system (5 minute system ANR)	Wedge 在 5 分钟内无响应

习题

1. Android 根据应用程序的组件以及组件当前运行状态将所有的进程按重要性程度从高到低划分为几个。

2. 简述 Android 开发中常用的四大基本组件是什么，各自有哪些作用。

3. 举例说明 Activity 生命周期中下列回调方法的调用时刻 onCreate()、onStart()、onResume()、onPause()、onStop()和 onDestroy()各自的作用。

第6章 Android组件之间的通信

Android应用程序组件之间通信的核心机制是Intent，Intent是一种消息传递机制，用于组件之间数据交换和发送广播消息。通过Intent可以开启一个Activity或Service等其他组件。

本章主要学习内容：
- 掌握Intent的各种属性；
- 掌握系统标准ActivityAction应用；
- 掌握Intent过滤器；
- 了解广播消息机制。

6.1 Intent简介

Intent是一种轻量级的消息传递机制，可以在同一个应用程序内部的不同组件之间传递信息，也可以在不同应用程序的组件之间传递信息，还可以作为广播事件发布Android系统消息。由于Intent的存在，使得Android系统中互相独立的组件成为了一个可以互相通信的组件集合。因此，无论这些组件是否在同一个应用程序中，Intent都可以将一个组件的数据或动作传递给另一个组件。

Intent是一个动作的完整描述，包含了动作的产生组件、接收组件和传递的数据信息。Android则根据Intent的描述，在不同组件间传递消息，负责找到对应的组件，将Intent传递给调用的组件，组件接收到传递的消息，执行相关动作，完成组件的调用。

Intent不仅可用于应用程序之间，也可用于应用程序内部的Activity/Service之间的交互。Intent为Activity、Service和BroadcastReceiver等组件提供交互能力，还可以启动Activity和Service，在Android系统上发布广播消息。这里的广播消息是指可以接收到的特定数据或消息，也可以是手机的信号变化或电池的电量过低等信息。

因此，Intent在这里起着一个媒体中介的作用，专门提供组件互相调用的相关信息，实现调用者与被调用者之间的解耦。在Android SDK中给出了Intent作用的表现形式。

（1）通过Context.startActivity()or Activity.startActivityForResult()启动一个Activity。

（2）通过Context.startService()启动一个服务，或者通过Context.bindService()和后台服务交互。

(3) 通过广播方法(例如 Context.sendBroadcast(),Context.sendOrderedBroadcast(),Context.sendStickyBroadcast())发给 broadcast receivers。

一般情况下,Intent 对某操作的抽象描述包含下面几个部分。

- 对执行动作的描述:操作(action)。
- 对这次动作相关联的数据进行描述:数据(data)。
- 对数据类型的描述:数据类型(type)。
- 对执行动作的附加信息进行描述:类别(category)。
- 其他一切附加信息的描述:附件信息(extras)。
- 对目标组件的描述:目标组件(component)。

6.1.1 Intent 的 action 属性

action 是要执行的动作,也可以是在广播 Intent 中已发生且正被报告的动作。action 部分是一个字符串对象。它描述了 Intent 会触发的动作。Android 系统中已经预定义了一些 action 常量,可以参看 SDK 帮助文档,表 6-1 给出了一些标准的 action 常量。

表 6-1　Intent 标准动作说明

常量	目标组件	描述
ACTION_CALL	activity	初始化一个电话呼叫
ACTION_EDIT	activity	显示可供用户编辑的数据
ACTION_MAIN	activity	将该 Activity 作为 task 的第一个 Activity,没有数据输入,也没有数据返回
ACTION_SYNC	activity	使服务器上的数据与移动设备上数据同步
ACTION_BATTERY_LOW	broadcast receiver	提示电池电量低
ACTION_HEADSET_PLUG	broadcast receiver	提示耳机塞入或拔出
ACTION_SCREEN_ON	broadcast receiver	屏幕已点亮
ACTION_TIMEZONE_CHANGED	broadcast receiver	时区设置改变

除上表介绍的 action 常量外,开发者也可以定义自己的 action 描述。一般来讲,定义自己的 action 字符串应该以应用程序的包名为前缀(防止重复定义)。由于 action 部分很大程度上决定了一个 Intent 的内容,特别是数据(data)和附加(extras)字段,就像一个方法名决定了参数和返回值。正是这个原因,应该尽可能地明确指定动作,并紧密关联到其他 Intent 字段。即应该定义组件能够处理的 Intent 对象的整个协议,而不仅仅是单独地定义一个动作。一个 Intent 对象的动作通过 setAction()方法设置,通过 getAction()方法读取。

6.1.2 Intent 的 data 属性

data,即执行动作要操作的数据。

data 描述了 Intent 的动作所能操作数据的 MIME 类型和 URL,不同的 Action 用不同的操作数据。例如,如果 Activity 字段是 ACTION_EDIT,data 字段将显示包含用于编辑的文档的 URI;如果 Activity 是 ACTION_CALL,data 字段是一个 tel://URI 和将拨打的号码；如果 Activity 是 ACTION_VIEW，data 字段是一个 http://URI，接收活动将被调用去下载和显示 URI 指向的数据。在许多情况下，数据类型能够从 URI 中推测出来，特别是 content://URIs，它表示位于设备上的数据被内容提供者(Content Provider)控制。但是类型也能够显示设置，setData()方法指定数据的 URI，setType()指定 MIME 类型，setDataAndType()指定数据的 URI 和 MIME 类型。通过 getData()读取 URI，getType()读取类型。

匹配一个 Intent 到一个能够处理 data 的组件，知道 data 的类型(它的 MIME 类型)和它的 URI 很重要。例如，一个组件能够显示图像数据就不应该被调用去播放音频文件。

6.1.3 Intent 的 type 属性

数据类型(type)，显式指定 Intent 的数据类型(MIME)。一般 Intent 的数据类型能够根据数据本身进行判定，但是通过设置这个属性，可以强制采用显式指定的类型而不再进行推导。

6.1.4 Intent 的 category 属性

category(类别)，被执行动作的附加信息。例如，LAUNCHER_CATEGORY 表示 Intent 的接受者应该在 Launcher 中作为顶级应用出现；而 ALTERNATIVE_CATEGORY 表示当前的 Intent 是一系列的可选动作中的一个，这些动作可以在同一块数据上执行。其他的如表 6-2 所示。

表 6-2 category 属性说明

常量	描述
CATEGORY_BROWSABLE	目标 Activity 可通过浏览器安全启动以显示一个链接相关的数据，如图片或邮件信息
CATEGORY_GADGET	Activity 可被嵌入另外一个拥有 gadget 的 Activity 中
CATEGORY_HOME	Activity 显示主页，即设备打开时用户看到的第一个界面或是用户按 Home 键时的界面
CATEGORY_LAUNCHER	Activity 是一个 task 的初始 Activity，是程序启动的高优先级 Activity
CATEGORY_PREFERENCE	目标 Activity 为 preference panel

通过 addCategory()方法添加一个种类到 Intent 对象中；通过 removeCategory()方法删除一个之前添加的种类；通过 getCategories()方法获取 Intent 对象中的所有种类。

6.1.5 Intent 的 extras 属性

extras(附加信息)是一组键值对，包含了需要传递给目标组件并有其处理的一些附加信息。就像动作关联的特定种类的数据 URIs，也关联到某些特定的附加信息。例如，一个 ACTION_TIMEZONE_CHANGE intent 有一个 time-zone 的附加信息，标识新的时区，ACTION_HEADSET_PLUG 有一个 state 附加信息，标识头部现在是否塞满或未塞满；有一个 name 附加信息，标识头部的类型。如果自定义了一个 SHOW_COLOR 动作，颜色值将可以设置在附加的键值对中。例如,如果要执行"发送电子邮件"这个动作,可以将电子邮件的标题、正文等保存在 extras 里,传给电子邮件发送组件。

Intent 有一系列 putXXX()方法用于插入各种附加数据,有一系列 getXXX()方法可以取出一系列数据。使用 Extras 可以为组件提供扩展信息。

6.1.6 Intent 的 component 属性

ComponentName(组件),指定 Intent 的目标组件的类名称。ComponentName 包含两个 String 成员,分别代表组件的全称类名和包名,包名必须和 AndroidManifest.xml 文件标记中的对应信息一致。ComponentName 通过 setComponent()、setClass()或 setClassName()设置,通过 getComponent()读取。

通常 Android 会根据 Intent 中包含的其他属性的信息(如 action、data/type、category)进行查找,最终找到一个与之匹配的目标组件。但是,如果 ComponentName 这个属性有指定,将直接使用指定的组件,而不再执行上述查找过程。指定了这个属性以后,Intent 的其他所有属性都是可选的。对于 Intent,组件名并不是必需的。如果一个 Intent 对象添加了组件名,则称该 Intent 为显式 Intent,这样的 Intent 在传递时会直接根据组件名去寻找目标组件。如果没有添加组件名,则称为隐式 Intent,Android 会根据 Intent 中的其他信息来确定响应该 Intent 的组件。

总之,action、data/type、category 和 extras 一起使系统能够理解诸如"查看某联系人的详细信息"或"给某人打电话"之类的短语。随着应用不断地加入系统中,Android 系统可以添加新的 action、data/type、category 来扩展功能。当然,最受益的还是应用本身,可以利用这套语言机制来处理不同的动作和数据。

6.2 系统标准 ActivityAction 应用

6.2.1 Activity 的启动

在 Android 系统中,应用程序一般都有多个 Activity,Intent 可以实现不同 Activity 之间的切换和数据传递。

Intent 启动 Activity 有以下两种方式。
- 显示启动,必须在 Intent 中指明启动的 Activity 所在的类。
- 隐式启动,Android 系统,根据 Intent 的动作和数据来决定启动哪一个 Activity,即在隐式启动时,Intent 中只包含需要执行的动作和所包含的数据,而无须指明具体启动哪一个 Activity,选择权由 Android 系统和最终用户来决定。

1. 显式启动

使用 Intent 来显式启动 Activity,首先需要创建一个 Intent,并为它指定当前的应用程序上下文以及要启动的 Activity,把创建好的这个 Intent 作为参数传递给 startActivity()方法。

```
1    Intent intent = new Intent(IntentDemo.this, ActivityToStart.class);
2    startActivity(intent);
```

下面用 IntentDemo 示例说明如何使用 Intent 启动新的 Activity。IntentDemo 示例包含两个 Activity,分别是 IntentDemoActivity 和 NewActivity。程序默认启动的 Activity 是 IntentDemo,在用户单击"启动 Activity"按钮后,程序启动的 Activity 是 NewActivity,如图 6-1 所示。

(a) IntentDemoActivity (b) NewActivity

图 6-1　IntentDemo 示例用户界面

在 IntentDemo 示例中使用了两个 Activity,因此需要在 AndroidManifest.xml 文件中注册这两个 Activity。注册 Activity 应使用＜activity＞标签,嵌套在＜application＞标签内部。

AndroidManifest.xml 文件代码如下:

```
1    <?xml version = "1.0" encoding = "utf-8"?>
2    <manifest xmlns:android = "http://schemas.android.com/apk/res/android"
3        package = "edu.hrbeu.IntentDemo"
4        android:versionCode = "1"
5        android:versionName = "1.0">
6        <application android:icon = "@drawable/icon" android:label = "@string/app_name">
7            <activity android:name = ".IntentDemo"
8                android:label = "@string/app_name">
9                <intent-filter>
10                   <action android:name = "android.intent.action.MAIN" />
11                   <category android:name = "android.intent.category.LAUNCHER" />
12               </intent-filter>
13           </activity>
14           <activity android:name = ".NewActivity"
15               android:label = "@string/app_name">
16           </activity>
17       </application>
18       <uses-sdk android:minSdkVersion = "14" />
19   </manifest>
```

Android 应用程序中,用户使用的每个组件都必须在 AndroidManifest.xml 文件中的 ＜application＞结点内定义。在上面的代码中,＜application＞结点下共有两个＜activity＞结点,分别代表应用程序中所使用的两个 Activity,IntentDemoActivity 和 NewActivity。

在 IntentDemoActivity.java 文件中,包含了使用 Intent 启动 Activity 的核心代码:

```
1   Button button = (Button)findViewById(R.id.btn);
2   button.setOnClickListener(new OnClickListener(){
3       public void onClick(View view){
4           Intent intent = new Intent(IntentDemoActivity.this, NewActivity.class);
5           startActivity(intent);
6       }
7   });
```

在单击事件的处理函数中,Intent 构造函数的第 1 个参数是应用程序上下文,在这里就是 IntentDemoActivity;第 2 个参数是接收 Intent 的目标组件,这里使用的是显式启动方式,直接指明了需要启动的 Activity。

2. 隐式启动

隐式启动的好处在于不需要指明需要启动哪一个 Activity,而由 Android 系统来决定,这样有利于降低组件之间的耦合度。

选择隐式启动 Activity,Android 系统会在程序运行时解析 Intent,并根据一定的规则对 Intent 和 Activity 进行匹配,使 Intent 上的动作、数据与 Activity 完全吻合。匹配的组件可以是程序本身的 Activity,也可以是 Android 系统内置的 Activity,还可以是第三方应用程序提供的 Activity。因此,这种方式强调了 Android 组件的可复用性。

例如,如果程序开发人员希望启动一个浏览器,查看指定的网页内容,却不能确定具体应该启动哪一个 Activity,此时则可以使用 Intent 的隐式启动方式,由 Android 系统在程序运行时决定具体启动哪一个应用程序的 Activity 来接收这个 Intent。程序开发人员可以将浏览动作和 Web 地址作为参数传递给 Intent,Android 系统则通过匹配动作和数据格式,找到最适合于此动作和数据格式的组件。

```
1   Intent intent = new Intent(Intent.ACTION_VIEW, Uri.parse("http://www.google.com.hk"));
2   startActivity(intent);
```

Intent 的动作是 Intent.ACTION_VIEW,数据是 Web 地址,使用 Uri.parse(urlString)方法,可以简单地把一个字符串解释成 Uri 对象。Android 系统在匹配 Intent 时,首先根据动作 Intent.ACTION_VIEW,得知需要启动具备浏览功能的 Activity,但具体是浏览电话号码还是浏览网页,还需要根据 URI 的数据类型来做最后判断。因为数据提供的是 Web 地址 http://www.google.com,所以最终可以判定 Intent 需要启动具有网页浏览功能的 Activity。在默认情况下,Android 系统会调用内置的 Web 浏览器。

Intent 的语法如下:

```
1   Intent intent = new Intent(Intent.ACTION_VIEW, Uri.parse(urlString));
```

Intent 构造函数的第 1 个参数是 Intent 需要执行的动作,Android 系统支持的常见动作字符串常量如表 6-3 所示。第 2 个参数是 URI,表示需要传递的数据。

表 6-3　Intent 常用动作

动　作	说　明
ACTION_ANSWER	打开接听电话的 Activity，默认为 Android 内置的拨号界面
ACTION_CALL	打开拨号盘界面并拨打电话，使用 Uri 中的数字部分作为电话号码
ACTION_DELETE	打开一个 Activity，对所提供的数据进行删除操作
ACTION_DIAL	打开内置拨号界面，显示 Uri 中提供的电话号码
ACTION_EDIT	打开一个 Activity，对所提供的数据进行编辑操作
ACTION_INSERT	打开一个 Activity，在提供数据的当前位置插入新项
ACTION_PICK	启动一个子 Activity，从提供的数据列表中选取一项
ACTION_SEARCH	启动一个 Activity，执行搜索动作
ACTION_SENDTO	启动一个 Activity，向数据提供的联系人发送信息
ACTION_SEND	启动一个可以发送数据的 Activity
ACTION_VIEW	最常用的动作，对以 Uri 方式传送的数据，根据 Uri 协议部分以最佳方式启动相应的 Activity 进行处理。对于 http:address 将打开浏览器查看；对于 tel:address 将打开拨号界面并呼叫指定的电话号码
ACTION_WEB_SEARCH	打开一个 Activity，对提供的数据进行 Web 搜索

WebViewIntentDemo 示例说明了如何隐式启动 Activity，用户界面如图 6-2(a)所示。

(a) 输入网址界面　　　　(b) 打开 Web 后的界面

图 6-2　WebViewIntentDemo 用户界面

当用户在文本框中输入 Web 地址后，通过单击"浏览此 URL"按钮，程序根据用户输入的 Web 地址生成一个 Intent，并以隐式启动的方式调用 Android 内置的 Web 浏览器，并打开指定的 Web 页面。本例输入的 Web 地址 http://www.google.com.hk，打开页面后的效果如图 6-2(b)所示。

6.2.2　获取 Activity 返回值

在上一小节 IntentDemo 示例中，通过 startActivity(Intent)方法启动 Activity，启动后的两个 Activity 之间相互独立，没有任何的关联。在很多情况下，后启动的 Activity 是为了让用户对特定信息进行选择，在后启动的 Activity 关闭时，这些信息是需要返回给先前启动的 Activity。后启动的 Activity 称为为"子 Activity"，先启动的 Activity 称为"父 Activity"。如果需要将子 Activity 的信息返回给父 Activity，则可以使用 Sub-Activity 的方式去启动子 Activity。

获取子 Activity 的返回值，一般可以分为以下 3 个步骤：

(1) 以 Sub-Activity 的方式启动子 Activity；
(2) 设置子 Activity 的返回值；
(3) 在父 Activity 中获取返回值。

下面详细介绍每一个步骤的过程和代码实现。

1. 以 Sub-Activity 的方式启动子 Activity

以 Sub-Activity 方式启动子 Activity，需要调用 startActivityForResult（Intent，requestCode)函数，参数 Intent 用于决定启动哪个 Activity，参数 requestCode 是请求码。因为所有子 Activity 返回时，父 Activity 都调用相同的处理函数，因此父 Activity 使用 requestCode 来确定数据是哪一个子 Activity 返回的。

显式启动子 Activity 的代码如下：

```
1  int SUBACTIVITY1 = 1;
2  Intent intent = new Intent(this, SubActivity1.class);
3  startActivityForResult(intent, SUBACTIVITY1);
```

隐式启动子 Activity 的代码如下：

```
1  int SUBACTIVITY2 = 2;
2  Uri uri = Uri.parse("content://contacts/people");
3  Intent intent = new Intent(Intent.ACTION_PICK, uri);
4  startActivityForResult(intent, SUBACTIVITY2);
```

2. 设置子 Activity 的返回值

在子 Activity 调用 finish()函数关闭前，调用 setResult()函数设定需要返回给父 Activity 的数据。setResult()函数有两个参数，一个是结果码，一个是返回值。结果码表明了子 Activity 的返回状态，通常为 Activity.RESULT_OK（正常返回数据）或者 Activity.RESULT_CANCELED（取消返回数据），也可以是自定义的结果码，结果码均为整数类型。返回值封装在 Intent 中，也就是说子 Activity 通过 Intent 将需要返回的数据传递给父 Activity。数据主要以 Uri 形式返回给父 Activity，此外还可以附加一些额外信息，这些额外信息用 Extra 的集合表示。

以下代码说明如何在子 Activity 中设置返回值：

```
1  Uri data = Uri.parse("tel:" + tel_number);
2  Intent result = new Intent(null, data);
3  result.putExtra("address", "JD Street");
4  setResult(RESULT_OK, result);
5  finish();
```

3. 在父 Activity 中获取返回值

当子 Activity 关闭后，父 Activity 会调用 onActivityResult()函数，用于获取子 Activity

的返回值。如果需要在父 Activity 中处理子 Activity 的返回值，则重载此函数即可。onActivityResult()函数的语法如下：

```
1  public void onActivityResult(int requestCode, int resultCode, Intent data);
```

其中第 1 个参数 requestCode 是请求码，用于判断第 3 个参数是哪一个子 Activity 的返回值；resultCode 用于表示子 Activity 的数据返回状态；Data 是子 Activity 的返回数据，返回数据类型是 Intent。根据返回数据的用途不同，Uri 数据的协议则不同，也可以使用 Extra 方法返回一些原始类型的数据。

以下代码说明如何在父 Activity 中处理子 Activity 的返回值：

```
1   private static final int SUBACTIVITY1 = 1;
2   private static final int SUBACTIVITY2 = 2;
3
4   @Override
5   public void onActivityResult(int requestCode, int resultCode, Intent data){
6       Super.onActivityResult(requestCode, resultCode, data);
7       switch(requestCode){
8           case SUBACTIVITY1:
9               if (resultCode == Activity.RESULT_OK){
10                  Uri uriData = data.getData();
11              }else if (resultCode == Activity.RESULT_CANCEL){
12              }
13          break;
14          case SUBACTIVITY2:
15              if (resultCode == Activity.RESULT_OK){
16                  Uri uriData = data.getData();
17              }
18          break;
19      }
20  }
```

代码的第 1 行和第 2 行是两个子 Activity 的请求码，在第 7 行对请求码进行匹配。代码第 9 行和第 11 行对结果码进行判断，如果返回的结果码是 Activity.RESULT_OK，则在代码的第 10 行使用 getData()函数获取 Intent 中的 Uri 数据；如果返回的结果码是 Activity.RESULT_CANCELED，则放弃所有操作。

ActivityCommunication 示例说明了如何以 Sub-Activity 方式启动子 Activity，以及如何使用 Intent 进行组件间通信。

该示例的主界面如图 6-3 所示。当用户单击"启动 Activity1"和"启动 Activity2"按钮时，程序将分别启动子 SubActivity1 和 SubActivity2，如图 6-4 所示。SubActivity1 提供了一个输入框，以及"接收"和"撤销"两个按钮。如果在输入框中输入信息后单击"接受"按钮，程序会把输入框中的信息传递给其父 Activity，并在父 Activity 的界面上显示。而如果用户单击"撤销"按钮，则程序不会向父 Activity 传递任何信息。SubActivity2 主要是为了说明如何在父 Activity 中处理多个子 Activity，因此仅提供了用于关闭 SubActivity2 的"关闭"按钮。

图 6-3　ActivityCommunication 用户界面

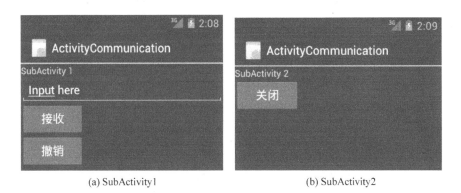

(a) SubActivity1　　　　　　　　　(b) SubActivity2

图 6-4　ActivityCommunication 的两个子 Activity

ActivityCommunication 示例的文件结构如图 6-5 所示，父 Activity 的代码在 ActivityCommunication.java 文件中，界面布局在 main.xml 中；两个子 Activity 的代码分别在 SubActivity1.java 和 SubActivity2.java 文件中，界面布局分别在 subactivity1.xml 和 subactivity2.xml 中。

图 6-5　ActivityCommunication 文件结构

ActivityCommunicationActivity.java 文件的核心代码如下：

```java
1   public class ActivityCommunicationActivity extends Activity {
2       private static final int SUBACTIVITY1 = 1;
3       private static final int SUBACTIVITY2 = 2;
4       TextView textView;
5       @Override
6       public void onCreate(Bundle savedInstanceState) {
7           super.onCreate(savedInstanceState);
8           setContentView(R.layout.main);
9           textView = (TextView)findViewById(R.id.textShow);
10          final Button btn1 = (Button)findViewById(R.id.btn1);
11          final Button btn2 = (Button)findViewById(R.id.btn2);
12  
13          btn1.setOnClickListener(new OnClickListener(){
14              public void onClick(View view){
15              Intent intent = new Intent(ActivityCommunication.this, SubActivity1.class);
16              startActivityForResult(intent, SUBACTIVITY1);
17              }
18          });
19  
20          btn2.setOnClickListener(new OnClickListener(){
21              public void onClick(View view){
22              Intent intent = new Intent(ActivityCommunication.this, SubActivity2.class);
23              startActivityForResult(intent, SUBACTIVITY2);
24              }
25          });
26      }
27  
28      @Override
29      protected void onActivityResult(int requestCode, int resultCode, Intent data) {
30          super.onActivityResult(requestCode, resultCode, data);
31  
32          switch(requestCode){
33          case SUBACTIVITY1:
34              if (resultCode == RESULT_OK){
35                  Uri uriData = data.getData();
36                  textView.setText(uriData.toString());
37              }
38              break;
39          case SUBACTIVITY2:
40              break;
41          }
42      }
43  }
```

在代码的第 2 行和第 3 行代码分别定义了两个子 Activity 的请求码。在代码的第 16 行和第 23 行代码以 Sub-Activity 的方式分别启动两个子 Activity。代码第 29 行是子 Activity 关闭后的返回值处理函数，其中 requestCode 是子 Activity 返回的请求码，与第 2

行和第 3 行定义的两个请求码相匹配；resultCode 是结果码，在代码第 32 行对结果码进行判断，如果等于 RESULT_OK，在第 35 行代码获取子 Activity 返回值中的数据；data 是返回值，子 Activity 需要返回的数据就保存在 data 中。

SubActivity1.java 的核心代码如下：

```
1   public class SubActivity1 extends Activity {
2       @Override
3       public void onCreate(Bundle savedInstanceState) {
4           super.onCreate(savedInstanceState);
5           setContentView(R.layout.subactivity1);
6           final EditText editText = (EditText)findViewById(R.id.edit);
7           Button btnOK = (Button)findViewById(R.id.btn_ok);
8           Button btnCancel = (Button)findViewById(R.id.btn_cancel);
9   
10          btnOK.setOnClickListener(new OnClickListener(){
11              public void onClick(View view){
12                  String uriString = editText.getText().toString();
13                  Uri data = Uri.parse(uriString);
14                  Intent result = new Intent(null, data);
15                  setResult(RESULT_OK, result);
16                  finish();
17              }
18          });
19  
20          btnCancel.setOnClickListener(new OnClickListener(){
21              public void onClick(View view){
22                  setResult(RESULT_CANCELED, null);
23                  finish();
24              }
25          });
26      }
27  }
```

代码第 13 行将 EditText 控件的内容作为数据保存在 Uri 中，并在第 14 行代码中构造 Intent。在第 15 行代码中，RESUIT_OK 作为结果码，通过调用 setResult() 函数，将 result 设定为返回值。最后在代码第 16 行调用 finish() 函数关闭当前的子 Activity。

SubActivity2.java 的核心代码：

```
1   public class SubActivity2 extends Activity {
2       @Override
3       public void onCreate(Bundle savedInstanceState) {
4           super.onCreate(savedInstanceState);
5           setContentView(R.layout.subactivity2);
6   
7           Button btnReturn = (Button)findViewById(R.id.btn_return);
8           btnReturn.setOnClickListener(new OnClickListener(){
9               public void onClick(View view){
```

```
10                    setResult(RESULT_CANCELED, null);
11                    finish();
12                }
13            });
14    }
15 }
```

在 SubActivity2 的代码中,第 10 行的 setResult()函数仅设置了结果码,第 2 个参数为 null,表示没有数据需要传递给父 Activity。

6.3　Intent 过滤器

Intent 过滤器是一种根据 Intent 中的动作(Action)、类别(Categorie)和数据(Data)等内容,对适合接收该 Intent 的组件进行匹配和筛选的机制。Intent 过滤器可以匹配数据类型、路径和协议,还可以确定多个匹配项顺序的优先级(Priority)。

应用程序的 Activity、Service 和 BroadcastReceiver 组件都可以注册 Intent 过滤器。这样,这些组件在特定的数据格式上则可以产生相应的动作。

隐式启动 Activity 时,并没有在 Intent 中指明 Activity 所在的类,因此,Android 系统一定存在某种匹配机制,使 Android 系统能够根据 Intent 中的数据信息,找到需要启动的 Activity。这种匹配机制是依靠 Android 系统中的 Intent 过滤器(Intent Filter)来实现的。

6.3.1　注册 Intent 过滤器

注册 Intent 过滤器的方法如下。

在 AndroidManifest.xml 文件的各个组件下定义＜intent-filter＞结点,然后在＜intent-filter＞结点中声明该组件所支持的动作、执行的环境和数据格式等信息。当然,也可以在程序代码中动态地为组件设置 Intent 过滤器。

＜intent-filter＞结点支持＜action＞标签、＜category＞标签和＜data＞标签,分别用于定义 Intent 过滤器的"动作"、"类别"和"数据"。

＜intent-filter＞节点支持的标签和属性说明如表 6-4 所示。

表 6-4　＜intent-filter＞结点属性

标　　签	属　　性	说　　明
＜action＞	android:name	指定组件所能响应的动作,用字符串表示,通常由 Java 类名和包的完全限定名构成
＜category＞	android:category	指定以何种方式去服务 Intent 请求的动作
＜data＞	Android:host	指定一个有效的主机名
	android:mimetype	指定组件能处理的数据类型
	android:path	有效的 URI 路径名
	android:port	主机的有效端口号
	android:scheme	所需要的特定协议

<category>标签用于指定 Intent 过滤器的服务方式，每个 Intent 过滤器可以定义多个<category>标签，程序开发人员可以使用自定义的类别，或使用 Android 系统提供的类别。Android 系统提供的类别如表 6-5 所示。

表 6-5　Android 系统提供的类别

值	说　明
ALTERNATIVE	Intent 数据默认动作的一个可替换的执行方法
SELECTED_ALTERNATIVE	和 ALTERNATIVE 类似，但替换的执行方法不是指定的，而是被解析出来的
BROWSABLE	声明 Activity 可以由浏览器启动
DEFAULT	为 Intent 过滤器中定义的数据提供默认动作
HOME	设备启动后显示的第一个 Activity
LAUNCHER	在应用程序启动时首先被显示

AndroidManifest.xml 文件中每个组件的<intent-filter>都被解析成一个 Intent 过滤器对象。当应用程序安装到 Android 系统时，所有的组件和 Intent 过滤器都会注册到 Android 系统中。这样，Android 系统便可以将任何一个 Intent 请求通过 Intent 过滤器映射到相应的组件上。

6.3.2　Intent 解析

Intent 到 Intent 过滤器的映射过程称为"Intent 解析"。Intent 解析可以在所有的组件中，找到一个可以与请求的 Intent 达成最佳匹配的 Intent 过滤器。Android 系统中 Intent 解析的匹配规则如下。

（1）Android 系统把所有应用程序包中的 Intent 过滤器集合在一起，形成一个完整的 Intent 过滤器列表。

（2）在 Intent 与 Intent 过滤器进行匹配时，Android 系统会将列表中所有 Intent 过滤器的"动作"和"类别"与 Intent 进行匹配，任何不匹配的 Intent 过滤器都将被过滤掉。没有指定"动作"的 Intent 过滤器可以匹配任何的 Intent，但是没有指定"类别"的 Intent 过滤器只能匹配没有"类别"的 Intent。

（3）把 Intent 数据 Uri 的每个子部与 Intent 过滤器的<data>标签中的属性进行匹配，如果<data>标签指定了协议、主机名、路径名或 MIME 类型，那么这些属性都要与 Intent 的 Uri 数据部分进行匹配，任何不匹配的 Intent 过滤器均被过滤掉。

（4）如果 Intent 过滤器的匹配结果多于一个，则可以根据在<intent-filter>标签中定义的优先级标签来对 Intent 过滤器进行排序，优先级最高的 Intent 过滤器将被选择。

IntentResolutionDemo 示例说明了如何在 AndroidManifest.xml 文件中注册 Intent 过滤器，以及如何设置<intent-filter>结点属性来捕获指定的 Intent。

AndroidManifest.xml 的完整代码如下：

```
1   <?xml version = "1.0" encoding = "utf - 8"?>
2   < manifest xmlns:android = "http://schemas.android.com/apk/res/android"
3       package = "edu.hrbeu.IntentResolutionDemo"
```

```
4       android:versionCode = "1"
5       android:versionName = "1.0">
6   <application android:icon = "@drawable/icon" android:label = "@string/app_name">
7       <activity android:name = ".IntentResolutionDemo"
8               android:label = "@string/app_name">
9           <intent-filter>
10              <action android:name = "android.intent.action.MAIN" />
11              <category android:name = "android.intent.category.LAUNCHER" />
12          </intent-filter>
13      </activity>
14      <activity android:name = ".ActivityToStart"
15              android:label = "@string/app_name">
16          <intent-filter>
17              <action android:name = "android.intent.action.VIEW" />
18              <category android:name = "android.intent.category.DEFAULT" />
19              <data android:scheme = "schemodemo" android:host = "edu.hrbeu" />
20          </intent-filter>
21      </activity>
22  </application>
23  <uses-sdk android:minSdkVersion = "14" />
24 </manifest>
```

在代码的第 7 行和第 14 行分别定义了两个 Activity。第 9 行到第 12 行代码是第 1 个 Activity 的 Intent 过滤器,动作是 android.intent.action.MAIN,类别是 android.intent.category.LAUNCHER,由此可知,这个 Activity 是应用程序启动后显示的默认用户界面。

第 16 行到第 20 行代码是第 2 个 Activity 的 Intent 过滤器,过滤器的动作是 android.intent.action.VIEW,表示根据 Uri 协议,以浏览的方式启动相应的 Activity;类别是 android.intent.category.DEFAULT,表示数据的默认动作;数据的协议部分是 android:scheme = "schemodemo",数据的主机名称部分是 android:host = "edu.hrbeu"。

在 IntentResolutionDemo.java 文件中,定义了一个 Intent 用于启动另一个 Activity,这个 Intent 与 Activity 设置的 Intent 过滤器是完全匹配的。IntentResolutionDemo.java 文件中 Intent 实例化和启动 Activity 的代码如下:

```
1   Intent intent = new Intent(Intent.ACTION_VIEW, Uri.parse("schemodemo://edu.hrbeu/path"));
2   startActivity(intent);
```

代码第 1 行所定义的 Intent,动作为 Intent.ACTION_VIEW,与 Intent 过滤器的动作 android.intent.action.VIEW 匹配;Uri 是 "schemodemo://edu.hrbeu/path",其中的协议部分为 "schemodemo",主机名部分为 "edu.hrbeu",也与 Intent 过滤器定义的数据要求完全匹配。因此,代码第 1 行定义的 Intent,在 Android 系统与 Intent 过滤器列表进行匹配时,会与 AndroidManifest.xml 文件中 ActivityToStart 定义的 Intent 过滤器完全匹配。

6.4 广播消息实例

Intent 的另一种用途是发送广播消息，应用程序和 Android 系统都可以使用 Intent 发送广播消息，广播消息的内容可以是与应用程序密切相关的数据信息，也可以是 Android 的系统信息，例如网络连接变化、电池电量变化、接收到短信或系统设置变化等。如果应用程序注册了 BroadcastReceiver，则可以接收到指定的广播消息。

使用 Intent 发送广播消息非常简单，只需要创建一个 Intent，并调用 sendBroadcast() 函数就可把 Intent 携带的信息广播出去。但需要注意的是，在构造 Intent 时必须定义一个全局唯一的字符串，用于标识其要执行的动作，通常使用应用程序包的名称。如果要在 Intent 传递额外数据，可以用 Intent 的 putExtra() 方法。下面的代码构造用于广播消息的 Intent，并添加了额外的数据，然后调用 sendBroadcast() 发送广播消息：

```
1  String UNIQUE_STRING = "edu.hrbeu.BroadcastReceiverDemo";
2  Intent intent = new Intent(UNIQUE_STRING);
3  intent.putExtra("key1", "value1");
4  intent.putExtra("key2", "value2");
5  sendBroadcast(intent);
```

BroadcastReceiver 用于监听广播消息，可以在 AndroidManifest.xml 文件或在代码中注册一个 BroadcastReceiver，并使用 Intent 过滤器指定要处理的广播消息。创建 BroadcastReceiver 需要继承 BroadcastReceiver 类，并重载 onReceive() 方法。示例代码如下：

```
1  public class MyBroadcastReceiver extends BroadcastReceiver {
2      @Override
3      public void onReceive(Context context, Intent intent) {
4          //TODO: React to the Intent received.
5      }
6  }
```

当 Android 系统接收到与注册 BroadcastReceiver 匹配的广播消息时，Android 系统会自动调用这个 BroadcastReceiver 接收广播消息。在 BroadcastReceiver 接收到与之匹配的广播消息后，onReceive() 方法会被调用，但 onReceive() 方法必须要在 5 秒钟执行完毕，否则 Android 系统会认为该组件失去响应，并提示用户强行关闭该组件。

BroadcastReceiverDemo 示例说明了如何在应用程序中注册 BroadcastReceiver 组件，并指定接收广播消息的类型。BroadcastReceiverDemo 示例的界面如图 6-6 所示，在单击"发生广播消息"按钮后，EditText 控件中内容将以广播消息的形式发生出去，示例内部的 BroadcastReceiver 将接收这个广播消息，并显示在用户界面的下方。

BroadcastReceiverDemo.java 文件中包含发送广播消息的代码，其关键代码如下：

图 6-6　BroadcastReceiverDemo 主界面

```
1    button.setOnClickListener(new OnClickListener(){
2        public void onClick(View view){
3            Intent intent = new Intent("edu.hrbeu.BroadcastReceiverDemo");
4            intent.putExtra("message", entryText.getText().toString());
5            sendBroadcast(intent);
6        }
7    });
```

代码第 3 行创建 Intent 时，将 edu.hrbeu.BroadcastReceiverDemo 作为识别广播消息的字符串标识，并在代码第 4 行将添加了额外信息，最后在代码第 5 行调用 sendBroadcast()函数发送广播消息。

为了能够使应用程序中的 BroadcastReceiver 接收指定的广播消息，首先要在 AndroidManifest.xml 文件中 BroadcastReceiver 结点下添加 Intent 过滤器，声明 BroadcastReceiver 可以接收的广播消息类型。AndroidManifest.xml 文件的完整代码如下：

```
1    <?xml version = "1.0" encoding = "utf - 8"?>
2    < manifest xmlns:android = "http://schemas.android.com/apk/res/android"
3        package = "edu.hrbeu.BroadcastReceiverDemo"
4        android:versionCode = "1"
```

```
5         android:versionName = "1.0">
6     <application android:icon = "@drawable/icon" android:label = "@string/app_name">
7         <activity android:name = ".BroadcastReceiverDemo"
8                   android:label = "@string/app_name">
9             <intent-filter>
10                <action android:name = "android.intent.action.MAIN" />
11                <category android:name = "android.intent.category.LAUNCHER" />
12            </intent-filter>
13        </activity>
14        <receiver android:name = ".MyBroadcastReceiver">
15            <intent-filter>
16                <action android:name = "edu.hrbeu.BroadcastReceiverDemo" />
17            </intent-filter>
18        </receiver>
19    </application>
20    <uses-sdk android:minSdkVersion = "14" />
21 </manifest>
```

在代码的第14行中创建了一个＜receiver＞结点,在第15行中声明了Intent过滤器的动作为"edu. hrbeu. BroadcastReceiverDemo",这与BroadcastReceiverDemo. java文件中Intent的动作相一致,表明这个BroadcastReceiver可以接收动作为"edu. hrbeu. BroadcastReceiverDemo"的广播消息。

MyBroadcastReceiver. java文件创建了一个自定义的BroadcastReceiver,其核心代码如下:

```
1  public class MyBroadcastReceiver extends BroadcastReceiver {
2      @Override
3      public void onReceive(Context context, Intent intent) {
4          String msg = intent.getStringExtra("message");
5          Toast.makeText(context, msg, Toast.LENGTH_SHORT).show();
6      }
7  }
```

代码第1行首先继承了BroadcastReceiver类,并在第3行重载了onReveive()函数。当接收到AndroidManifest. xml文件定义的广播消息后,程序将自动调用onReveive()函数进行消息处理。代码第4行通过调用getStringExtra()函数,从Intent中获取标识为message的字符串数据,并使用Toast()函数将信息显示在界面上。

习题

1. 依据创建Intent对象时是否明确指定接收组件名称,Intent可分为哪两种类型。
2. 简要说明Intent中6个主要属性名称及功能。
3. BroadcastReceiver对象的主要功能是什么。
4. BroadcastReceiver作为应用级组件必须经过注册才能处理广播消息,注册有哪两种方式。

第7章 后台服务

Service用于后台完成用户指定的操作,是Android的四大组件之一。Service没有可以操作的用户界面,Service是在后台运行的,可以用于音乐播放器、文件下载工具等应用程序。如果有某个组件需要在运行时向用户呈现可操作的信息就应该选择Activity,否则应该选择Service。

本章主要学习内容:
- 掌握Service的启动方式和基础;
- 掌握本地服务应用;
- 了解Service的生命周期。

7.1 Service 介绍

Service是没有用户界面,在后台执行长时间操作的应用程序组件之一。其他的应用程序组件可以启动服务,而且启动的服务即便是用户切换到其他应用程序也可以一直在后台保持运行状态。应用程序组件还可以绑定到服务上面,甚至可以通过IPC(Interprocess Communication)机制来提供可供其他应用程序远程访问的服务。

7.1.1 Service 启动方式

1. startService()

应用程序组件(如Activity),可以通过调用Context.startService()来启动一个服务。服务一旦启动,便在后台执行,即便启动服务的应用程序组件被销毁了,也不影响服务的继续运行。通过startService()方式启动的服务,通常来讲都会去执行一个比较单一的任务,且不会返回结果给调用者。

2. bindService()

应用程序组件可以通过调用Context.bindService()绑定一个服务。通过bindService()启动的服务可以返回结果给调用者,甚至可以通过IPC机制进行跨进程的通信。一旦应用程序组件与之绑定,服务开始运行。多个组件可以同时绑定到同一个服务上面,直到所有的组件都与服务解绑之后,服务才会被销毁。

上述实际上介绍了两种不同种类的服务,但是可以让某一个服务同时具备两种功能。

是否具备这两种功能仅仅取决于是否实现了 Service 的一些不同的回调方法。实现了 onStartCommand()方法就允许其他组件通过调用 startService()方法来启动这个服务；实现了 onBind()方法就允许其他组件绑定到这个服务上面，意味着可以接收到服务返回的结果。

在默认情况下，Service 是运行在应用程序进程的 UI 主线程当中的。这也就意味着，如果需要在 Service 当中执行一些比较耗时的操作，应该在服务当中创建子线程去做这些比较耗时，耗 CPU 的操作。通过在 Service 中创建子线程的方式，可以有效地避免应用程序产生 ANR 的错误。

7.1.2 Service 基础

如果需要在应用程序当中创建服务，则必须创建一个继承 android.app.Service 的子类。在创建的子类当中，需要去实现一些处理 Service 生命周期的重要的回调方法：

- onStartCommand：当其他组件，例如 Activity，调用 startService()方法的时候，系统就会回调 onStartCommand()这个方法。一旦这个方法被调用过后，服务就会启动并开始在后台运行。操作完成过后，可以通过调用 stopSelf()或者调用 Context.stopService()用于停止服务的运行。
- onBind：当其他组件调用 bindService()方法的时候，系统就会回调 onBind()这个方法。在这个方法里面，需要给绑定到此服务的组件返回了一个实现了 android.os.IBinder 接口的对象。如果服务不支持其他组件的绑定，则此方法返回 null 就可以了。
- onCreate：当服务第一次创建的时候，系统就会回调 onCreate()这个方法。系统会在 onStartCommand()或 onBind()方法之前调用这个方法，且仅仅调用一次。如果服务已经在运行了，则不会调用这个方法。
- onDestory：当服务不再被使用的时候，系统就会回调 onDestory()这个方法。在这个方法里面，一般需要去做一些释放资源的操作。对于 Service 来说，这是系统回调的最后一个方法。

如果 Service 是通过其他组件调用 startService()方法而运行的，那么 Service 会一直运行直到调用自己的 stopSelf()方法，或者是其他组件显示的调用 stopService()方法。

如果 Service 是通过其他组件调用 bindService()方法而运行的，那么只有当绑定到此服务的所有组件都与之解绑，这时服务才会被销毁。

Android 系统在内存不足的情况下，为了保证与用户正在交互的应用程序的正常使用，会强制地停止一些 Service 的运行。如果 Service 被绑定到正在用户交互的 Activity 组件上，那么这种被强制停止运行的概率就会小很多。

7.2 本地服务

Android Service 事实上又分为两种类型，一种是本地服务，另外一种是远程服务。本地服务指的是只用于当前应用程序内部的服务，而远程服务指的是通过 IPC 机制进行进程间通信的服务。

本地服务，大体上又可以分为两种，第一种是不需要跟组件进行交互的服务，也就是通过调用 Context.startService()方法启动的服务；另外一种是需要跟组件进行交互的服务，也就是通过调用 Context.bindService()方法启动的服务。

通过调用 Context.startService()方法启动的服务，与调用者之间没有任何关联，即使调用者退出了，服务也可以继续运行。在服务未被创建的时候，系统会首先调用服务的 onCreate()方法，接着调用服务的 onStartCommand()方法。如果在调用 Context.startService()方法前服务已经被创建，多次调用 Context.startService()方法并不会导致服务的多次创建，但会导致 onStartCommand()方法多次被调用。可以通过调用 Context.stopService()方法或调用服务本身的 stopSelf()方法停止服务的运行，服务结束时会调用 onDestory()方法。

通过调用 Context.bindService()方法启动的服务，调用者与服务绑定在了一起。调用者一旦退出，服务也会相应地停止运行。在服务未被创建的时候，系统会首先调用服务的 onCreate()方法，接着调用服务的 onBind()方法，如果调用者与服务已经处于绑定的状态，多次调用 Context.bindService()方法并不会导致 onBind()方法被调用多次。采用 Context.bindService()方法启动的服务，只能通过调用 Context.unbindService()方法解除调用者与服务的绑定，服务结束时会调用 onDestory()方法。

7.2.1 不需要与组件交互本地服务

下面通过实例介绍如何使用不需要与组件交互的本地服务。

（1）创建一个新的 Android 工程，工程名为 ServiceDemo，应用程序名为 ServiceDemo，包名为 hlju.edu.cn，创建的 Activity 的名字为 MainActivity，最小 SDK 版本根据选择的目标 API 会自动添加。

（2）修改 res 目录下 layout 文件夹中的 activity_main.xml 文件，设置线性布局，添加两个 Button 控件，对相应控件进行描述，并设置相关属性，代码如下：

```
1   <LinearLayout xmlns:android = "http://schemas.android.com/apk/res/android"
2       xmlns:tools = "http://schemas.android.com/tools"
3       android:layout_width = "fill_parent"
4       android:layout_height = "fill_parent"
5       android:orientation = "vertical">
6       <Button
7           android:id = "@ + id/startService"
8           android:layout_width = "fill_parent"
9           android:layout_height = "wrap_content"
10          android:text = "启动本地服务"/>
11      <Button
12          android:id = "@ + id/endService"
13          android:layout_width = "fill_parent"
14          android:layout_height = "wrap_content"
15          android:text = "停止本地服务"/>
16  </LinearLayout>
```

（3）在 AndroidManifest.xml 文件中声明 Service，代码如下：

```
<service android:name=".MyLocalService"></service>
```

（4）在 src 目录中 hlju.edu.cn 包下创建 MyLocalService.java 文件，代码如下：

```
1   package hlju.edu.cn;
2
3   import android.app.Notification;
4   import android.app.NotificationManager;
5   import android.app.PendingIntent;
6   import android.app.Service;
7   import android.content.Context;
8   import android.content.Intent;
9   import android.os.IBinder;
10  import android.os.SystemClock;
11  import android.util.Log;
12  public class MyLocalService extends Service {
13      private static final String TAG = "MyLocalService";
14      private static int NOTIFICATION_ID = 1;
15      private boolean isRunning;
16      private NotificationManager notificationManager;
17      @Override
18      public void onCreate() {
19          super.onCreate();
20          Log.d(TAG, "onCreate() method was invoked");
21          notificationManager = (NotificationManager)getSystemService(Context.NOTIFICATION_SERVICE);
22      }
23
24      @SuppressWarnings("deprecation")
25      private void displayNotification(String message){
26          Notification notification = new Notification(R.drawable.ic_launcher, message, System.currentTimeMillis());
27          PendingIntent contentIntent = PendingIntent.getActivity(this, 0, new Intent(this, MainActivity.class), 0);
28          notification.setLatestEventInfo(this, "Background Service", message, contentIntent);
29          notificationManager.notify(NOTIFICATION_ID ++, notification);
30      }
31
32      @Override
33      public int onStartCommand(Intent intent, int flags, int startId) {
34          Log.d(TAG, "onStartCommand() method was invoked");
35          displayNotification("Service onStartCommand");
36
37          isRunning = true;
38          Thread serviceThread = new Thread(new Runnable() {
39              @Override
```

```
40          public void run() {
41              while(isRunning){
42                  Log.d(TAG, "service Thread executes again");
43                  SystemClock.sleep(10 * 1000);
44          } } });
45      serviceThread.start();
46      return super.onStartCommand(intent, flags, startId);
47   }
48
49   @Override
50   public IBinder onBind(Intent arg0) {
51       return null;
52   }
53
54   @Override
55   public void onDestroy() {
56       super.onDestroy();
57       displayNotification("Service onDestory");
58       isRunning = false;
59       Log.d(TAG, "onDestory() method was invoked");
60   }
61 }
```

第 14 行代码表示线程是否可以继续运。第 18 行代码表示服务第一次创建时调用。第 21 行代码表示获取通知管理器。第 35 行代码表示提示通知消息。第 43 行代码表示休眠 10 秒钟。第 45 行代码表示启动子线程。第 50 行代码表示不需要与组件绑定时，onBind()方法返回 null。第 55 行代码表示当服务停止运行时调用。第 58 行代码表示停止子线程的运行。

（5）修改 src 目录中 hlju.edu.cn 包下的 MainActivity.java 文件，代码如下：

```
1  package hlju.edu.cn;
2
3  import android.app.Activity;
4  import android.content.Intent;
5  import android.os.Bundle;
6  import android.view.Menu;
7  import android.view.MenuItem;
8  import android.view.View;
9  import android.view.View.OnClickListener;
10 import android.widget.Button;
11 public class MainActivity extends Activity implements OnClickListener{
12     Button startServiceBtn;
13     Button endServiceBtn;
14     @Override
15     public void onCreate(Bundle savedInstanceState) {
16         super.onCreate(savedInstanceState);
17         setContentView(R.layout.activity_main);
18         startServiceBtn = (Button)findViewById(R.id.startService);
19         endServiceBtn = (Button)findViewById(R.id.endService);
20         startServiceBtn.setOnClickListener(this);
21         endServiceBtn.setOnClickListener(this);
```

```
22      }
23
24      @Override
25      public void onClick(View v) {
26       int viewId = v.getId();
27       if(R.id.startService == viewId){
28           tartService(new Intent(this, MyLocalService.class));
29       }else if(R.id.endService == viewId){
30           stopService(new Intent(this, MyLocalService.class));
31       }
32      }
33      @Override
34      public boolean onCreateOptionsMenu(Menu menu) {
35          // Inflate the menu; this adds items to the action bar if it is present.
36          getMenuInflater().inflate(R.menu.main, menu);
37          return true;
38      }
39
40      @Override
41      public boolean onOptionsItemSelected(MenuItem item) {
42          // Handle action bar item clicks here. The action bar will
43          // automatically handle clicks on the Home/Up button, so long
44          // as you specify a parent activity in AndroidManifest.xml.
45          int id = item.getItemId();
46          if (id == R.id.action_settings) {
47              return true;
48          }
49          return super.onOptionsItemSelected(item);
50      }
51 }
```

第 12 行代码表示声明启动服务按钮。第 13 行代码表示声明停止服务按钮。第 18 行代码表示找到启动服务按钮。第 19 行代码表示找到停止服务按钮。第 20 行代码表示绑定单击事件监听器。第 27 行代码表示如果启动服务按钮单击为真。第 28 行代码表示启动服务。第 29 行代码表示如果停止服务按钮单击为真。

（6）部署 ServiceDemo 工程，程序运行结果如图 7-1 所示。单击"启动本地服务"按钮，程序运行如图 7-2 所示。此时通过调用 Context.startService() 的方式来启动一个服务，在模拟器通知栏中可以看到服务运行，如图 7-3 所示。当单击"启动本地服务"按钮时，MyLocalService 里面的 onCreate() 及 onStartCommand() 函数依次被调动，LogCat 输出如图 7-4 所示。

当单击"停止本地服务"按钮时，如图 7-5 所示。此时在通知栏可以看到 1 个通知，如图 7-6 所示。

图 7-1　程序运行窗口

图 7-2　启动本地服务

图 7-3　模拟器通知栏

图 7-4　单击"启动本地服务"按钮时 LogCat 输出图

图 7-5　停止本地服务

图 7-6　模拟器通知栏

同时 MyLocalService 里面的 onDestroy() 函数会被调用。通过 LogCat 输出,如图 7-7 所示。

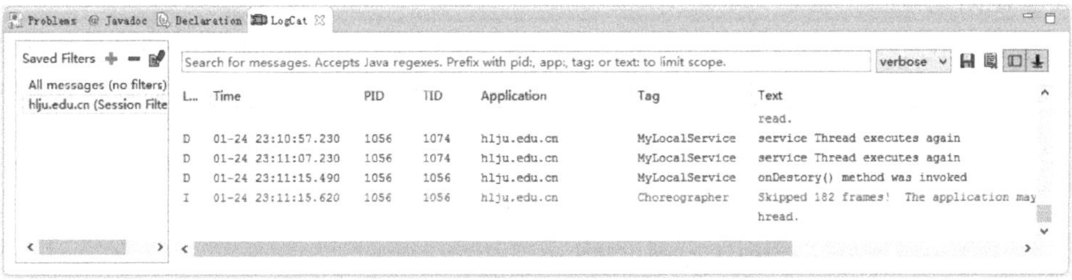

图 7-7 单击"停止本地服务"按钮时 LogCat 输出

连续单击"启动本地服务"和"停止本地服务"按钮时,LogCat 输出,如图 7-8 所示。

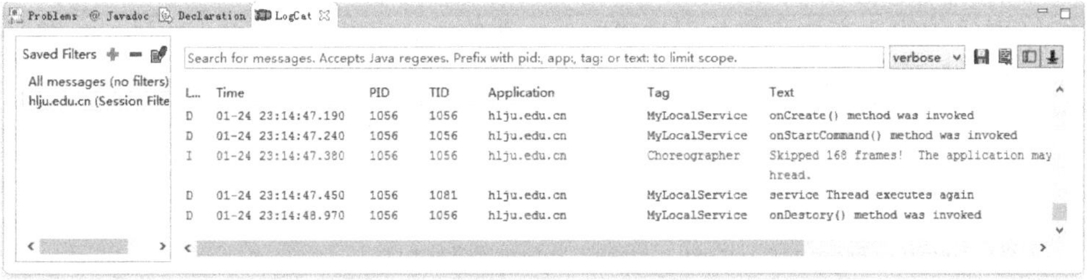

图 7-8 连续单击"启动"和"停止"按钮时,LogCat 输出

7.2.2 本地服务结合广播接收器

Android 应用程序可以使用广播接收器来接收有兴趣的广播,或自行发送广播让其他应用程序知道该应用程序状态有改变。

Android 系统本身就会常常发出广播,例如接到来电、收到短信、启动照相设备、系统开机和电池剩余电量过低等,Android 系统都会发出广播。

广播接收器本身并没有任何使用界面,它是一个继承 android.content.BroadcastReceiver 抽象类的子类,等接收到指定的广播而触发时,即在实现的 onReceive() 抽象方法回应广播来执行所需操作。代码如下:

```
1   public class BroadcastRecvDemo extends BroadcastReceiver {
2       public void onReceive(Context context, Intent intent) {
3           …
4       }
5   }
```

第 1 行代码表示继承 BroadcastReceiver 类。第 2 行代码表示实现 onReceive() 抽象方法来处理接收的广播。

Android 应用程序也可以发送自定义广播,从一个活动到另一个活动,或者完全不同的 Android 应用程序。例如,使用广播让其他应用程序知道已经完成文件下载,在 Java 程序是使用 Intent 对象的 sendBroadcast()方法来发送自定义广播。因为广播接收器类本身并没有程序进入点,在移动设备安装广播接收器后,用户不能直接执行,需要在活动通过 Intent 意图对象发出广播启动。

一般的情况下可以从活动发送广播,通常都是从一个 Android 应用程序发生广播,让另一个广播接收器的 Android 应用程序接收广播,而较少在同一个应用程序使用广播接收器。在活动发送广播同样是使用 Intent,代码如下:

```
1   public void btn1_Click(View view){
2       Intent I = new Intent("android.broadcast.TOAST");
3       sendBroadcast(I);
4   }
```

第1行代码表示 btn1_Click()方法是事件处理方法。第2行代码表示创建 Intent 对象,Intent 对象的构造方法参数是广播接收器注册的动作名称 android.broadcast.TOAST。第3行代码表示使用 sendBroadcast()方法发送广播。

等接收到指定广播而触发时,广播接收器可以启动活动、服务、显示通知或信息文字,即在实现的 onReceive()抽象方法回应广播来执行所需要的操作,代码如下:

```
1   public void onReceive(Context context,Intent intent){
2       Toast.makeText(context, "收到 Toast 广播!", Toast.LENGTH_SHORT).show;
3   }
```

第1行代码表示 ToastBroadcastReceiver 类的 onReceive 方法。第2行代码表示在接收到广播后使用 Toast 类显示收到广播的信息文字。

在 AndroidManifest.xml 需要注册此广播接收器,代码如下:

```
1   < receiver android:name = ".ToastBroadcastReceiver">
2       < intent - filter >
3       < action android:name = "android.broadcast.TOAST"/>
4       </intent - filter >
5   </receiver >
```

下面通过实例介绍如何使用广播接收器。

(1) 创建一个新的 Android 工程,工程名为 BroadcastDemo,应用程序名为 BroadcastDemo,包名为 hlju.edu.cn,创建的 Activity 的名字为 MainActivity,最小 SDK 版本根据选择的目标 API 会自动添加。

(2) 修改 res 目录下 layout 文件夹中的 activity_main.xml 文件,设置线性布局,添加3个 Button 控件,对相应控件进行描述,并设置相关属性,代码如下:

```
1   <?xml version = "1.0" encoding = "utf - 8"?>
2   < LinearLayout xmlns:android = "http://schemas.android.com/apk/res/android"
3       android:orientation = "vertical"
4       android:layout_width = "fill_parent"
```

```
5       android:layout_height = "fill_parent">
6       <Button android:id = "@ + id/btn1"
7         android:text = "发送 Toast 广播"
8         android:layout_width = "fill_parent"
9         android:layout_height = "wrap_content"
10        android:onClick = "btn1_Click"/>
11      <Button android:id = "@ + id/btn2"
12        android:text = "发送通知广播"
13        android:layout_width = "fill_parent"
14        android:layout_height = "wrap_content"
15        android:onClick = "btn2_Click"/>
16      <Button android:id = "@ + id/btn3"
17        android:text = "发送活动广播"
18        android:layout_width = "fill_parent"
19        android:layout_height = "wrap_content"
20        android:onClick = "btn3_Click"/>
21    </LinearLayout>
```

（3）在 AndroidManifest.xml 文件注册广播接收器，代码如下：

```
1    <activity android:name = ".TargetActivity"/>
2        <receiver android:name = ".ToastBroadcastReceiver">
3            <intent-filter>
4                <action android:name = "android.broadcast.TOAST"/>
5            </intent-filter>
6        </receiver>
7        <receiver android:name = ".NotificationBroadcastReceiver">
8            <intent-filter>
9                <action android:name = "android.broadcast.NOTIFICATION"/>
10           </intent-filter>
11       </receiver>
12       <receiver android:name = ".ActivityBroadcastReceiver">
13           <intent-filter>
14               <action android:name = "android.broadcast.ACTIVITY"/>
15           </intent-filter>
16       </receiver>
```

（4）在 src 目录中 hlju.edu.cn 包下创建 ToastBroadcastReceiver.java 文件，代码如下：

```
1    package hlju.edu.cn;
2    import android.content.BroadcastReceiver;
3    import android.content.Context;
4    import android.content.Intent;
5    import android.widget.Toast;
6    public class ToastBroadcastReceiver extends BroadcastReceiver {
7        @Override
8        public void onReceive(Context context, Intent intent) {
9            Toast.makeText(context, "收到 Toast 广播!",
```

```
10                     Toast.LENGTH_SHORT).show();
11           }
12 }
```

第 6 行代码表示继承 BroadcastReceiver 类来创建广播接收器 ToastBroadcastReceiver。第 8 行代码表示实现 onReceive()抽象方法。第 9 行代码表示使用 Toast 类显示收到广播的信息文字。

（5）在 src 目录中 hlju.edu.cn 包下创建 NotificationBroadcastReceiver.java 文件，代码如下：

```
1  package hlju.edu.cn;
2  import android.app.Notification;
3  import android.app.NotificationManager;
4  import android.app.PendingIntent;
5  import android.content.BroadcastReceiver;
6  import android.content.Context;
7  import android.content.Intent;
8  public class NotificationBroadcastReceiver extends BroadcastReceiver {
9      private static final int NOTIF_ID = 1;
10     @Override
11     public void onReceive(Context context, Intent intent) {
12         NotificationManager notifMgr = (NotificationManager)
13             context.getSystemService(Context.NOTIFICATION_SERVICE);
14         Notification note = new Notification(R.drawable.ic_launcher, "收到通知广播!", System.currentTimeMillis());
15         Intent i = new Intent(context, MainActivity.class);
16         PendingIntent pi = PendingIntent.getActivity(context, 0, i, PendingIntent.FLAG_UPDATE_CURRENT);
17         note.setLatestEventInfo(context, "这是广播发送的通知", null, pi);
18         notifMgr.notify(NOTIF_ID, note);
19     } }
```

第 8 行代码表示继承 BroadcastReceiver 类来创建广播接收器 NotificationBroadcastReceiver。第 11 行代码表示实现 onReceive()抽象方法。

（6）在 src 目录中 hlju.edu.cn 包下创建 TargetActivity.java 文件，代码如下：

```
1  package hlju.edu.cn;
2  import android.app.Activity;
3  import android.app.AlertDialog;
4  import android.os.Bundle;
5
6  public class TargetActivity extends Activity {
7      @Override
8      protected void onCreate(Bundle savedInstanceState) {
9          super.onCreate(savedInstanceState);
10         new AlertDialog.Builder(this).setMessage("收到活动广播!").show();
11     }
12 }
```

第 10 行代码使用 AlertDialog.Builder 类创建对话框，显示收到活动广播。

（7）在 src 目录中 hlju.edu.cn 包下创建 ActivityBroadcastReceiver.java 文件，代码如下：

```
1   package hlju.edu.cn;
2   import android.content.BroadcastReceiver;
3   import android.content.Context;
4   import android.content.Intent;
5   public class ActivityBroadcastReceiver extends BroadcastReceiver {
6       @Override
7       public void onReceive(Context context, Intent intent) {
8           Intent i = new Intent(context, TargetActivity.class);
9           i.addFlags(Intent.FLAG_ACTIVITY_NEW_TASK);
10          context.startActivity(i);
11      }
12  }
```

第 5 行代码表示继承 BroadcastReceiver 类来创建广播接收器 ActivityBroadcastReceiver。第 7 行代码表示实现 onReceive() 抽象方法。第 8 行代码表示创建 Intent 对象，启动 TargetActivity 活动。第 9 行代码表示使用 addFlags() 方法添加参数的标记值，表示启动的活动是一个新任务。

（8）修改 src 目录中 hlju.edu.cn 包下的 MainActivity.java 文件，代码如下：

```
1   package hlju.edu.cn;
2   import android.app.Activity;
3   import android.content.Intent;
4   import android.os.Bundle;
5   import android.view.View;
6   import android.view.Menu;
7   import android.view.MenuItem;
8
9   public class MainActivity extends Activity {
10      @Override
11      protected void onCreate(Bundle savedInstanceState) {
12          super.onCreate(savedInstanceState);
13          setContentView(R.layout.activity_main);
14      }
15      public void btn1_Click(View view) {
16          Intent i = new Intent("android.broadcast.TOAST");
17          sendBroadcast(i);
18      }
19      public void btn2_Click(View view) {
20          Intent i = new Intent("android.broadcast.NOTIFICATION");
21          sendBroadcast(i);
22      }
```

```
23      public void btn3_Click(View view) {
24      Intent i = new Intent("android.broadcast.ACTIVITY");
25          sendBroadcast(i);
26      }
27  }
```

第 15 行代码表示 Button 组件的 bnt1 的事件处理方法。第 16 行代码表示创建 Intent 对象 android.broadcast.TOAST。第 17 行代码表示调用 Context 类的 sendBroadcast。第 19 行代码表示 Button 组件的 bnt2 的事件处理方法。第 20 行代码表示创建 Intent 对象 android.broadcast.NOTIFICATION。第 21 行代码表示调用 Context 类的 sendBroadcast。第 23 行代码表示 Button 组件的 bnt3 的事件处理方法。第 24 行代码表示创建 Intent 对象 android.broadcast.ACTIVITY。第 25 行代码表示调用 Context 类的 sendBroadcast。

(9) 部署 BroadcastDemo 工程,程序运行结果,如图 7-9 所示。单击"发送 Toast 广播"按钮,程序运行如图 7-10 所示。

单击"发送通知广播"按钮,程序运行如图 7-11 所示。在通知窗口可以看到通知广播,如图 7-12 所示。当单击"发送活动广播"按钮时,程序运行如图 7-13 所示。

图 7-9　程序运行结果

图 7-10　发送 Toast 广播　　　　　　图 7-11　发送通知广播

图 7-12　通知窗口　　　　　　　图 7-13　发送活动广播

7.2.3　与组件交互本地服务

7.2.1节和7.2.2节两个实例分别演示了Activity组件以及Broadcast Receiver组件是如何调用本地服务的,但是并没有与本地服务进行交互。

下面通过实例介绍通过Activity如何与本地服务进行交互。

(1) 创建一个新的Android工程,工程名为LocalServiceDemo,应用程序名为LocalServiceDemo,包名为hlju.edu.cn,创建的Activity的名字为MainActivity,最小SDK版本根据选择的目标API会自动添加。

(2) 修改res目录下layout文件夹中的activity_main.xml文件,设置线性布局,添加两个Button控件,对相应控件进行描述,并设置相关属性,代码如下:

```
1  < LinearLayout xmlns:android = "http://schemas.android.com/apk/res/android"
2      xmlns:tools = "http://schemas.android.com/tools"
3      android:layout_width = "fill_parent"
4      android:layout_height = "fill_parent"
5      android:orientation = "vertical">
6  < Button android:id = "@ + id/bind_button"
7          android:layout_width = "fill_parent"
8          android:layout_height = "wrap_content"
9          android:text = "绑定服务"/>
10 < Button android:id = "@ + id/unbind_button"
11         android:layout_width = "fill_parent"
12         android:layout_height = "wrap_content"
13         android:text = "解除绑定服务"/>
14 </LinearLayout >
```

（3）在 AndroidManifest.xml 文件当中声明 Service，代码如下：

```
1   <application
2       android:allowBackup = "true"
3       android:icon = "@drawable/ic_launcher"
4       android:label = "@string/app_name"
5       android:theme = "@style/AppTheme" >
6       <activity
7           android:name = ".MainActivity"
8           android:label = "@string/app_name" >
9           <intent-filter>
10              <action android:name = "android.intent.action.MAIN" />
11              <category android:name = "android.intent.category.LAUNCHER" />
12          </intent-filter>
13      </activity>
14      <service android:name = ".LocalService"></service>
15  </application>
```

（4）在 src 目录中 hlju.edu.cn 包下创建 LocalService.java 文件，代码如下：

```
1   package hlju.edu.cn;
2
3   import java.util.Random;
4   import android.app.Notification;
5   import android.app.NotificationManager;
6   import android.app.PendingIntent;
7   import android.app.Service;
8   import android.content.Context;
9   import android.content.Intent;
10  import android.os.Binder;
11  import android.os.IBinder;
12  import android.util.Log;
13
14  public class LocalService extends Service {
15      private static final String TAG = "LocalService";
16      private final IBinder mBinder = new LocalBinder();
17      private final Random mGenerator = new Random();
18
19      private static int NOTIFICATION_ID = 1;
20      private NotificationManager notificationManager;
21      public class LocalBinder extends Binder {
22          LocalService getService() {
23              return LocalService.this;
24          }
25      }
26
27      @Override
28      public void onCreate() {
29          super.onCreate();
```

```
30          notificationManager = (NotificationManager) getSystemService (Context.NOTIFICATION_
   SERVICE);
31          displayNotification("Bind Service Created");
32          Log.d(TAG, "onCreate() method was invoked");
33      }
34
35      @Override
36      public IBinder onBind(Intent intent) {
37        Log.d(TAG, "onBind() method was invoked");
38            return mBinder;
39      }
40
41      public int getRandomNumber() {
42        Log.d(TAG, "getRandomNumber() method was invoked");
43          return mGenerator.nextInt(100);
44      }
45      @Override
46      public void onDestroy() {
47        super.onDestroy();
48        displayNotification("Bind Service Destoryed");
49        Log.d(TAG, "onDestroy() method was invoked");
50      }
51      @Override
52      public boolean onUnbind(Intent intent) {
53        Log.d(TAG, "onUnbind() method was invoked");
54        return super.onUnbind(intent);
55      }
56          private void displayNotification(String message){
57        Notification notification = new Notification(R.drawable.ic_launcher, message,
   System.currentTimeMillis());
58            PendingIntent contentIntent = PendingIntent.getActivity(this, 0, new Intent(this,
   MainActivity.class), 0);
59            notification.setLatestEventInfo(this, "Background Service", message, contentIntent);
60        notificationManager.notify(NOTIFICATION_ID ++, notification);
61      }
62 }
```

第 16 行代码表示将 IBinder 返回给调用的客户端。第 17 行代码表示随机数生成器。第 20 行代码表示通知管理器。第 23 行代码表示返回当前 Service 的实例,使客户端可以调用 Service 里面声明的公共方法。第 41 行代码表示调用客户端的方法。第 56 行代码表示显示通知。

(5) 修改 src 目录中 hlju.edu.cn 包下的 MainActivity.java 文件,代码如下:

```
1   package hlju.edu.cn;
2
3   import android.app.Activity;
4   import android.content.ComponentName;
```

```java
5  import android.content.Context;
6  import android.content.Intent;
7  import android.content.ServiceConnection;
8  import android.os.Bundle;
9  import android.os.IBinder;
10 import android.view.Menu;
11 import android.view.MenuItem;
12 import android.view.View;
13 import android.view.View.OnClickListener;
14 import android.widget.Button;
15 import android.widget.Toast;
16 import hlju.edu.cn.LocalService.LocalBinder;
17 public class MainActivity extends Activity implements OnClickListener{
18     Button bindServiceBtn;
19     Button unbindServiceBtn;
20     boolean mBound = false;
21     LocalService mService = null;
22 private ServiceConnection mConnection = new ServiceConnection() {
23        @Override
24        public void onServiceDisconnected(ComponentName name) {
25            mBound = false;
26        }
27        @Override
28        public void onServiceConnected(ComponentName name, IBinder service) {
29            LocalBinder binder = (LocalBinder)service;
30            mService = binder.getService();
31            mBound = true;
32            int num = mService.getRandomNumber();
33            Toast.makeText(MainActivity.this, "随机数:" + num , Toast.LENGTH_LONG).show();
34        }
35    };
36    @Override
37    public void onCreate(Bundle savedInstanceState) {
38        super.onCreate(savedInstanceState);
39        setContentView(R.layout.activity_main);
40        bindServiceBtn = (Button)findViewById(R.id.bind_button);
41        unbindServiceBtn = (Button)findViewById(R.id.unbind_button);
42        bindServiceBtn.setOnClickListener(this);
43        unbindServiceBtn.setOnClickListener(this);
44    }
45    @Override
46    public void onClick(View v) {
47     int viewId = v.getId();
48     if(R.id.bind_button == viewId){
49         Intent intent = new Intent(this, LocalService.class);
50         bindService(intent, mConnection, Context.BIND_AUTO_CREATE);
51     }else if(R.id.unbind_button == viewId){
52         if(mBound){
53             unbindService(mConnection);
54             mBound = false;
55         }
56    } }
```

第 18 行代码声明绑定服务按钮。第 19 行代码声明解绑服务按钮。第 20 行代码表示是否绑定进行标示。第 25 行代码表示解绑服务。第 31 行代码表示成功绑定服务。第 32 行代码表示获取随机数。第 40 行代码表示找到绑定服务按钮。第 41 行代码表示未找到绑定服务按钮。第 42 行代码表示绑定单击事件监听器。第 48 行代码表示判断是否为绑定按钮。第 51 行代码表示判断是否为解绑按钮。

（6）部署 LocalServiceDemo 工程，程序运行结果，如图 7-14 所示。单击"绑定服务"按钮，程序运行如图 7-15 所示。在通知栏可以看到绑定服务的创建通知，如图 7-16 所示。

图 7-14　程序运行结果

图 7-15　绑定服务

图 7-16　通知栏

此时 LogCat 的输出效果如图 7-17 所示。

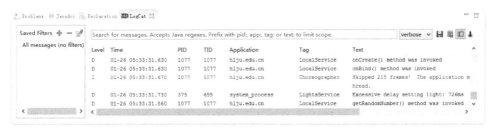

图 7-17　单击"绑定服务"按钮时的 LogCat

单击"解除绑定服务"按钮，程序运行如图 7-18 所示。在通知栏可以看到绑定服务的创建通知，如图 7-19 所示。

此时 LogCat 的输出效果如图 7-20 所示。

图 7-18　解除绑定服务　　　　　　　　图 7-19　通知栏

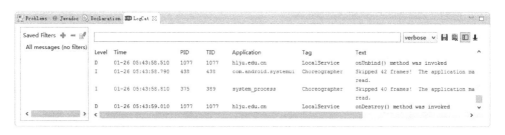

图 7-20　单击"解除绑定服务"按钮时的 LogCat

7.2.4　Service 与 Thread 的区别

　　Thread 是程序执行的最小单元,它是分配 CPU 的基本单位,可以用 Thread 来执行一些异步的操作。子线程当中的 run()方法体,是在子线程当中运行的。Service 是 Android 的四大组件之一,主要用于在后台重复执行一些比较耗时的操作。Service 分为本地服务以及远程服务两种,对于本地服务来说,是在当前应用程序进程的 UI 主线程当中运行的,对于远程服务来说,是在被调用的远程服务所在的应用程序进程的 UI 主线程当中运行的,为了防止对 UI 线程的阻塞,一般在 Service 当中也要新建 Thread 来使得后台比较耗时的操作能够在非 UI 主线程当中执行。

　　如果在一个 Activity 组件当中启动一个子线程,由于线程是程序执行的不同线索,所以子线程的执行是独立于 UI 主线程的,即便是启动子线程的 Activity 组件已经被销毁了,子线程仍可以继续执行下去。而且没有办法在不同的 Activity 当中对同一个子线程进行控制,因为 Android 当中的子线程一般都是通过匿名内部类的方式进行声明的。

　　对于一个 Service 来说,系统只会创建它的一个实例,也就是 Service 的 onCreate()方法只会在第一次创建的时候才会被调用。多次调用 Context.startService()方法不会启动多个 Service,只是相应 Service 对象的 onStartCommand()方法会被调用多次,所以可以在不同组件中调用 Context.stopService()来销毁系统当中通过该方法启动的 Service。多次调用 Context.bindService()方法也不会启动多个 Service,只是相应 Service 对象的 onBind()方法会被调用多次,也可以在不同的组件当中通过调用 Context.unbindService()来销毁系统当中通过该方法启动的 Service。

总体来说,Thread 不在 UI 主线程当中运行,不易于管理,而 Service 默认是在主线程当中运行,易于管理,适合重复执行比较耗时的后台操作。

7.3 管理 Service 的生命周期

服务的生命周期比 Activity 简单很多,但是开发人员需要更加关注服务如何创建和销毁,因为服务在用户不知情时就可以在后台运行。服务的生命周期可以分为两个不同的路径。

当其他组件调用 startService()方法时,服务被创建。接着服务无限期运行,服务必须调用 stopSelf()方法或者由其他组件调用 stop Service()方法来停止。当服务停止时,系统将其销毁。

当其他组件调用 bindService()方法时,服务被创建。接着客户端通过 IBinder 接口与服务通信。客户端通过 unbindService()方法关闭连接。多个客户端能绑定到同一个服务,并且当它们都被解除绑定时,系统销毁服务。

这两条路径并非完全独立,即开发人员可以绑定已经使用 startService()方法启动的服务,例如,后天音乐服务能通过包含音乐信息的 Intent 调用 startService()方法启动。然后,当用户需要控制播放器或者获得当前音乐信息时,可以调用 bindService()方法绑定 Activity 到服务。此时,stopService()和 stopSelf()方法直到全部客户端解除绑定时才能停止服务。两类服务的生命周期如图 7-21 所示。

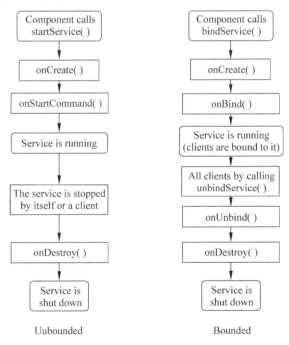

图 7-21 服务生命周期

下面创建一个综合实例。

（1）创建一个新的 Android 工程，工程名为 TimeServiceDemo，应用程序名为 TimeServiceDemo，包名为 hlju.edu.cn，创建的 Activity 的名字为 MainActivity，最小 SDK 版本根据选择的目标 API 会自动添加。

（2）修改 res 目录下 layout 文件夹中的 activity_main.xml 文件，设置线性布局，添加 1 个 TextView 控件，对相应控件进行描述，并设置相关属性，代码如下：

```xml
1  <?xml version = "1.0" encoding = "utf-8"?>
2  <LinearLayout xmlns:android = "http://schemas.android.com/apk/res/android"
3      android:layout_width = "fill_parent"
4      android:layout_height = "fill_parent"
5      android:orientation = "vertical" >
6      <TextView
7          android:id = "@+id/textView"
8          android:layout_width = "fill_parent"
9          android:layout_height = "wrap_content"
10         android:gravity = "center"
11         android:text = "@string/activity_title"
12         android:textColor = "@android:color/black"
13         android:textSize = "25dp" />
```

（3）在 AndroidManifest.xml 文件当中声明 Service，代码如下：

```xml
1  <?xml version = "1.0" encoding = "utf-8"?>
2  <manifest xmlns:android = "http://schemas.android.com/apk/res/android"
3      package = "hlju.edu.cn"
4      android:versionCode = "1"
5      android:versionName = "1.0" >
6      <uses-sdk
7          android:minSdkVersion = "14"
8          android:targetSdkVersion = "21" />
9      <application
10         android:allowBackup = "true"
11         android:icon = "@drawable/ic_launcher"
12         android:label = "@string/app_name"
13         android:theme = "@style/AppTheme" >
14         <activity
15             android:name = ".MainActivity"
16             android:label = "@string/app_name" >
17             <intent-filter>
18                 <action android:name = "android.intent.action.MAIN" />
19                 <category android:name = "android.intent.category.LAUNCHER" />
20             </intent-filter>
21         </activity>
22         <service android:name = ".TimeService" >
23         </service>
24     </application>
25 </manifest>
```

（4）在 src 目录中 hlju.edu.cn 包下创建 TimeService.java 文件，代码如下：

```
1   package hlju.edu.cn;
2
3   import java.util.Timer;
4   import java.util.TimerTask;
5   import android.app.Notification;
6   import android.app.NotificationManager;
7   import android.app.PendingIntent;
8   import android.app.Service;
9   import android.content.Context;
10  import android.content.Intent;
11  import android.os.IBinder;
12
13  public class TimeService extends Service {
14
15      private Timer timer;
16
17      @Override
18      public IBinder onBind(Intent intent) {
19          return null;
20      }
21
22      @Override
23      public void onCreate() {
24          super.onCreate();
25          timer = new Timer(true);
26      }
27
28      @Override
29      public void onStart(Intent intent, int startId) {
30          super.onStart(intent, startId);
31          timer.schedule(new TimerTask() {
32
33              @Override
34              public void run() {
35                  String ns = Context.NOTIFICATION_SERVICE;
36                  NotificationManager manager = (NotificationManager) getSystemService(ns);
37                  Notification notification = new Notification(R.drawable.ic_launcher, getText(R.string.ticker_text), System.currentTimeMillis());
38                  CharSequence contentTitle = getText(R.string.content_title);
39                  CharSequence contentText = getText(R.string.content_text);
40                  Intent intent = new Intent(TimeService.this, MainActivity.class);
41                  PendingIntent contentIntent = PendingIntent.getActivity(TimeService.this, 0, intent, Intent.FLAG_ACTIVITY_NEW_TASK);
42                  notification.setLatestEventInfo(TimeService.this, contentTitle, contentText, contentIntent);
43                  manager.notify(0, notification);
44                  TimeService.this.stopSelf();
45              }
46          }, 60000);        }        }
47
```

第 25 行代码表示创建 Timer 对象。第 36 行代码表示获得通知管理器。第 37 行代码表示创建通知。第 38 行代码表示定义通知的标题。第 39 行代码表示定义通知的内容。第 40 行代码表示创建 Intent 对象。第 41 行代码表示创建 PendingIntent 对象。第 42 行代码表示定义通知行为。第 43 行代码表示显示通知。第 44 行代码表示停止服务。第 47 行代码表示设定的时间。

(5) 修改 src 目录中 hlju.edu.cn 包下的 MainActivity.java 文件,代码如下:

```java
1   package hlju.edu.cn;
2
3   import android.app.Activity;
4   import android.content.Intent;
5   import android.os.Bundle;
6   import android.view.Menu;
7   import android.view.MenuItem;
8
9   public class MainActivity extends Activity {
10
11      @Override
12      protected void onCreate(Bundle savedInstanceState) {
13          super.onCreate(savedInstanceState);
14          setContentView(R.layout.activity_main);
15
16          startService(new Intent(this, TimeService.class));
17      }
18      @Override
19      public boolean onCreateOptionsMenu(Menu menu) {
20          // Inflate the menu; this adds items to the action bar if it is present.
21          getMenuInflater().inflate(R.menu.main, menu);
22          return true;
23      }
24      @Override
25      public boolean onOptionsItemSelected(MenuItem item) {
26          // Handle action bar item clicks here. The action bar will
27          // automatically handle clicks on the Home/Up button, so long
28          // as you specify a parent activity in AndroidManifest.xml.
29          int id = item.getItemId();
30          if (id == R.id.action_settings) {
31              return true;
32          }
33          return super.onOptionsItemSelected(item);
34      }
35  }
```

(6) 部署 LocalServiceDemo 工程,程序运行结果如图 7-22 所示。1 分钟后程序运行结果如图 7-23 所示。在通知栏可以看到通知,如图 7-24 所示。

图 7-22　程序运行结果　　　　图 7-23　1 分钟后程序运行结果

图 7-24　通知栏

习题

1. 简述 Android Service 的本地服务。
2. Service 与 Thread 的区别是什么。
3. 简述两类服务的生命周期。

第8章 数据存储与访问

在 Android 系统中,数据存储和使用与通常的数据操作有很大的不同。Android 中所有的应用程序数据都为自己应用程序所有,其他应用程序如果共享、访问别的应用程序数据,必须通过 Android 系统提供的方式才能访问或者暴露自己的私有数据供其他应用程序使用。Android 平台中实现数据存储的方式有 5 种,分别是使用 SharedPreferences 存储数据、文件存储数据、SQLite 数据库存储数据、使用 ContentProvider 存储数据和网络存储数据。

本章主要学习内容:
- 掌握 SharedPreferences 应用;
- 掌握文件存储的各种应用;
- 掌握 SQLite 数据库的创建、操作和管理;
- 了解数据共享。

8.1 SharedPreferences

8.1.1 SharedPreferences 简介

前面已经讲述了在 Android 系统中可以使用一个 SharedPreferences 类来保存一些系统的配置信息、窗口的状态等。SharedPreferences 接口位于 android.cotent 包下,它是一个轻量级的存储类,特别适合用于保存软件配置参数。

使用 SharedPreferences 保存数据,最终是以 XML 文件存储数据,是基于 XML 文件存储键值对(Name/Value Pair,NVP)数据。XML 处理时 Dalvik 会通过自带底层的本地 XML Parser 解析,例如 XMLpull 方式。SharedPreferences 保存数据的文件存储在目录/data/data/<package name>/shared_prefs 下。SharedPreferences 不仅能够保存数据,还能够实现不同应用程序间的数据共享。

由于 SharedPreferences 完全对用户屏蔽对文件系统的操作过程,在开发中 SharedPreferences 对象本身只能获取数据而不支持存储和修改,存储修改是通过 Editor 对象实现的。

SharedPreferences 支持各种基本数据类型,包括整型、布尔型、浮点型和长型等。

1. SharedPreferences 访问模式

在 Android 系统中，SharedPreferences 分为许多权限，其支持的访问模式有 3 种：私有、全局读和全局写。

私有(Context.MODE_PRIVATE)：为默认操作模式，代表该文件是私有数据，只能被应用本身访问，在该模式下写入的内容会覆盖原文件的内容。

全局读(Context.MODE_WORLD_READABLE)：不仅创建程序可以对其进行读取或写入，其他应用程序也有读取操作的权限，但没有写入操作的权限。

全局写(Context.MODE_WORLD_WRITEABLE)：创建程序和其他程序都可以对其进行写入操作，但没有读取的权限。

2. 使用 SharedPreferences 实现存储

访问自身程序数据的通常用法如下。

1) 定义 SharedPreferences 的访问模式

在使用 SharedPreferences 前，先定义 SharedPreferences 的访问模式。

如将访问模式定义为私有模式：

```
public static int MODE = Context.MODE_PRIVATE;
```

也可以将 SharedPreferences 的访问模式设定为既可以全局读，也可以全局写。设定如下：

```
public static int MODE = Context.MODE_WORLD_READABLE + Context.MODE_WORLD_WRITEABLE;
```

2) 定义 SharedPreferences 的名称

SharedPreferences 的名称与在 Android 文件系统中保存的文件同名。因此，只要具有相同的 SharedPreferences 名称的 NVP 内容都会保存在同一个文件中，例如：

```
public static finial String PR_NAME = "SaveFile";
```

3) 获取 SharedPreferences 对象

使用 SharedPreferences，需要将上述定义的访问模式和 SharedPreferences 名称作为参数，传递到 getSharedPreferences 方法并获取到 SharedPreferences 对象。

```
SharedPreferences SharedPreferences = getSharedPreferences(PR_NAME,MODE);
```

4) 利用 edit()方法获取 Editor 对象

在获取到 SharedPreferences 对象后，可以通过 SharedPreferences.Editor 类对 SharedPreferences 进行修改。

```
Editor editor = sharedPreferences.edit();
```

5）通过 Editor 对象存储 key-value 键值对数据

```
1   editor.putString("Name","Jim");
2   editor.putInt("Age","17");
3   editor.putFloat("Height","1.81");
```

6）通过 commit()方法提交数据

```
editor.commit();
```

完成上述步骤后，如果需要从已经保存的 SharedPreferences 中读取数据，同样是调用 getSharedPreferences()方法，并在方法的第 1 个参数中指明需要访问的 SharedPreferences 名称，然后通过 get<Type>()方法获取保存在 SharedPreferences 中的 NVP。

```
1   SharedPreferences sharedPreferences = getSharedPreferences(PR_NAME,MODE);
2   String name = sharedPreferences.getString("NAME","name");
3   int age = sharedPreferences.getInt("Age",18);
4   float height = sharedPreferences.getFloat("Height",);
```

上述代码中，get<Type>()方法中的第 1 个参数是 NVP 的名称。

第 2 个参数是在无法获取到数值的时候使用的默认值。如第 4 行中的 getFloat()的第 2 个参数为默认值，如果 preference 中不存在该 key，将返回默认值。

3. 访问其他应用程序数据的 SharedPreferences

如果需要创建访问其他应用程序数据的 SharedPreferences，其前提条件如下。

在 SharedPreferences 对象创建时，为其指定 Context.MODE_WORLD_READABLE 或者 Context.MODE_WORLD_WRITEABLE 权限。

```
1   Context otherApps = creatPackageContext ("com.hissoft.sharedpreferences ", Context.CONTEXT_IGNORE_SECURITY);
2   SharedPreferences sharedPreferences = otherApps.getSharedPreferences ("testApps", Context.MODE_WORLD_READABLE);
3   String name = sharedPreferences.getString("name"," ");
4   int age = sharedPreferences.getInt("Age",20);
```

如果想采用读取 XML 文件方式，直接访问其他应用 SharedPreferences 对应的 XML 文档，代码如下：

```
File sfx = new File("/data/data/<package name>/shared_prefs/mypreferences.xml");
```

代码中的<package name>应替换成应用的包名。

4. 访问资源文件

（1）访问存储在 res 目录下的文件，如 res/raw 目录下：

```
InputStream ismp3 = getResource().openRawResource(R.raw.testVideo);
```

代码的作用是存储声音文件。

（2）访问存储在 assets 目录下的文件。

```
InputStream anyFile = getAssets().open(name);
```

代码的作用是存储数据文件。

注意：存储文件的大小有限制。

SharedPreferences 对象与后续讲解的 SQLite 数据库相比，省略了创建数据库、创建表和写 SQL 语句等诸多操作，相对而言更加方便、简洁。但是 SharedPreferences 也有其自身缺陷，例如其只能存储 boolean、int、float、long 和 String 这 5 种简单的数据类型，无法进行条件查询等。所以不论 SharedPreferences 的数据存储操作是如何简单，它也只能是存储方式的一种补充，而无法完全替代如 SQLite 数据库这样的其他数据存储方式。

8.1.2 存储应用程序数据实例

上面简单介绍了 SharedPreferences 的基础知识和存储访问应用方法，下面通过一个案例详细介绍 SharedPreferences 存储程序数据的应用。

（1）创建一个新的 Android 工程，工程名为 SharedPreferencesDemo，应用程序名为 SharedPreferencesDemo，包名为 hlju.edu.cn，创建的 Activity 的名字为 MainActivity，最小 SDK 版本根据选择的目标 API 会自动添加。

（2）修改 res 目录下 layout 文件夹中的 activity_main.xml 文件，设置线性布局，添加两个 TextView 控件、两个 EditText 和 1 个 Button 控件，对相应控件进行描述，并设置相关属性，代码如下：

```
1  <?xml version = "1.0" encoding = "utf - 8"?>
2  < LinearLayout xmlns:android = "http://schemas.android.com/apk/res/android"
3      android:layout_width = "fill_parent"
4      android:layout_height = "fill_parent"
5      android:orientation = "vertical" >
6      < TextView
7          android:layout_width = "fill_parent"
8          android:layout_height = "wrap_content"
9          android:text = "用户名" />
10     < EditText
11         android:id = "@ + id/username"
12         android:layout_width = "fill_parent"
13         android:layout_height = "wrap_content" />
```

```
14      <TextView
15          android:layout_width = "fill_parent"
16          android:layout_height = "wrap_content"
17          android:text = "年龄" />
18      <EditText
19          android:id = "@ + id/age"
20          android:layout_width = "fill_parent"
21          android:layout_height = "wrap_content"
22          android:numeric = "integer" />
23      <Button
24          android:id = "@ + id/saveBtn"
25          android:layout_width = "fill_parent"
26          android:layout_height = "wrap_content"
27          android:text = "保存个人信息" />
28  </LinearLayout>
```

(3) 修改 src 目录中 hlju.edu.cn 包下的 MainActivity.java 文件,代码如下:

```
1   package hlju.edu.cn;
2
3   import android.app.Activity;
4   import android.content.Context;
5   import android.content.SharedPreferences;
6   import android.content.SharedPreferences.Editor;
7   import android.os.Bundle;
8   import android.view.Menu;
9   import android.view.MenuItem;
10  import android.view.View.OnClickListener;
11  import android.widget.Button;
12  import android.widget.EditText;
13  import android.widget.Toast;
14  public class MainActivity extends Activity implements OnClickListener {
15      private  static final String FILE_NAME = "info";
16      EditText usernameEdit;
17      EditText ageEdit;
18      Button saveBtn;
19  @Override
20  public void onCreate(Bundle savedInstanceState) {
21          super.onCreate(savedInstanceState);
22          setContentView(R.layout.activity_main);
23  usernameEdit = (EditText)findViewById(R.id.username);
24          ageEdit = (EditText)findViewById(R.id.age);
25          saveBtn = (Button)findViewById(R.id.saveBtn);
26          saveBtn.setOnClickListener(this);
27  }
28  @Override
29  public void onClick(View v) {
30      if(R.id.saveBtn == v.getId()){
```

```
31        saveInfo();
32      }
33   }
34   private void saveInfo(){
35       SharedPreferences infoPref = getSharedPreferences(FILE_NAME, Context.MODE_PRIVATE);
36       Editor editor = infoPref.edit();
37       String username = usernameEdit.getText().toString();
38       int   age = Integer.parseInt(ageEdit.getText().toString());
39       editor.putString("username", username);
40       editor.putInt("age", age);
41       editor.commit();
42       Toast.makeText(this, "保存个人信息成功", Toast.LENGTH_LONG ).show();
43   }
44     @Override
45     public boolean onCreateOptionsMenu(Menu menu) {
46       // Inflate the menu; this adds items to the action bar if it is present.
47       getMenuInflater().inflate(R.menu.main, menu);
48       return true;
49      }
50
51     @Override
52     public boolean onOptionsItemSelected(MenuItem item) {
53       // Handle action bar item clicks here. The action bar will
54       // automatically handle clicks on the Home/Up button, so long
55       // as you specify a parent activity in AndroidManifest.xml.
56       int id = item.getItemId();
57       if (id == R.id.action_settings) {
58         return true;
59       }
60       return super.onOptionsItemSelected(item);
61     }
62   }
```

（4）右击工程 SharedPreferencesDemo，在弹出的快捷菜单中选择 Run As|Android Application 命令，工程运行结果如图 8-1 所示。在界面当中输入用户名和年龄，如图 8-2 所示。单击"保存个人信息"按钮后运行效果如图 8-3 所示。

数据存储在路径 data/data/hlju.edu.cn/shared_prefs/目录下，通过选择 Eclipse 菜单中的 Window→Show View→Other 菜单项，在对话框中展开 android 文件夹，选择下面的 File Explorer 视图，然后在 File Explorer 视图中展开 data/data/hlju.edu.cn/shared_prefs/，可以找到名称为 info.xml 文件，单击 Pull a file from a device 按钮导出文件，文件内容如下面代码所示。

图 8-1　工程运行图

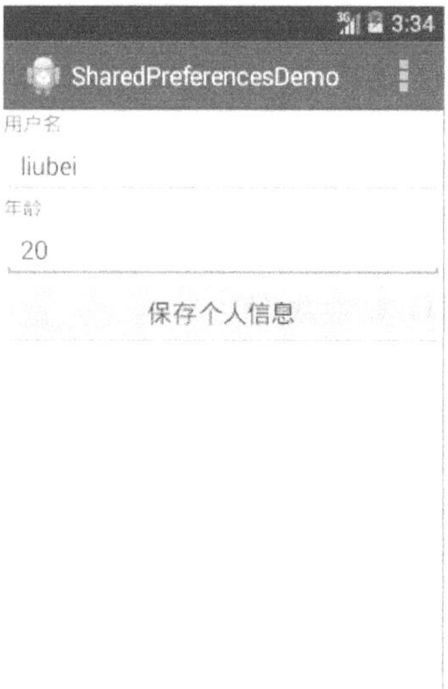

图 8-2　输入用户名和年龄　　　　　图 8-3　保存个人信息效果

```
1    <?xml version = '1.0' encoding = 'utf-8' standalone = 'yes' ?>
2    <map>
3        <string name = "username">liubei</string>
4        <int name = "age" value = "20" />
5    </map>
```

也可以使用 adb shell 命令，进入设备或者模拟器的 Shell 环境中，在这个 Linux Shell 中，允许执行各种 Linux 命令。由于创建的 info.xml 文件位于 data/data/hlju.edu.cn/shared_prefs/目录下，在控制台窗口输入 adb shell 命令过后，通过 Linux 的 cd 命令，进入 info.xml 文件所在的目录后，在输入 Linux 查看文件内容的命令 cat info.xml，控制台窗口的输出如图 8-4 所示。

图 8-4　控制台窗口的输出

8.1.3　读取其他应用程序数据实例

上述介绍了 SharedPreferences 存储应用程序数据的方法，下面通过一个案例详细介绍 SharedPreferences 访问其他应用程序数据的应用。

(1) 创建一个新的 Android 工程,工程名为 SharedPreferencesDemo1,应用程序名为 SharedPreferencesDemo1,包名为 hlju.edu.cn,创建的 Activity 的名字为 MainActivity,最小 SDK 版本根据选择的目标 API 会自动添加。

(2) 修改 res 目录下 layout 文件夹中的 activity_main.xml 文件,设置线性布局,添加一个 Button 控件,并设置相关属性,代码如下:

```xml
1   <?xml version = "1.0" encoding = "utf-8"?>
2   <LinearLayout xmlns:android = "http://schemas.android.com/apk/res/android"
3       android:layout_width = "fill_parent"
4       android:layout_height = "fill_parent"
5       android:orientation = "vertical" >
6     <Button
7         android:id = "@+id/readBtn"
8         android:layout_width = "fill_parent"
9         android:layout_height = "wrap_content"
10        android:text = "@string/information" />
11  </LinearLayout>
```

(3) 修改 src 目录中 hlju.edu.cn 包下的 MainActivity.java 文件,代码如下:

```java
1   package hlju.edu.cn;
2
3   import android.app.Activity;
4   import android.content.Context;
5   import android.content.SharedPreferences;
6   import android.os.Bundle;
7   import android.view.Menu;
8   import android.view.MenuItem;
9   import android.view.View;
10  import android.view.View.OnClickListener;
11  import android.widget.Button;
12  import android.widget.Toast;
13
14  public class MainActivity extends Activity implements OnClickListener {
15      private  static final String FILE_NAME = "info";
16      Button readBtn;
17  @Override
18  public void onCreate(Bundle savedInstanceState) {
19      super.onCreate(savedInstanceState);
20      setContentView(R.layout.activity_main);
21      readBtn = (Button)findViewById(R.id.readBtn);
22      readBtn.setOnClickListener(this);
23  }
24  @Override
25  public void onClick(View v) {
26  if(R.id.readBtn == v.getId()){
27          readInfo();
28      }
29  }
30
```

```
31    private void readInfo(){
32        SharedPreferences infoPref = getSharedPreferences(FILE_NAME, Context.MODE_PRIVATE);
33        String username = infoPref.getString("username", "");
34        int age = infoPref.getInt("age", 0);
35        Toast.makeText(this, "用户名: " + username + " 年龄: " + age, Toast.LENGTH_LONG).show();
36    }
37
38    @Override
39
40    public boolean onCreateOptionsMenu(Menu menu) {
41        // Inflate the menu; this adds items to the action bar if it is present.
42        getMenuInflater().inflate(R.menu.main, menu);
43        return true;
44    }
45
46    @Override
47    public boolean onOptionsItemSelected(MenuItem item) {
48        // Handle action bar item clicks here. The action bar will
49        // automatically handle clicks on the Home/Up button, so long
50        // as you specify a parent activity in AndroidManifest.xml.
51        int id = item.getItemId();
52        if (id == R.id.action_settings) {
53            return true;
54        }
55        return super.onOptionsItemSelected(item);
56    }
57 }
```

（4）部署运行 SharedPreferencesDemo1 工程，读取上一案例存储的 username 和 age 值，程序运行窗口如图 8-5 所示。单击"读取信息"按钮后的运行效果如图 8-6 所示。

图 8-5　运行窗口　　　　　　　　图 8-6　读取信息效果

8.2 文件存储

Android 系统使用的是基于 Linux 的文件系统,应用程序开发人员可以建立和访问程序自身的私有文件,也可以访问保存在资源目录中的原始文件和 XML 文件。此外,还可以在 SD 卡等外部存储设备中保存文件信息等。

8.2.1 文件存储简介(内部存储)

在 Android 系统中,允许应用程序创建仅能够自身访问的私有文件,文件保存在设备的内部存储器上,文件默认保存路径位置是/data/data/<package name>/files/目录下。

Android 系统不仅支持标准 Java 的 IO 类和方法,还提供了能够简化读写流式文件过程的方法。关于文件存储,Activity 提供了 openFileOutput()方法和 openFileInput()方法。

openFileOutput()方法可以用于把数据输出到文件中。

openFileInput()方法为打开应用程序私有文件读取数据。

具体的实现过程在 J2SE 环境中保存数据到文件中是一样的。文件可用于存储大量数据,如文本、图片和音频等。

下述是 openFileOutput()方法和 openFileInput()方法的用法。

1) openFileOutput()方法的用法

openFileOutput()方法为打开应用程序私有文件写入数据,如果指定的文件不存在,则创建一个新的文件。

openFileOutput()方法的语法声明:

```
public FileOutputStream openFileOutput(String name, int mode)
```

第 1 个参数是文件名称,这个参数不能包含路径分隔符"/"。

第 2 个参数是文件操作模式。

使用 openFileOutput()方法创建新文件,代码如下:

```
1   Sting filename = "data.txt";
2   FileOutputStream fos = openFileOutput(filename, Context.MODE_PRIVATE);
3   Sting ts = "What is this?";
4   fos.write(ts.getBytes());
5   fos.flush;
6   fos.close;
```

第 1 行定义了创建文件的名称"data.txt",第 2 行使用 openFileOutput()方法以私有模式建立文件,第 4 行将数据写入文件,第 5 行将缓存中所有剩余的数据写入文件,第 6 行关闭流。

2) openFilInput()方法的用法

如果要打开存储在/data/data/<package name>/files 目录下的应用程序私有的文件,可以使用 openFilInput()方法。

openFileInput()方法的语法声明:

```
public FileInputStream openFileInput(String name)
```

方法参数也是文件名称,字符串中不能包含路径分隔符"/"。

如果想直接使用文件的绝对路径,可以使用如下代码:

```
1   File filename = new File("/data/data/hlju.edu.cn/files/data.txt");
2   FileInputStream inStream = new FileInputStream(filename);
```

上面第1行文件路径中的 hlju.edu.cn 为应用所在包。

对于私有文件只能被创建该文件的应用访问。如果希望文件能被其他应用读和写,可以在创建文件时指定 Context.MODE_WORLD_READABLE 和 Context.MODE_WORLD_WRITEABLE 权限。

Activity 还提供了 getCacheDir()和 getFilesDir()方法。

getCacheDir()方法用于获取/data/data/<package name>/cache 目录。

getFilesDir()方法用于获取/data/data/<package name>/files 目录。

注意:

(1) openFileOutput()方法和 openFileInput()方法使用时必须使用 try{} catch{}捕获异常。

(2) 创建的文件保存在/data/data/<package name>/files 目录下,如 data/data/hlju.edu.cn/files/data.txt。

同上节讲述的一样,通过 File Explorer 视图,在 File Explorer 视图中展开/data/data/<package name>/files 目录即可看到该文件。Android 系统支持 4 种文件操作模式,如表 8-1 所示。

表 8-1　Android 系统支持的 4 种文件操作模式

文件操作模式	值	描 述
MODE_PRIVATE	0	私有模式,文件仅能够被文件创建程序访问,或具有相同 UID 的程序访问。为默认操作模式,代表该文件是私有数据,只能被应用本身访问。在该模式下写入的内容会覆盖原文件的内容,如果想把新写入的内容追加到原文件中,可以使用 Context.MODE_APPEND
MODE_APPEND	32768	追加模式,模式会检查文件是否存在,存在则追加内容,否则就创建新文件
MODE_WORLD_READABLE	1	全局读模式,允许任何程序读取私有文件
MODE_WORLD_WRITEABLE	2	全局写模式,允许任何程序写入私有文件

注意: 在使用上述模式时,可以用"+"来选择多种模式,例如 openFileOutput(FILENAME,Context.MODE_READABLE+Context.MODE_WORLD_WRITEABLE);。

8.2.2　文件存储应用实例

8.2.1节介绍了文件存储访问方式及访问方法,下面通过一个文件存储案例详细介绍访问 File 的应用。

（1）创建一个新的 Android 工程，工程名为 FileWriteAndDemo，应用程序名为 FileWriteAndDemo，包名为 hlju.edu.cn，创建的 Activity 的名字为 MainActivity，最小 SDK 版本根据选择的目标 API 会自动添加。

（2）修改 res 目录下 layout 文件夹中的 activity_main.xml 文件，设置线性布局，添加 1 个 EditText 和两个 Button 控件，代码如下：

```
1   <LinearLayout xmlns:android = "http://schemas.android.com/apk/res/android"
2       android:layout_width = "fill_parent"
3       android:layout_height = "fill_parent"
4       android:orientation = "vertical">
5   <!-- 接收用户输入的数据 -->
6       <EditText
7           android:id = "@ + id/myEditText"
8           android:layout_width = "fill_parent"
9           android:layout_height = "wrap_content" />
10  <!-- 保存数据 -->
11      <Button
12          android:id = "@ + id/saveInternal"
13          android:layout_width = "fill_parent"
14          android:layout_height = "wrap_content"
15          android:text = "保存数据"/>
16  <!-- 读取数据 -->
17      <Button
18          android:id = "@ + id/loadInternal"
19          android:layout_width = "fill_parent"
20          android:layout_height = "wrap_content"
21          android:text = "读取数据"/>
22  </LinearLayout>
```

（3）修改 src 目录下包 hlju.edu.cn 中的 MainActivity.java 文件，代码如下：

```
1   package hlju.edu.cn;
2
3   import java.io.ByteArrayOutputStream;
4   import java.io.FileInputStream;
5   import java.io.FileOutputStream;
6   import android.app.Activity;
7   import android.content.Context;
8   import android.os.Bundle;
9   import android.util.Log;
10  import android.view.Menu;
11  import android.view.MenuItem;
12  import android.view.View;
13  import android.view.View.OnClickListener;
14  import android.widget.Button;
15  import android.widget.EditText;
```

```java
16   import android.widget.Toast;
17
18   public class MainActivity extends Activity implements OnClickListener {
19       private static final String TAG = "MainActivity";
20   EditText myEditText;
21       Button saveBtn;
22       Button readBtn;
23
24       @Override
25       public void onCreate(Bundle savedInstanceState) {
26         super.onCreate(savedInstanceState);
27         setContentView(R.layout.activity_main);
28         myEditText = (EditText) findViewById(R.id.myEditText);
29         saveBtn = (Button) findViewById(R.id.saveInternal);
30         readBtn = (Button) findViewById(R.id.loadInternal);
31         saveBtn.setOnClickListener(this);
32         readBtn.setOnClickListener(this);
33       }
34
35       @Override
36       public void onClick(View view) {
37         int viewId = view.getId();
38         if (R.id.saveInternal == viewId) {
39           try {
40             writeToInternalStorage("data.txt", myEditText.getText().toString().getBytes());
41             Toast.makeText(this, "写入数据成功", Toast.LENGTH_LONG).show();
42           } catch (Exception e) {
43             Toast.makeText(this, "写入数据失败", Toast.LENGTH_LONG).show();
44           }
45         } else if (R.id.loadInternal == viewId) {// 读取数据按钮单击
46           try {
47             String readData = new String(readFormInternalStorage("data.txt"), "UTF-8");
48             Toast.makeText(this, "读取的数据: " + readData, Toast.LENGTH_LONG).show();
49           } catch (Exception e) {
50             Toast.makeText(this, "读取数据失败", Toast.LENGTH_LONG).show();
51           }
52         }
53       }
54   private  void writeToInternalStorage(String filename, byte[] content) {
55       if(null == content || 0 == content.length){//对写入数据合法性进行校验
56
57           return;
58       }
59       try {
60         FileOutputStream fos = openFileOutput(filename, Context.MODE_PRIVATE);
61         fos.write(content);
62         fos.close();
63       } catch (Exception e) {
64         Log.d(TAG, "写入失败: exception:" + e);
```

```
65          throw new RuntimeException("write to internal storage exception");
66      }
67  }
68  private byte[] readFormInternalStorage(String fileName){
69      int len = 1024;
70      byte[] buffer = new byte[len];
71      try {
72          FileInputStream fis = openFileInput(fileName);
73          ByteArrayOutputStream baos = new ByteArrayOutputStream();
74          int nrb = fis.read(buffer, 0, len);
75          while (nrb != -1) {
76              baos.write(buffer, 0, nrb);
77              nrb = fis.read(buffer, 0, len);
78          }
79          buffer = baos.toByteArray();
80          baos.close();
81          fis.close();
82      } catch (Exception e) {
83          Log.d(TAG, "读取失败: exception:" + e);
84          throw new RuntimeException("read from  internal storage exception");
85      }
86      return buffer;
87  }
88
89  @Override
90  public boolean onCreateOptionsMenu(Menu menu) {
91      // Inflate the menu; this adds items to the action bar if it is present.
92      getMenuInflater().inflate(R.menu.main, menu);
93      return true;
94  }
95
96  @Override
97  public boolean onOptionsItemSelected(MenuItem item) {
98      // Handle action bar item clicks here. The action bar will
99      // automatically handle clicks on the Home/Up button, so long
100     // as you specify a parent activity in AndroidManifest.xml.
101     int id = item.getItemId();
102     if (id == R.id.action_settings) {
103         return true;
104     }
105     return super.onOptionsItemSelected(item);
106     }
107 }
```

(4) 部署工程 FileWriteAndDemo,程序运行效果如图 8-7 所示。输入存储的文件内容,单击"保存数据"按钮,如果保存成功,会显示写入数据成功的提示信息,如图 8-8 所示。保存完成后,单击"读取数据"按钮会显示读取的数据,如图 8-9 所示。

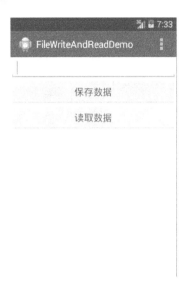 

　　图 8-7　运行效果　　　　　　　　　　图 8-8　保存数据

　　文件存储到目录/data/data/hlju.edu.cn/files/下,可以打开 DDMS 视图下的 File Explorer 面板进行查看。也可以进入设备的 Shell 环境当中,然后进入文件存储的指定目录,最后通过 Linux 的 Cat 命令,查看指定文件的内容,如图 8-10 所示。

　　图 8-9　读取数据　　　　　　　图 8-10　Shell 窗口查看文件的内容

　　(5) 如需要存入 SD 卡,需要在 AndroidManifest.xml 文件中的＜manifest＞中添加读写文件的权限,具体代码见 8.2.4 节的案例。

8.2.3　SD Card 存储简介

　　SD 卡(Secure Digital Memory Card)是 Android 的外部存储设备,广泛使用于数码设备上,Android 系统提供了对 SD 卡的便捷的访问方法。

上节讲述了使用 Activity 的 openFileOutput()方法保存文件,文件是存储在手机自身空间内,一般手机的自身存储空间不大,如果要存储像视频这样的大文件,人们通常把它放置在外部的存储设备 SDCard 上。

SD 卡适用于保存大尺寸的文件或者是一些无须设置访问权限的文件,可以保存录制的大容量的视频文件和音频文件等。

SD 卡使用的是 FAT(File Allocation Table)的文件系统,不支持访问模式和权限控制,但可以通过 Linux 文件系统的文件访问权限的控制保证文件的私密性。

1. SD 卡创建方式

Android 模拟器支持 SD 卡,但模拟器中没有默认的 SD 卡,应用程序开发人员必须在模拟器中手工添加 SD 卡的映像文件。

创建 SD 卡有两种方式:一种是通常在 Eclipse 创建模拟器时创建 SD 卡;另外一种是使用<Android SDK>/tools 目录下的 mksdcard 工具创建 SD 卡映像文件。

在控制台窗口中进入 Android SDK 安装路径的 tools 目录下,使用 mksdcard 工具,命令如下:

```
mksdcard  -1  SDCa  256M c:\android\sdcard_file
```

第 1 个参数-1 表示后面的字符串是 SD 卡的标签,这个新建立的 SD 卡的标签是 SDCa。
第 2 个参数 256M 表示 SD 卡的容量是 256 MB。
第 3 个参数表示 SD 卡映像文件的保存位置,上面的命令将映像保存在 c:\android 目录下的 sdcard_file 文件中。在 CMD 中执行该命令后,可在所指定的目录中找到生产的 SD 卡映像文件。

2. 访问 SD 卡

在编程访问 SD 卡,往 SD 卡存储文件之前,首先程序需要先判断手机是否装有 SD 卡(检测系统的/sdcard 目录是否可用),并且可以进行读写。如果不可用,则说明设备中的 SD 卡已经被移除(如用在 Android 模拟器,则表明 SD 卡映像没有被正确加载);如果可用,则直接通过使用标准的 Java.io.File 类进行访问。

注意:在处理中外字符时需要注意编码问题,发送和接收、保存和读取都采用相同的字符编码,一般采用 utf-8 编码,以防出现乱码。

8.2.4 SD 卡存储应用实例

上述介绍了 SD Card 创建方式及访问方法,下面通过一个案例详细介绍访问 SD Card 的应用。

(1) 创建一个新的 Android 工程,工程名为 SDCardDemo,应用程序名为 SDCardDemo,包名为 hlju.edu.cn,创建的 Activity 的名字为 MainActivity,最小 SDK 版本根据选择的目标 API 会自动添加。

(2) 修改 res 目录下 layout 文件夹中的 activity_main.xml 文件,设置线性布局,添加 1

个 EditText 控件和两个 Button 控件描述,并设置相关属性,代码如下:

```
1   <LinearLayout xmlns:android = "http://schemas.android.com/apk/res/android"
2       android:layout_width = "fill_parent"
3       android:layout_height = "fill_parent"
4       android:orientation = "vertical" >
5       <EditText
6           android:id = "@ + id/myEditText"
7           android:layout_width = "fill_parent"
8           android:layout_height = "wrap_content"
9           android:minLines = "3"/>
10      <LinearLayout
11          android:layout_width = "fill_parent"
12          android:layout_height = "wrap_content"
13          android:orientation = "horizontal"
14          android:weightSum = "2" >
15          <Button
16              android:id = "@ + id/saveSDCard"
17              android:layout_width = "fill_parent"
18              android:layout_height = "wrap_content"
19              android:layout_weight = "1"
20              android:text = "保存数据" />
21          <Button
22              android:id = "@ + id/loadSDCard"
23              android:layout_width = "fill_parent"
24              android:layout_height = "wrap_content"
25              android:layout_weight = "1"
26              android:text = "读取数据" />
27      </LinearLayout>
28  </LinearLayout>
```

(3) 在 AndroidManifest.xml 文件中,<manifest>根结点下添加在 SD 卡中创建、删除和写入数据的权限,代码如下:

```
1   <LinearLayout xmlns:android = "http://schemas.android.com/apk/res/android"
2       android:layout_width = "fill_parent"
3       android:layout_height = "fill_parent"
4       android:orientation = "vertical" >
5   <?xml version = "1.0" encoding = "utf - 8"?>
6   <manifest xmlns:android = "http://schemas.android.com/apk/res/android"
7       package = "hlju.edu.cn"
8       android:versionCode = "1"
9       android:versionName = "1.0" >
10      <uses - sdk
11          android:minSdkVersion = "14"
12          android:targetSdkVersion = "21" />
13  <uses - permission android:name = "android.permission.MOUNT_UNMOUNT_FILESYSTEMS"/>
14  <uses - permission android:name = "android.permission.WRITE_EXTERNAL_STORAGE"/>
```

```
15      <application
16          android:allowBackup = "true"
17          android:icon = "@drawable/ic_launcher"
18          android:label = "@string/app_name"
19          android:theme = "@style/AppTheme" >
20          <activity
21              android:name = ".MainActivity"
22              android:label = "@string/app_name" >
23              <intent-filter>
24                  <action android:name = "android.intent.action.MAIN" />
25                  <category android:name = "android.intent.category.LAUNCHER" />
26              </intent-filter>
27          </activity>
28      </application>
29  </manifest>
```

(4) 修改 src 目录中 hlju.edu.cn 中的 MainActivity.java 文件,代码如下:

```
1   package hlju.edu.cn;
2
3   import java.io.ByteArrayOutputStream;
4   import java.io.File;
5   import java.io.FileInputStream;
6   import java.io.FileOutputStream;
7   import android.app.Activity;
8   import android.os.Bundle;
9   import android.os.Environment;
10  import android.util.Log;
11  import android.view.Menu;
12  import android.view.MenuItem;
13  import android.view.View;
14  import android.view.View.OnClickListener;
15  import android.widget.Button;
16  import android.widget.EditText;
17  import android.widget.Toast;
18  public class MainActivity extends Activity implements OnClickListener {
19      private static final String TAG = "MainActivity";
20      EditText dataEditText;
21      Button saveExternalBtn;
22      Button loadExternalBtn;
23      @Override
24      public void onCreate(Bundle savedInstanceState) {
25          super.onCreate(savedInstanceState);
26          setContentView(R.layout.activity_main);
27          dataEditText = (EditText) findViewById(R.id.myEditText);
28          saveExternalBtn = (Button) findViewById(R.id.saveSDCard);
29          loadExternalBtn = (Button) findViewById(R.id.loadSDCard);
30          saveExternalBtn.setOnClickListener(this);
```

```java
31        loadExternalBtn.setOnClickListener(this);
32    }
33    @Override
34    public void onClick(View v) {
35        int viewId = v.getId();
36        if (R.id.saveSDCard == viewId) {
37            try {
38                String content = dataEditText.getText().toString();
39                saveToSDCard("data.txt", content.getBytes());
40                Toast.makeText(this, "写入 SD Card 成功", Toast.LENGTH_LONG).show();
41            } catch (Exception e) {
42                Toast.makeText(this, "写入 SD Card 失败", Toast.LENGTH_LONG).show();
43            }
44        } else if (R.id.loadSDCard == viewId) {
45            try {
46    Toast.makeText(this, new String(loadFromSDCard("data.txt"), "UTF-8"), Toast.LENGTH_LONG).show();
48            } catch (Exception e) {
49                Toast.makeText(this, "读取 SD Card 失败", Toast.LENGTH_LONG).show();
50            }
51        }
52    }
53    private boolean isExternalStorageWriteable() {
54        return Environment.MEDIA_MOUNTED.equals(Environment.getExternalStorageState()) ? true : false;
55    }
56    private boolean isExternalStorageReadable() {
57        return Environment.MEDIA_MOUNTED.equals(Environment.getExternalStorageState()) || Environment.MEDIA_MOUNTED_READ_ONLY.equals(Environment.getExternalStorageState()) ? true : false;
58    }
59    private void saveToSDCard(String fileName, byte[] content) {
60        if (isExternalStorageWriteable()) {
61            File sdcardDir = Environment.getExternalStorageDirectory();
62            File baseDir = new File(sdcardDir, "chapter8_2_4");
63            File file = new File(baseDir, fileName);
64            try {
65                if(!baseDir.exists()){
66                    baseDir.mkdir();
67                }
68                if(file.exists()){
69                    file.createNewFile();
70                }
71                FileOutputStream fos = new FileOutputStream(file);
72                fos.write(content);
73                fos.close();
74            } catch (Exception e) {
75                Log.d(TAG, "写入 SD Card 出错,exception:" + e);
76                throw new RuntimeException("write to sdcard exception");
```

```java
77       }
78     }
79   }
80   private byte[] loadFromSDCard(String fileName) {
81     if (isExternalStorageReadable()) {
82       int len = 1024;
83       byte[] buffer = new byte[len];
85       File dir = Environment.getExternalStorageDirectory();
86       File file = new File(dir, "chapter8_2_4/" + fileName);
87       try {
88         FileInputStream fis = new FileInputStream(file);
89         ByteArrayOutputStream baos = new ByteArrayOutputStream();
90         int nrb = fis.read(buffer, 0, len);
91         while (nrb != -1) {
92           baos.write(buffer, 0, nrb);
93           nrb = fis.read(buffer, 0, len);
94         }
95         buffer = baos.toByteArray();
96         fis.close();
97       } catch (Exception e) {
98         Log.d(TAG, "读取 SD Card 出错: exception:" + e);
99     throw new RuntimeException("read from sd card exception");
100       }
101       return buffer;
102     }
103     throw new RuntimeException("sd card is not readable");
104   }
105   @Override
106   public boolean onCreateOptionsMenu(Menu menu) {
107     getMenuInflater().inflate(R.menu.main, menu);
108     return true;
109   }
110   @Override
111   public boolean onOptionsItemSelected(MenuItem item) {
112       // Handle action bar item clicks here. The action bar will
113       // automatically handle clicks on the Home/Up button, so long
114       // as you specify a parent activity in AndroidManifest.xml.
115     int id = item.getItemId();
116     if (id == R.id.action_settings) {
117       return true;
118     }
119     return super.onOptionsItemSelected(item);
120   }
121 }
```

（5）部署运行 SDCardDemo 工程，如图 8-11 所示。然后在文本框中输入"Please store the data in the SDCard"，单击"保存数据"按钮，显示"写入 SDCard 成功"说明数据成功写入 SD 卡，如图 8-12 所示。单击"读取数据"按钮，显示 SD 卡中存取的数据，如图 8-13 所示。

图 8-11　运行效果　　　　　　　　　图 8-12　写入 SD Card

图 8-13　读取数据

文件存储到目录/mnt/media_rw/sdcard/chapter8_2_4/下,可以打开 DDMS 视图下的 File Explorer 面板进行查看。也可以进入设备的 Shell 环境当中,然后进入文件存储的指定目录,最后通过 Linux 的 Cat 命令,查看指定文件的内容,如图 8-14 所示。

图 8-14　Shell 环境查看文件内容

8.3　SQLite 数据库存储

8.3.1　SQLite 数据库简介

SQLite 是在 2000 年由 D. Richard Hipp 发布的轻量级嵌入式关系型数据库，它支持 SQL 语言，是开源的项目，在 Android 系统平台中集成了嵌入式关系型数据库（SQLite）。

1. SQLite 数据库体系结构

SQLite 数据库由 SQL 编译器、内核、后端以及附件 4 部分组成。SQLite 利用虚拟机和虚拟数据库引擎（VDBE）操作，使调试、修改和扩展 SQLite 的内核变得更加方便。

1）Interface

接口由 SQLite C API 组成，SQLite 类库大部分的公共接口程序是由 main.c、legacy.c 和 vdbeapi.c 源文件中的功能执行的。但有些程序是分散在其他文件夹的，因为在其他文件夹里它们可以访问有文件作用域的数据结构。例如：

sqlite3_get_table() 在 table.c 中执行。

sqlite3_mprintf() 在 printf.c 中执行。

sqlite3_complete() 在 tokenize.c 中执行。

Tcl 接口程序用 tclsqlite.c 执行。

因此，无论是应用程序、脚本还是库文件，最终都是通过接口与 SQLite 交互。

为了避免和其他软件在名字上有冲突，SQLite 类库中所有的外部符合都是以 sqlite3 为前缀来命名的。这些被用于做外部使用的符号是以 sqlite3_ 开头来命名的。

2）Tokenizer

当执行一个包含 SQL 语句的字符串时，接口程序要把这个字符串传递给 tokenizer。Tokenizer 的任务是把原有字符串分成一个个标识符，并把这些标识符传递给剖析器。Tokenizer 是在 C 文件夹 tokenizer.c 中用手编译的。

在这个设计中需要注意的一点是 tokenizer 调用 parser。即用 tokenizer 调用 parser 会使程序运行得更顺利。

3）Parser

Parser 是 Lemon（LALA(1) 文法分析器生成工具）生成的分析器的核心例程，在分析器调用 ParserAlloc 后，分析器就可以把切分的词传递给 Parser 进行语法分析。Tokenizer 和 Parser 对 SQL 语句进行语法检查，然后把 SQL 语句转化为底层能更方便处理的分层的数据结构，这种分层的数据结构称为"语法树"，把语法树传给 Code Generator 进行处理。

4）Code Generator

在 Parser 收集完符号并转换成完全的 SQL 语句时，它调用 Code Generator 来产生虚拟的机器代码，这些机器代码将按照 SQL 语句的要求来工作。在代码产生器中有许多文件，如 attach.c、auth.c、build.c、delete.c、expr.c、insert.c、pragma.c、select.c、trigger.c、update.c、vacuum.c 和 where.c。在这些文件中，expr.c 处理表达式代码的生成。where.c 处理 SELECT、UPDATE 和 DELETE 语句中 WHERE 子句代码的生成。文件 attach.c、 delete.c、insert.c、select.c、trigger.c、update.c 和 vacuum.c 处理 SQL 语句中具有同样名字的语句代码的生成(每个文件调用 expr.c 和 where.c 中的程序)。所有 SQL 的其他语句的代码都是由 build.c 生成的。文件 auth.c 执行 sqlite3_set_authorizer()功能。

5) Virtual Machine

由 Code Generator(代码生成器)产生的程序由 Virtual Machine(虚拟机器)来运行。总而言之,虚拟机器主要用于执行一个为操作数据库而设计的抽象的计算引擎。机器有一个用于存储中间数据的存储栈。每个指令包含一个操作代码和 3 个额外的操作数。

虚拟机器本身是被包含在一个单独的文件 vdbe.c 中的。虚拟机器也有它自己的标题文件:vdbe.h 在虚拟机器和剩下的 SQLite 类库之间定义了一个接口程序,bdbeInt.h 定义了虚拟机器的结构。文件 vdbeaux.c 包含了虚拟机器所使用的实用程序和一些被其他类库用于建立 VM 程序的接口程序模块。文件 vdbeapi.c 包含虚拟机器的外部接口。单独的值(字符串、整数、浮动点数值、BLOBS)被存储在一个叫作 Mem 的内部目标程序里,Mem 是由 vdbemem.c 执行的。

6) B-Tree

SQLite 数据库在磁盘里维护,使用源文件 btree.c 中的 B-Tree(B-树)执行。数据库中的每个表格和目录使用一个单独的 B-tree。所有的 B-trees 被存储在同样的磁盘文件里。文件格式的细节被记录在 btree.c 开头的备注里。B-tree 子系统的接口程序被标题文件 btree.h 所定义。主动功能就是索引,它维护着各个页面之间复杂的关系,便于快速找到所需要的数据。

7) Pager

B-tree 模块要求信息来源于磁盘上固定规模的程序块。默认程序块的大小是 1024 个字节,但是可以在 512~65 536 个字节间变化。Pager(页面高速缓存)负责读、写和高速缓存这些程序块。页面高速缓存还提供重新运算和提交抽象命令,它还管理关闭数据库文件夹。B-tree 驱动器要求页面高速缓存器中的特别的页,当它想修改页或重新运行改变的时候会通报页面高速缓存。为了保证所有的需求被快速、安全和有效地处理,页面高速缓存处理所有微小的细节。运行页面高速缓存的代码在专门的 C 源文件的 pager.c 中。页面高速缓存的子系统的接口程序被目标文件 pager.h 所定义。页缓存的主要作用就是通过操作系统接口在 B-树和磁盘之间传递页面。

8) OS Interface

为了在 POSIX 和 Win32 之间提供一些可移植性,SQLite 操作系统的接口程序使用了一个提取层。OS 提取层的接口程序被定义在 os.h 中。每个支持的操作系统都有自己的执行文件:UNIX 使用 os_unix.c,Windows 使用 os_win.c。每个具体的操作器都具有自己的标题文件,如 os_unix.c、os_win.c 等。

9) Utilities

内存分配和字符串比较程序位于 util.c。Parser 使用的表格符号被 hash.c 中的无用信息表格维护。源文件 utf.c 包含 UNICODE 转换子程序。SQLite 有自己的执行文件 printf()(有一些扩展)。在 printf.c 中,还有它自己随机数量产生器在 random.c 中。

10) Test Code

如果计算回归测试脚本,多于一半的 SQLite 代码数据库的代码将被测试。在主要代码文件中有许多assert()语句。另外,源文件test1.c通过test5.c和md5.c执行只为测试用的扩展名。os_test.c向后的接口程序通过模拟断电来验证页面调度程序中的系统性事故恢复机制。

2. SQLite数据库特点

- 更加适用于嵌入式系统,嵌入到使用它的应用程序中;
- 占用非常少,运行高效可靠,可移植性好;
- 提供了零配置(zero-configuration)运行模式;
- SQLite数据库不仅提高了运行效率,而且屏蔽了数据库使用和管理的复杂性,程序仅需要进行最基本的数据操作,其他操作可以交给进程内部的数据库引擎完成;
- SQLite数据库具有很强的移植性,可以运行在Windows、Linux、BSD、Mac OS X和一些商用UNIX系统,例如Sun的Solaris,IBM的AIX;SQLite数据库也可以工作在许多嵌入式操作系统下,如QNX、VxWorks、Palm OS、Symbin和Windows CE。

3. SQLite数据库和其他数据库的区别

SQLite和其他数据库最大的不同就是对数据类型的支持,SQLite3支持NULL、INTEGER、REAL(浮点数字)、TEXT(字符串文本)和BLOB(二进制对象)数据类型。虽然它支持的类型只有5种,但实际上sqlite3也接受varchar(n)、char(n)、decimal(p,s)等数据类型,只不过在运算或保存时会转成对应的5种数据类型。

8.3.2 创建SQLite数据库方式

创建SQLite数据库的方式有两种,分别是使用sqlite3工具命令行方式和使用程序编码方式,下面就分别介绍它们创建数据库的过程。

1. sqlite3工具命令行方式(适合调试用)

sqlite3是SQLite数据库自带的一个基于命令行的SQL命令执行工具,并可以显示命令执行结果,sqlite3工具被集成在Android系统中,用户在Linux的命令行界面中输入sqlite3可启动sqlite3工具,并得到工具的版本信息,在CMD中输入adb shell命令可以启动Linux的命令行界面,过程如下。

(1)首先用命令或启动Eclipse中启动模拟器,然后在CMD下输入命令"adb shell"进入设备Linux控制台,出现提示符"#"后输入命令"sqlite3",如图8-15所示。

图8-15 进入sqlite

在启动 sqlite3 工具后，提示符从"♯"变为"sqlite＞"，表示命令行界面进入与 SQLite 数据库的交互模式，此时可以输入命令建立、删除或修改数据库的内容。正确退出 sqlite3 工具的方法是使用命令.exit。如图 8-16 所示。

图 8-16　sqlite3 退出命令

（2）命令行方式手动创建 sqlite 数据库，步骤如下。

① CMD 下输入命令 adb shell 进入设备 Linux 控制台。

② ♯ cd　/data/data，进入应用 data 目录，如图 8-17 所示。

图 8-17　进入 data 目录

③ ♯ ls，列表目录，查看文件，如图 8-18 所示。找到项目包目录并进入，如图 8-19 所示。

图 8-18　查看目录

图 8-19　进入项目包目录

④ 使用 ls 命令查看有无 databases 目录，如果没有，则创建一个，如图 8-20 所示。

⑤ 使用 ls 命令查看列表目录会看到有一个文件为 mydb.db，即 sqlite 数据库，如图 8-21 所示。

图 8-20　创建数据库

图 8-21　查看 sqlite 数据库文件

2．使用程序编码方式（通常使用）

在程序代码中动态建立 sqlite 数据库是比较常用的方法。在程序运行过程中，当需要进行数据库操作时，应用程序会首先尝试打开数据库，此时如果数据库并不存在，程序会自动建立数据库，然后再打开数据库。

在 Android 应用程序中创建使用 SQLite 数据库有两种方式：一种是自定义类继承 SQLiteOpenHelper；另一种是调用 openOrCreateDatabases()方法创建数据库。下面分别进行介绍。

1）自定义类继承 SQLiteOpenHelper，创建数据库

在 Android 应用程序中使用 SQLite，必须自己创建数据库，然后创建表、索引，填充数据。Android 提供了 SQLiteOpenHelper 帮助创建一个数据库，只要继承 SQLiteOpenHelper 类就可以轻松地创建数据库。SQLiteOpenHelper 类根据开发应用程序的需要，封装了创建和更新数据库使用的逻辑。

创建 SQLiteOpenHelper 的子类，至少需要实现 3 个方法。

（1）构造函数，调用父类 SQLiteOpenHelper 的构造函数。这个方法需要 4 个参数：上下文环境（例如一个 Activity）、数据库名字、一个可选的游标工厂（通常是 Null）和一个代表正在使用的数据库模型版本的整数。

（2）onCreate()方法，它需要一个 SQLiteDatabase 对象作为参数，根据需要对这个对象填充表和初始化数据。

（3）onUpgrage()方法，它需要 3 个参数，一个 SQLiteDatabase 对象、一个旧的版本号和一个新的版本号，这样就可以知道如何把一个数据库从旧的模型转变到新的模型了。

2）调用 openOrCreateDatabases()方法创建数据库

在 Android 中创建和打开一个数据库都可以使用 openOrCreateDatabase 方法来实现，

因为它会自动去检测是否存在这个数据库,如果存在则打开,如果不存在则创建一个数据库,创建成功则返回一个 SQLiteDatebase 对象,否则抛出异常 FileNotFoundException。

8.3.3 SQLite 数据库操作

在编程实现时,一般将所有对数据库的操作都封装在一个类中,因此只要调用这个类,就可以完成对数据库的添加、更新、删除和查询等操作。下面就在数据库中对创建表、索引,给表添加数据等操作进行介绍。

1. 创建表和索引

为了创建表和索引,需要调用 SQLiteDatabase 的 execSQL()方法来执行 DDL 语句。如果没有异常,这个方法没有返回值。

```
db.execSQL("CREATE TABLE mytable(_id INTEGER PRIMARY KEY AUTOINCREMENT , title TEXT ,value REAL);");
```

上述语句创建表名为 mytable,表有一个列名为_id,并且是主键,列值是会自动增长的整数。另外还有两列:title(字符)和 value(浮点数)。SQLite 会自动为主键列创建索引。通常情况下,第一次创建数据库时便创建了表和索引。

另外,SQLiteDatabase 类提供了一个重载后的 execSQL(String sql ,Object[] bindArgs)方法。

使用这个方法支持使用占位符参数(?)。使用例子如下:

```
1  SQLiteDatabase db = ...;
2  db.execSQL("insert into person(name,age) values(?,?)",new Object[]{"liubei",20});
3  db.close();
```

第 2 行的第 1 个参数为 SQL 语句,第 2 个参数为 SQL 语句中占位符参数的值,参数值在数组中的顺序要和占位符的位置对应。

如果不需要改变表的 schema,则不需要删除表和索引。删除表和索引需要使用 execSQL()方法调用 DROP INDEX 和 DROP TABLE 语句。

2. 给表添加数据

给数据库中表添加数据有两种方法。

(1) 使用 execSQL()方法执行 INSERT、UPDATE、DELETE 等语句来更新表的数据。execSQL()方法适用于所有不返回结果的 SQL 语句。如:

```
db.execSQL("INSERT INTO widgets (name, inventory)" + "VALUES ('Sprocket', 5)");
```

(2) 使用 SQLiteDatabase 对象的 insert()、update()、delete()方法。这些方法把 SQL 语句的一部分作为参数。

1) insert()方法

insert()方法用于添加数据,各个字段的数据使用ContentValues进行存储。ContentValues类似于MAP,相对于MAP,它提供了存取数据对应的put(String key, Xxx value)和getAsXxx t(String key)方法,key为字段名称,value为字段值。例如：

```
1  SQLiteDatabase db = databaseHelp.getWritableDatabase();
2  ContentValues values = new ContentValues;
3  values.put("name","liubei");
4  values.put("age","20");
5  long rowid = db.insert("person",null,values);
```

第5行返回新添记录的行号,与主键id无关。不管第3个参数是否包含数据,执行insert()方法必然会添加一条记录。如果第3个参数为空,会添加一条除主键之外其他字段值为Null的记录。

2) update()方法

update()方法有4个参数,分别是表名、表示列名和值的ContentValues对象,可选的WHERE条件和可选的填充WHERE语句的字符串,这些字符串会替换WHERE条件中的"?"标记。

update()根据条件更新指定列的值,所以用execSQL()方法可以达到同样的目的。WHERE条件及其参数和用过的SQL APIs类似。例如：

```
1  String[] parms = new String[]{"this is a string"};
2  db.update("widgets",replacements, "name = ?",parms);
```

3) delete()方法

delete()方法的使用和update()类似,使用表名,可选的WHERE条件和相应的填充WHERE条件的字符串。例如：

```
1  db. delete ("person"," person id <?",new String[],{ "2"});
2  db.close();
```

3. 查询数据库

在Android系统中,数据库查询结果的返回值并不是数据集合的完整备份,而是返回数据集的指针,这个指针就是Cursor类。

Cursor类支持在查询的数据集合中多种方式移动,并能够获取数据集合的属性名称和序号。

查询数据库使用SELECT从SQLite数据库检索数据有两种方法：一种是使用rawQuery()直接调用SELECT语句;另一种是使用query()方法构建一个查询。

(1) 使用rawQuery()直接调用SELECT语句。

调用SQLiteDatabase类的rawQuery()方法用于执行select语句。

例如：

```
Cursor c = db.rawQuery("SELECT name FROM sqlite_master WHERE type = 'table' AND name = 'mytable'",null);
```

rawQuery()方法的第 1 个参数为 select 语句；第 2 个参数为 select 语句中占位符参数的值，如果 select 语句没有使用占位符，该参数可以设置为 null。

带占位符参数的 select 语句使用例子如下：

```
Cursor cursor = db.rawQuery("select * from person where name like ? and age = ?",new String[]
{"%liubei%","20"};
```

在上面例子中，查询 SQLite 系统表（sqlite_master）检查 table 表是否存在。返回值是一个 cursor 对象。这个对象的方法可以迭代查询结果。如果检查是动态的，使用这个方法就会非常复杂。

例如，当需要查询的列在程序编译的时候不能确定，这时使用 query()方法会方便很多。

（2）使用 query()方法构建一个查询。

调用 SQLiteDatabase 类的 query()，query()的语法如下：

```
Cursor android.database.sqlite.SQLiteDatabase.query(String table,String[] columns,String
selection,String[] selectionArgs,String groupBy,String having,string orderBy,String limit)
```

query()的参数说明如表 8-2 所示。

表 8-2 query()的参数说明

位置	类型+名称	说明
1	String table	表名称
2	String[] columns	返回的属性列名称
3	String selection	查询条件子句
4	String[] selectionArgs	如果在查询条件中使用的是问号，则需要定义替换符的具体内容
5	String groupBy	分组方式
6	String having	定义组的过滤器
7	String limit	制定偏移量和获取的记录数

例如：

```
1  SQLiteDatabase db = databaseHelper.getWritableDatabase();
2  Cursor cursor = db.query("person",new String[] {"personid,name,age"},"name like?",new
String[]{"%Tom%"},null,null, "personid desc","1,2");
3  while(cursor.moveToNext()){
4  int personid = cursor.getInt(0);
5  String name = cursor.getString(1);
6  int age = cursor.getInt(2);
7  }
8      cursor.close();
9  db.close();
```

第 4 行是获取第 1 列的值，第 1 列的索引从 0 开始。第 5 行是获取第 2 列的值。第 6

行是获取第 3 列的值。

在 Android 的 SQLite 数据库使用游标,不论如何执行查询,都会返回一个 Cursor 对象。Cursor 类的常用方法和说明如表 8-3 所示。

表 8-3 Cursor 类的常用方法和说明

方 法	说 明
moveToFirst	将指针移动到第一条数据上
moveToNext	将指针移动到下一条数据上
moveToPrevious	将指针移动到上一条数据上
getCount	获取集合的数据数量
getColumnIndexOrThrow	返回指定属性名称的序号。如果属性不存在,则产生异常
getColumnName	返回指定序号的属性名称
getColumnNames	返回属性名称的字符串数组
getColumnIndex	根据属性名称返回序号
moveToPosition	将指针移动到指定的数据上
getPosition	返回当前指针的位置
getString、getInt 等	获取给定字段当前记录的值
requery	重新执行查询得到游标
close	释放游标资源

8.3.4 SQLite 数据库管理

在 Android 系统中,针对 sqlite 数据库的查看和管理有两种方式,一种是使用 Eclipse 插件 DDMS 查看和管理;另一种是使用 Android 工具包中的 adb 工具来查看和管理。Android 项目中 sqlite 数据库位置:/data/data/<package-name>/databases/。

1. 使用 Eclipse 插件 DDMS 查看和管理 SQLite 数据库

(1) 在 Eclipse 中打开 DDMS 视图,如图 8-22 所示。如果是模拟器进行项目调试,必须先启动模拟器,打开 DDMS 视图才能有内容。

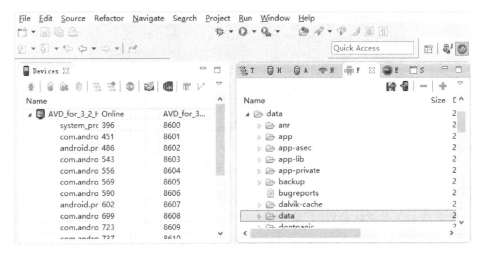

图 8-22 DDMS 视图下查看 data

（2）选择 File Explorer 窗口，然后在 /data/data/＜package-name＞/databases/ 目录下打开 database 文件即可看见 sqlite 数据库文件，如图 8-23 所示。选择 Pull a file from the device 按钮，可以导出 sqlite 数据库文件，如图 8-24 所示。

图 8-23　sqlite 数据库文件路径

图 8-24　导出数据库文件

2．使用 adb 工具管理 sqlite 数据库

（1）在控制台窗口中输入命令"adb shell"，进入设备 Linux 控制台，出现提示符"♯"后，输入命令"cd /data/data/＜package-name＞/databases"进入目录。使用"ls"命令查看数据库文件是否存在，如图 8-25 所示。

图 8-25　命令查看数据库

（2）输入命令"sqlite3"，进入 sqlite 管理模式命令，如图 8-26 所示。

sqlite 命令行工具默认是以"；"结束语句的。所以如果只是一行语句，要在末尾加"；"，或者在下一行中输入，这样 sqlite 命令才会被执行。具体 sqlite3 的命令如表 8-4 所示。

图 8-26　sqlite 命令模式

表 8-4　sqlite3 的命令表

命　　　令	说　　　明
. bail ON|OFF	遇到错误时不再继续,默认为 OFF
. databases	列出附加到数据库的数据库和文件
. dump ? TABLE? …	保存表到 SQL 格式的文件中,没有指表名,则保存所有,如果要保存到磁盘上需要结合. output 命令
. echo ON|OFF	打开/关闭命令行回显
. exit	退出该命令行
. explain ON|OFF	以合适的方式显示表头,不带参数则为开启
. header(s) ON|OFF	是否显示表头,和. explain 差别不是很大
. help	显示帮助信息
. import FILE TABLE	从文件中导入表
. indices TABLE	显示索引
. mode MODE ? TABLE?	设置输出模式
. nullvalue STRING	以 STRING 代替 NULL 值的输出
. output FILENAME	输出到文件,而不是显示在屏幕上
. output stdout	输出到屏幕上
. prompt MAIN CONTINUE	替换默认的命令提示信息,默认就是 sqlite>
. quit	退出命令行
. read FILENAME	执行 FILENAME 中的 SQL 语句
. schema ? TABLE?	显示 CREATE 语句
. show	显示各种设置
. tables ? PATTERN?	查看数据库的表列表
. timeout MS	在 MS 时间内尝试打开被锁定的表
. widthNUM NUM …	设置 column 模式中的列的宽度
. timer ON|OFF	显示 CPU 时间

8.3.5　SQLite 数据库应用案例

通过一个案例,详细介绍访问 SQLite 数据库的应用。

（1）创建一个新的 Android 工程,工程名为 SQLiteDemo,应用程序名为 SQLiteDemo,包名为 hlju. edu. cn,创建的 Activity 的名字为 MainActivity,最小 SDK 版本根据选择的目标 API 会自动添加。

（2）修改 res 目录下 layout 文件夹中的 activity_main. xml 文件,设置线性布局,添加 4

个 Button 控件和 1 个 ListView 控件描述,并设置相关属性,代码如下:

```
1   <?xml version = "1.0" encoding = "utf-8"?>
2   <LinearLayout xmlns:android = "http://schemas.android.com/apk/res/android"
3       android:orientation = "vertical"
4       android:layout_width = "fill_parent"
5       android:layout_height = "fill_parent">
6       <Button
7           android:id = "@ + id/addBtn"
8           android:layout_width = "fill_parent"
9           android:layout_height = "wrap_content"
10          android:text = "添加记录"/>
11      <Button
12          android:id = "@ + id/updateBtn"
13          android:layout_width = "fill_parent"
14          android:layout_height = "wrap_content"
15          android:text = "修改记录"/>
16      <Button
17          android:id = "@ + id/deleteBtn"
18          android:layout_width = "fill_parent"
19          android:layout_height = "wrap_content"
20          android:text = "删除记录"/>
21      <Button
22          android:id = "@ + id/queryBtn"
23          android:layout_width = "fill_parent"
24          android:layout_height = "wrap_content"
25          android:text = "查询记录"/>
26      <ListView
27          android:id = "@ + id/listView"
28          android:layout_width = "fill_parent"
29          android:layout_height = "wrap_content"/>
30  </LinearLayout>
```

(3) 新建一个继承于 android.database.sqlite.SQLiteOpenHelper 类的子类: MySQLiteOpenHelper,在 src 目录中 hlju.edu.cn 包下创建 MyContentProvider.java 文件,代码如下:

```
1   public class MySQLiteOpenHelper extends SQLiteOpenHelper {
2       private static final String DATABASE_NAME = "test.db";
3       private static final int DATABASE_VERSION = 1;
4       public MySQLiteOpenHelper(Context context) {
5           super(context, DATABASE_NAME, null, DATABASE_VERSION);
6       }
7       @Override
8       public void onCreate(SQLiteDatabase db) {
9           db.execSQL("CREATE TABLE IF NOT EXISTS person" +
10              "(_id INTEGER PRIMARY KEY AUTOINCREMENT, name VARCHAR, age INTEGER, info TEXT)");
11      }
```

```
12      @Override
13      public void onUpgrade(SQLiteDatabase db, int oldVersion, int newVersion) {
14          db.execSQL("ALTER TABLE person ADD COLUMN other STRING");
15      } }
```

第 8 行代码表示数据库第一次被创建时 onCreate 会被调用。第 9 行代码表示在 onCreate()方法当中调用 SQLiteDatabase 对象的 execSQL()方法来执行 DDL 语句,如果没有异常发生,这个方法没有返回值。数据库第一次创建时 onCreate()方法会被调用,可以执行创建表的语句,当系统发现版本变化之后会调用第 13 行代码。第 13 行代码表示如果 DATABASE_VERSION 值被改为 2,系统发现现有数据库版本不同,即会调用 onUpgrade 方法,可以执行修改表结构等语句。

(4) 在 src 目录中 hlju.edu.cn 包下创建 Person.java 文件,代码如下:

```
1   public class Person {
2       public int _id;
3       public String name;
4       public int age;
5       public String info;
6       public Person() {
7       }
8       public Person(String name, int age, String info) {
9           this.name = name;
10          this.age = age;
11          this.info = info;
12      }
13  }
```

第 2 行代码定义 id,第 3 行代码定义姓名,第 4 行代码定义年龄,第 5 行代码定义信息。

(5) 在 src 目录中 hlju.edu.cn 包下创建 DBManager.java 文件,代码如下:

```
1   public class DBManager {
2       private MySQLiteOpenHelper helper;
3       private SQLiteDatabase db;
4
5       public DBManager(Context context) {
6           helper = new MySQLiteOpenHelper(context);
7           db = helper.getWritableDatabase();
8       }
9       public void add(List<Person> persons) {
10          db.beginTransaction();
11          try {
12              for (Person person : persons) {
13                  db.execSQL("INSERT INTO person VALUES(null, ?, ?, ?)", new Object[]{person.name, person.age, person.info});
14              }
```

```
15          db.setTransactionSuccessful();
16       } finally {
17          db.endTransaction();
18       }
19    }
20    public void updateAge(Person person) {
21       ContentValues cv = new ContentValues();
22       cv.put("age", person.age);
23       db.update("person", cv, "name = ?", new String[]{person.name});
24    }
25    public void deleteOldPerson(Person person) {
26       db.delete("person", "age >= ?", new String[]{String.valueOf(person.age)});
27    }
28    public List<Person> query() {
29       ArrayList<Person> persons = new ArrayList<Person>();
30       Cursor c = queryTheCursor();
31     while (c.moveToNext()) {
32       Person person = new Person();
33       person._id = c.getInt(c.getColumnIndex("_id"));
34       person.name = c.getString(c.getColumnIndex("name"));
35       person.age = c.getInt(c.getColumnIndex("age"));
36       person.info = c.getString(c.getColumnIndex("info"));
37       persons.add(person);
38     }
39     c.close();
40     return persons;
41    }
42    public Cursor queryTheCursor() {
43      Cursor c = db.rawQuery("SELECT * FROM person", null);
44      return c;
45    }   public void closeDB() {
46       db.close();
47    }
48    }
```

第 7 行代码表示 getWritableDatabase 内部调用了 mContext.openOrCreateDatabase（mName，0，mFactory）；所以要确保 context 已初始化，可以把实例化 DBManager 的步骤放在 Activity 的 onCreate 里。第 9 行代码表示添加 Person 列表至数据库 person 表当中。第 20 行代码表示根据名称更改数据。第 26 行代码表示根据年龄的比对删除数据。第 28 行代码表示以列表方式返回所有的 Person。

（6）修改 src 目录中 hlju.edu.cn 包下的 MainActivity.java 文件，代码如下：

```
1  package hlju.edu.cn;
2
3  import java.util.ArrayList;
4  import java.util.HashMap;
5  import java.util.List;
```

```java
6   import java.util.Map;
7   import android.app.Activity;
8   import android.os.Bundle;
9   import android.view.Menu;
10  import android.view.MenuItem;
11  import android.view.View;
12  import android.view.View.OnClickListener;
13  import android.widget.Button;
14  import android.widget.ListView;
15  import android.widget.SimpleAdapter;
16
17  public class MainActivity extends Activity implements OnClickListener{
18      private DBManager mgr;
19      private ListView listView;
20      private Button addBtn, updateBtn, deleteBtn, queryBtn;
21        @Override
22        public void onCreate(Bundle savedInstanceState) {
23            super.onCreate(savedInstanceState);
24            setContentView(R.layout.activity_main);
25            mgr = new DBManager(this);
26            initUI();
27      }
28      addBtn = (Button)findViewById(R.id.addBtn);
29      updateBtn = (Button)findViewById(R.id.updateBtn);
30      deleteBtn = (Button)findViewById(R.id.deleteBtn);
31      queryBtn = (Button)findViewById(R.id.queryBtn);
32      addBtn.setOnClickListener(this);
33      updateBtn.setOnClickListener(this);
34      deleteBtn.setOnClickListener(this);
35      queryBtn.setOnClickListener(this);
36      listView = (ListView) findViewById(R.id.listView);
37      }   @Override
38      public void onClick(View v) {
39          int viewId = v.getId();
40      switch(viewId){
41      case R.id.addBtn:
42          add();
43          break;
44      case R.id.updateBtn:
45          update();
46          break;
47      case R.id.deleteBtn:
48          delete();
49          break;
50      case R.id.queryBtn:
51          query();
52          break;
53      }
54      }
55
56      public void add() {
57      ArrayList<Person> persons = new ArrayList<Person>();
```

```
58
59        Person person1 = new Person("张飞", 22, "猛将");
60        Person person2 = new Person("赵云", 20, "常胜将军");
61        Person person3 = new Person("关羽", 23, "武神");
62        Person person4 = new Person("诸葛亮", 18, "军师");
63        Person person5 = new Person("刘备", 30, "君主");
64        persons.add(person1);
65        persons.add(person2);
66        persons.add(person3);
67        persons.add(person4);
68        persons.add(person5);
69
70        mgr.add(persons);
71      }
72
73      public void update() {
74        Person person = new Person();
75        person.name = "张飞";
76        person.age = 21;
77        mgr.updateAge(person);
78      }
79
80      public void delete() {
81        Person person = new Person();
82          person.age = 18;
83        mgr.deleteOldPerson(person);
84      }
85
86      public void query() {
87        List<Person> persons = mgr.query();
88        ArrayList<Map<String, String>> list = new ArrayList<Map<String, String>>();
89        for (Person person : persons) {
90          HashMap<String, String> map = new HashMap<String, String>();
91          map.put("name", person.name);
92          map.put("info", person.age + " years old, " + person.info);
93          list.add(map);
94        }
95        SimpleAdapter adapter = new SimpleAdapter(this, list, android.R.layout.simple_list_item_2,
96          new String[]{"name", "info"}, new int[]{android.R.id.text1, android.R.id.text2});
97        listView.setAdapter(adapter);
98      }
99      protected void onDestroy() {
100       super.onDestroy();
101       mgr.closeDB();
102     }
103     @Override
104     public boolean onCreateOptionsMenu(Menu menu) {
```

```
105         // Inflate the menu; this adds items to the action bar if it is present.
106         getMenuInflater().inflate(R.menu.main, menu);
107         return true;
108     }
109     @Override
110      public boolean onOptionsItemSelected(MenuItem item) {
111             // Handle action bar item clicks here. The action bar will
112         // automatically handle clicks on the Home/Up button, so long
113         // as you specify a parent activity in AndroidManifest.xml.
114         int id = item.getItemId();
115         if (id == R.id.action_settings) {
116            return true;
117         }
118         return super.onOptionsItemSelected(item);
119     }
120 }
```

（7）部署 SQLiteDemo 工程，程序运行结果，如图 8-27 所示。单击"添加记录"按钮，然后单击"查询记录"按钮，如图 8-28 所示。

图 8-27　运行界面　　　　　　　　图 8-28　查询记录

8.4　数据共享

8.4.1　ContentProvider 简介

ContentProvider 类位于 android.content 包下，ContentProvider（数据提供者）是在应用程序间共享数据的一种接口机制。

1. ContentProvider 作用

在 Android 系统中 ContentProvider 的作用是对外共享数据,也就是说 ContentProvider 提供了在多个应用程序之间统一的数据共享方法,将需要共享的数据封装起来,提供了一组供其他应用程序调用的接口,通过 ContentResolver 来操作数据。应用程序可以指定需要共享的数据,而其他应用程序则可以在不知数据来源、路径的情况下对共享数据进行查询、添加、删除和更新等操作。使用 ContentProvider 对外共享数据的好处是统一了数据的访问方式。如果用户不需要在多个应用程序之间共享数据,可以通过上节讲述 SQLiteDatabase 创建数据库的方式实现数据内部共享。

2. ContentProvider 调用原理

ContentProvider 创建和使用前,需要先通过数据库、文件系统或网络实现底层数据存储功能,然后自定义类继承 ContentProvider 类,并在其中实现基本数据操作的接口函数,包括添加、删除、查找和更新等功能。

ContentProvider 的接口函数不能直接使用,需要使用 ContentResolver 对象,通过 URI 间接调用 ContentProvider。

用户使用 ContentResolver 对象与 ContentProvider 进行交互,而 ContentResolver 则通过 URI 确定需要访问的 ContentProvider 的数据集。ContentProvider 负责组织应用程序的数据,向其他应用程序提供数据。ContentResolver 则负责获取 ContentProvider 提供的数据,修改、添加、删除和更新数据等。

8.4.2 URI、UriMatcher 和 ContentUris 简介

1. URI 简介

URI 代表了要操作的数据 Uri 的信息,主要有两部分。

(1) 需要操作的 ContentProvider。

(2) 对 ContentProvider 中的数据进行操作,通过 URI 来确定。

分别就上述 URI 包含的两部分进行介绍。

① ContentProvider 数据模式。ContentProvider 的数据模式类似于数据库的数据表,每行是一条记录,每列具有相同的数据类型,每条记录都包含一个长型的字段_ID,用于唯一标识每条记录。

ContentProvider 可以提供多个数据集,调用者使用 URI 对不同的数据集的数据进行操作。ContentProvider 数据模式如表 8-5 所示。

表 8-5　ContentProvider 数据模式

_ID	NAME	NUMBER	EMAIL
1	Alice	13913913900	alice@google.com
2	Black	12345678	black@126.com
3	Jack	1333333333	jack@qq.com

② URI(通用资源标志符，Uniform Resource Identifier)。URI 用于定位任何远程或本地的可用资源，在 ContentProvider 使用的 URI 通常由以下几部分组成，如图 8-29 所示。

图 8-29　URI 的组成结构

ContentProvider(数据提供者)的 scheme 已经由 Android 所规定，scheme 为：content://；content:// 是通用前缀，表示该 URI 用于 ContentProvider 定位资源，无须修改。

主机名或 <authority> 是授权者名称，用于确定具体由哪一个 ContentProvider 提供资源，外部调用者可以根据这个标识来找到它。因此，一般 <authority> 都由类的小写全称组成，以保证唯一性。

路径是数据路径(<data_path>)，用于确定请求的是哪个数据集。

如果 ContentProvider 仅提供一个数据集，数据路径则是可以省略的。

如果 ContentProvider 提供多个数据集，数据路径则必须指明具体是哪一个数据集。

数据集的数据路径可以写成多段格式，例如 person/name 和 person/age。<id> 是数据编号，用于唯一确定数据集中的一条记录，用于匹配数据集中_ID 字段的值。

如果请求的数据并不只限于一条数据，则 <id> 可以省略。

Android SDK 推荐的方法是：在提供数据表字段中包含一个 ID，在创建表时 INTEGER PRIMARY KEY AUTOINCREMENT 标识此 ID 字段。

例如：person/22 表示要操作 person 表中 id 为 22 的记录。person/22/age 表示要操作 person 表中 id 为 22 的记录的 age 字段。/person 表示要操作 person 表中的所有记录。

2. UriMatcher 类简介

上述 URI 代表了要操作的数据，需要解析 URI 并从 URI 中获取数据。

UriMatcher 类是 Android 系统提供了的用于操作 URI 的工具类。它用于匹配 URI，用法如下。

(1) 注册需要匹配 URI 路径，代码如下：

```
1   UriMatcher sMatcher = new UriMatcher(UriMatcher.NO_MATCH);
2   sMatcher.addURI("hlju.edu. provider.helloprovider ", "person", 1);
```

第 1 行中的常量 UriMatcher.NO_MATCH 表示不匹配任何路径的返回码，第 2 行表示添加需要匹配 URI，如果匹配就会返回匹配码。

上述代码中 addURI() 方法的声明语法：

```
public void addURI(String authority, String path, int code)
```

authority：表示匹配的授权者名称。
path：表示数据路径。
♯：可以代表任何数字。
code：表示返回代码。

(2) 使用 sMatcher.match(uri)方法对输入的 URI 进行匹配。如果匹配就返回匹配码,匹配码是调用 addURI()方法传入的第 3 个参数。假设匹配 content:// hlju.edu.provider.helloprovider/person 路径,返回的匹配码为 1。代码如下:

```
1    sMatcher.addURI("hlju.edu. provider.helloprovider ", "person /♯", 2);
2    switch (sMatcher.match(Uri.parse("content://hlju.edu.provider. helloprovider/person /
     1"))) {
3        case 1
4          break;
5        case 2
6          break;
7        default:
8          break;
9    }
```

第 1 行中的"♯"为通配符,第 7 行表示不匹配。

3. ContentUris 类简介

ContentUris 类也是 Android 系统提供了的用于操作 URI 的工具类,用于操作 URI 路径后面的 ID 部分,它有两个比较常用的方法:withAppendedId(uri, id)和 parseId(uri)方法。

8.4.3 创建 ContentProvider

ContentProvider 创建步骤分 3 步。

(1) 自定义类继承 ContentProvider,并重载 ContentProvider 的 6 个方法。

新创建的自定义类继承 ContentProvider 后,需要重载 6 个方法,代码如下:

```
1    public class ContentProviderDemo extends ContentProvider{
2      public boolean onCreate();
3      public Uri insert(Uri uri,ContentValues values);
4      public int delete(Uri uri,String selection, String[] selectionArgs);
5      public int update(Uri uri, ContentValues values,String selection, String[]
       selectionArgs);
6      public Cursor query(Uri uri, String[] projection, String selection, String[] selectionArgs,
       String sortOrder);
7      public String getType(Uri uri)
8    }
```

第 2 行表示初始化底层数据集和建立数据连接等工作。第 3 行表示添加数据集,第 4 行表示删除数据集,第 5 行表示更新数据集,第 6 行表示查询数据集,第 7 行表示返回指定 URI 的 MIME 数据类型。

(2) 实现 UriMatcher。在新创建的 ContentProvider 类中,通过创建一个 UriMatcher,用于判断 URI 是单条数据还是多条数据。通常为了便于判断和使用 URI,一般将 URI 的授权者名称和数据路径等内容声明为静态常量,并声明 CONTENT_URI。

（3）在AndroidManifest.xml文件中注册ContentProvider。实现完成上述ContentProvider类的代码后，需要在AndroidManifest.xml文件中进行注册，在<application>根结点下，添加<provider>标签，并设置属性。

8.4.4 ContentResolver 操作数据

使用ContentResolver类可以完成外部应用对ContentProvider中的数据进行添加、删除、修改和查询操作。ContentResolver对象的创建，可以使用Activity提供的getContentResolver()方法。ContentResolver类的方法如下。

public Uri insert(Uri uri, ContentValues values)：该方法用于往ContentProvider中添加数据。

public int delete(Uri uri, String selection, String[] selectionArgs)：该方法用于从ContentProvider删除数据。

public int update(Uri uri, ContentValues values, String selection, String[] selectionArgs)：该方法用于更新ContentProvider中的数据。

public Cursor query(Uri uri, String[] projection, String selection, String[] selectionArgs, String sortOrder)：该方法用于从ContentProvider中获取数据。

8.4.5 ContentProvider 应用实例

通过一个案例，详细介绍ContentProvider的应用。

（1）创建一个新的Android工程，工程名为ContentProviderDemo，应用程序名为ContentProviderDemo，包名为hlju.edu.cn，创建的Activity的名字为MainActivity，最小SDK版本根据选择的目标API会自动添加。

（2）修改res目录下layout文件夹中的activity_main.xml文件，设置线性布局中嵌套线性布局，添加一个TextView控件，并设置相关属性，代码如下：

```xml
1  <?xml version = "1.0" encoding = "utf-8"?>
2  <LinearLayout xmlns:android = "http://schemas.android.com/apk/res/android"
3      android:layout_width = "fill_parent"
4      android:layout_height = "fill_parent"
5      android:orientation = "vertical" >
6      <TextView
7          android:id = "@ + id/result"
8          android:layout_width = "wrap_content"
9          android:layout_height = "wrap_content"
10         android:textColor = "@android:color/black"
11         android:textSize = "25dp" />
12 </LinearLayout>
```

（3）在AndroidManifest.xml文件中，增加读取联系人的权限，代码如下：

```xml
<uses-permission android:name = "android.permission.READ_CONTACTS"/>
```

(4) 修改 src 目录中 hlju.edu.cn 包下的 MainActivity.java 文件,代码如下:

```java
package hlju.edu.cn;
import android.app.Activity;
import android.content.ContentResolver;
import android.database.Cursor;
import android.os.Bundle;
import android.provider.ContactsContract.Contacts;
import android.widget.TextView;
import android.view.Menu;
import android.view.MenuItem;
public class MainActivity extends Activity {
    private String[] columns = { Contacts._ID,
            Contacts.DISPLAY_NAME,
    };

    @Override
    public void onCreate(Bundle savedInstanceState) {
        super.onCreate(savedInstanceState);
        setContentView(R.layout.activity_main);
        TextView tv = (TextView) findViewById(R.id.result);
        tv.setText(getQueryData());
    }

    private String getQueryData() {
        StringBuilder sb = new StringBuilder();
        ContentResolver resolver = getContentResolver();
        Cursor cursor = resolver.query(Contacts.CONTENT_URI, columns, null, null, null);
        int idIndex = cursor.getColumnIndex(columns[0]);
        int displayNameIndex = cursor.getColumnIndex(columns[1]);
        sb.append("   ID   " + "姓名" + "\n");
        for (cursor.moveToFirst(); !cursor.isAfterLast(); cursor.moveToNext()) {
            int id = cursor.getInt(idIndex);
            String displayName = cursor.getString(displayNameIndex);
            sb.append("   " + id + "      " + displayName + "\n");
        }
        cursor.close();
        return sb.toString();
    }

    @Override
    public boolean onCreateOptionsMenu(Menu menu) {
        // Inflate the menu; this adds items to the action bar if it is present.
        getMenuInflater().inflate(R.menu.main, menu);
        return true;
    }
    @Override
    public boolean onOptionsItemSelected(MenuItem item) {
```

```
47         // Handle action bar item clicks here. The action bar will
48         // automatically handle clicks on the Home/Up button, so long
49         // as you specify a parent activity in AndroidManifest.xml.
50         int id = item.getItemId();
51         if (id == R.id.action_settings) {
52             return true;
53         }
54         return super.onOptionsItemSelected(item);
55     }
56 }
```

第 11 行代码表示获取通讯录的 ID 值。第 12 行代码表示获取通讯录的姓名。第 19 行代码表示获取布局文件中的标签。第 20 行代码表示为标签设置数据。第 25 行代码表示获取 ContentResolver 对象。第 27 行代码表示获取 ID 记录的索引值。第 28 行代码表示获取姓名记录的索引值。第 30 行代码表示迭代全部记录。

（5）部署 ContentProviderDemo 工程，程序运行后，如图 8-30 所示。

图 8-30　读取联系人

习题

1. 简述 SharedPreferences 的访问模式。
2. SDCard 存储的特点是什么。
3. SQLite 数据库特点是什么。
4. SQLite 数据库和其他数据库的区别是什么。
5. ContentProvider 作用是什么。

第 9 章 多媒体

Android 提供了常见媒体的编码、解码机制,因此可以非常容易地集成音频、视频和图片等多媒体文件到应用程序中。可以调用 Android 提供的现有 API,非常容易地实现相册、播放器、录音和摄像等应用程序。

本章主要学习内容：
- 掌握音频和视频播放的应用；
- 掌握音频和视频录制的应用；
- 了解 TTS 的使用。

9.1 音频播放

播放媒体的来源可以来自源文件、文件系统或网络的数据流。

9.1.1 MediaPlayer 的介绍

OpenCore 为 Android 用户提供了强大的多媒体开发运用功能,为了简化音频、视频系统的开发和播放,Android 提供了一个综合的 MediaPlayer 类以简化对多媒体的操作,通过 MediaPlayer 类可以对音视频文件进行播放。

使用 MediaPlayer 进行音频的播放,主要有以下几个步骤。

1. 获得 MediaPlayer 类的实例

可以直接使用 new 的方式,代码如下。

```
MediaPlayer mediaPlayer = new MediaPlayer();
```

也可以调用 MediaPlayer 类的 create 静态方法,代码如下。

```
MediaPlayer mediaPlayer = MediaPlayer.create(this,R.raw.test);
```

2. 如何设置要播放的文件

MediaPlayer 要播放的文件主要包括 3 个来源：应用中自带的 resource 资源、存储在

SDCard 卡或其他文件路径下的媒体文件和网络上的文件。

3. 对播放器的控制

Android 通过控制播放器状态的方式来控制媒体文件的播放。内容如下。

（1）prepare()和 prepareAsync()提供了同步和异步两种方式设置播放器进入 prepare 状态，需要注意的是，如果 MediaPlayer 实例是由 create 静态方法创建的，那么第一次启动播放前就不需要调用 prepare()方法了，因为在 create 静态方法中已经调用过 prepare()方法了。

（2）start()方法是真正启动文件播放的方法。

（3）pause()方法和 stop()方法比较简单，起到暂停播放和停止播放的作用。

（4）seekTo()方法是定位方法，可以让播放器从指定的位置开始播放，需要注意的是该方法是个异步方法，也就是说该方法返回时并不意味着定位完成。

（5）release()方法可以使播放器占用资源，一旦确定不再使用播放器时应当尽早调用它释放资源。

（6）reset()方法可以使播放器从 Error 状态中恢复过来，重新回到 Idle 状态。

MediaPlayer 的常用方法，如表 9-1 所示。

表 9-1　MediaPlayer 的常用方法

方　　法	描　　述
create(Context context,Uri uri)	静态方法，通过 URI 创建一个多媒体播放器
create(Context context,int resid)	静态方法，通过资源 ID 创建一个多媒体播放器
create(Context context, Uri uri, SurfaceHolder holder)	静态方法，通过 URI 和指定 SurfaceHolder 抽象类创建一个多媒体播放器
getCurrentPosition()	返回 Int，得到当前播放位置
getDuration()	返回 Int，得到文件的时间
getVideoHeight()	返回 Int，得到视频的高度
getVideoWidth()	返回 Int，得到视频的宽度
isLooping()	返回 boolean，是否循环播放
isPlaying()	返回 boolean，是否正在播放
pause()	无返回值，暂停
prepare()	无返回值，准备同步
prepareAsync()	无返回值，准备异步
release()	无返回值，释放 MediaPlayer 对象
reset()	无返回值，重置 MediaPlayer 对象
seekTo(int msec)	无返回值，指定播放的位置（以毫秒为单位的时间）
setAudioStreamType(int streamtype)	无返回值，指定流媒体的类型
setDataSource(String path)	无返回值，设置多媒体数据来源（根据路径）
setDataSource(FileDescriptor fd, long offiset, long length)	无返回值，设置多媒体数据来源（根据 FileDescriptor）
setDataSource(FileDescriptor fd)	无返回值，设置多媒体数据来源（根据 FileDescriptor）
setDataSource(Context context,Uri uri)	无返回值，设置多媒体数据来源（根据 URI）
setDisplay(SurfaceHolder sh)	无返回值，设置用 SurfaceHolder 来显示多媒体

续表

方　　法	描　　述
setLooping(Boolean looping)	无返回值,设置是否循环播放
setScreenOnWhilePlaying(Boolean screenOn)	无返回值,设置是否使用 SurfaceHolder 显示
setVolume(float leftVolume, float rightVolume)	无返回值,设置音量
start()	无返回值,开始播放
stop()	无返回值,停止播放

4. 设置播放器的监听器

MediaPlayer 提供了一些设置不同监听器的方法来更好地对播放器的工作状态进行监听,设置播放器时需要考虑到播放器可能出现的情况,设置好监听和处理逻辑,以保持播放器的健壮性。MediaPlayer 的监听事件如表 9-2 所示。

表 9-2　MediaPlayer 的监听事件

事　　件	描　　述
setOnBufferingUpdateListener(MediaPlayer.OnBufferingUpdateListener listener)	网络流媒体的缓冲监听
setOnCompletionListener(MediaPlayer.OnCompletionListener listener)	网络流媒体播放结束监听
setOnErrorListener(MediaPlayer.OnErrorListener listener)	设置错误信息监听
setOnVideoSizeChangedListener(MediaPlayer.OnVideoSizeChangedListener listener)	视频尺寸监听

9.1.2　MediaPlayer 播放音频

下面通过实例介绍如何使用 MediaPlayer 对象来播放音频文件。

(1) 创建一个新的 Android 工程,工程名为 MediaPlayerDemo,应用程序名为 MediaPlayerDemo,包名为 hlju.edu.cn,创建的 Activity 的名字为 MainActivity,最小 SDK 版本根据选择的目标 API 会自动添加。

(2) 修改 res 目录下 layout 文件夹中的 activity_main.xml 文件,设置线性布局,添加 3 个 Button 控件,对相应控件进行描述,并设置相关属性,代码如下:

```
1   <?xml version = "1.0" encoding = "utf - 8"?>
2   < LinearLayout xmlns:android = "http://schemas.android.com/apk/res/android"
3       android:layout_width = "fill_parent"
4       android:layout_height = "fill_parent"
5       android:orientation = "vertical" >
6       < Button
7           android:id = "@ + id/play"
8           android:layout_width = "fill_parent"
9           android:layout_height = "wrap_content"
10          android:text = "play" >
11      </Button >
12      < Button
13          android:id = "@ + id/pause"
```

```
14          android:layout_width = "fill_parent"
15          android:layout_height = "wrap_content"
16          android:text = "pause" >
17      </Button>
18      <Button
19          android:id = "@ + id/stop"
20          android:layout_width = "fill_parent"
21          android:layout_height = "wrap_content"
22          android:text = "stop" >
23      </Button>
24  </LinearLayout>
```

（3）修改 src 目录中 hlju.edu.cn 包下的 MainActivity.java 文件，代码如下：

```
1   package hlju.edu.cn;
2
3   import android.app.Activity;
4   import android.content.res.AssetFileDescriptor;
5   import android.content.res.AssetManager;
6   import android.media.MediaPlayer;
7   import android.media.MediaPlayer.OnCompletionListener;
8   import android.os.Bundle;
9   import android.util.Log;
10  import android.view.Menu;
11  import android.view.MenuItem;
12  import android.view.View;
13  import android.view.View.OnClickListener;
14  import android.widget.Button;
15
16  public class MainActivity extends Activity implements OnClickListener {
17      private static final String TAG = "MainActivity";
18      private Button playBtn;
19      private Button pauseBtn;
20      private Button stopBtn;
21      private MediaPlayer mediaPlayer;
22
23      @Override
24      public void onCreate(Bundle savedInstanceState) {
25          super.onCreate(savedInstanceState);
26          setContentView(R.layout.activity_main);
27          playBtn = (Button) findViewById(R.id.play);
28          pauseBtn = (Button) findViewById(R.id.pause);
29          stopBtn = (Button) findViewById(R.id.stop);
30          playBtn.setOnClickListener(this);
31          pauseBtn.setOnClickListener(this);
32          stopBtn.setOnClickListener(this);
33      }
34
```

```java
35      @Override
36      public void onClick(View v) {
37        int id = v.getId();
38        switch (id) {
39        case R.id.play:
40          try {
41            AssetManager assetManager = this.getAssets();
42            AssetFileDescriptor fileDescriptor = assetManager.openFd ("1.mp3");
43            mediaPlayer = new MediaPlayer();
44            mediaPlayer.setDataSource(fileDescriptor.getFileDescriptor(), fileDescriptor.
   getStartOffset(), fileDescriptor.getLength());
45            mediaPlayer.prepare();
46            mediaPlayer.start();
47          } catch (Exception e) {
48            Log.e(TAG, "play music occurs errors:" + e.getMessage());
49          }
50          mediaPlayer.setOnCompletionListener(new OnCompletionListener() {
51            @Override
52            public void onCompletion(MediaPlayer mp) {
53              mp.release();
54            }
55          });
56          break;
57        case R.id.pause:
58          if (mediaPlayer != null) {
59            mediaPlayer.pause();
60          }
61          break;
62        case R.id.stop:
63          if (mediaPlayer != null) {
64            mediaPlayer.stop();
65          }
66          break;
67        }
68      }
69
70      @Override
71      protected void onDestroy() {
72        super.onDestroy();
73        if(mediaPlayer != null){
74            mediaPlayer.release();
75        }
76      }
77      @Override
78      public boolean onCreateOptionsMenu(Menu menu) {
79          // Inflate the menu; this adds items to the action bar if it is present.
80          getMenuInflater().inflate(R.menu.main, menu);
81          return true;
82      }
```

```
83        @Override
84        public boolean onOptionsItemSelected(MenuItem item) {
85            // Handle action bar item clicks here. The action bar will
86            // automatically handle clicks on the Home/Up button, so long
87            // as you specify a parent activity in AndroidManifest.xml.
88            int id = item.getItemId();
89            if (id == R.id.action_settings) {
90                return true;
91            }
92            return super.onOptionsItemSelected(item);
93        }
94    }
```

第 18 行声明播放按钮,第 19 行声明暂停按钮,第 20 行声明停止按钮,第 27 行表示找到 id 为 play 的播放按钮,第 28 行表示找到 id 为 pause 的暂停按钮,第 29 行表示找到 id 为 stop 的停止按钮,第 30 行绑定单击事件监听器,第 39 行设置播放按钮,第 42 行要使用的音频文件"1.mp3"要先将它存储到"assets",第 50 行设置媒体播放器监听器,第 57 行为暂停按钮,第 62 行为停止按钮。

(4) 部署 MediaPlayerDemo 工程,程序运行结果如图 9-1 所示。

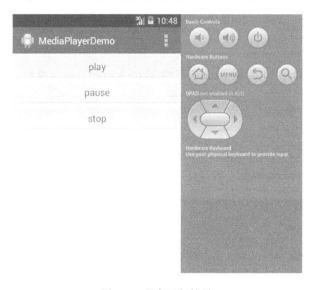

图 9-1　程序运行结果

9.2　视频播放

Android 提供了 3 种方式来实现视频的播放。

使用其自带的播放器。指定 Action 为 ACTION_VIEW,Data 为 Uri,Type 为其 MIME 类型。

使用 VideoView 来播放。在布局文件中使用 VideoView 结合 MediaController 来实现

对其控制。

使用 MediaPlayer 类和 SurfaceView 来实现。这种方式最灵活,也最复杂。

9.2.1 自带播放器播放视频

下面通过实例介绍演示如何使用自带的播放器来播放视频文件。

(1) 创建一个新的 Android 工程,工程名为 SelfDemo,应用程序名为 SelfDemo,包名为 hlju.edu.cn,创建的 Activity 的名字为 MainActivity,最小 SDK 版本根据选择的目标 API 会自动添加。

(2) 修改 res 目录下 layout 文件夹中的 activity_main.xml 文件,设置线性布局,添加 1 个 Button 控件,对相应控件进行描述,并设置相关属性,代码如下:

```
1   <LinearLayout xmlns:android = "http://schemas.android.com/apk/res/android"
2       xmlns:tools = "http://schemas.android.com/tools"
3       android:layout_width = "fill_parent"
4       android:layout_height = "fill_parent" >
5   <Button
6       android:id = "@ + id/play_video"
7       android:layout_width = "fill_parent"
8       android:layout_height = "wrap_content"
9       android:text = "播放视频"/>
10  </LinearLayout>
```

(3) 修改 src 目录中 hlju.edu.cn 包下的 MainActivity.java 文件,代码如下:

```
1   package hlju.edu.cn;
2
3   import android.app.Activity;
4   import android.content.Intent;
5   import android.net.Uri;
6   import android.os.Bundle;
7   import android.os.Environment;
8   import android.view.Menu;
9   import android.view.MenuItem;
10  import android.view.View;
11  import android.view.View.OnClickListener;
12  import android.widget.Button;
13  public class MainActivity extends Activity {
14      Button playVideoButton;
15      @Override
16      public void onCreate(Bundle savedInstanceState) {
17          super.onCreate(savedInstanceState);
18          setContentView(R.layout.activity_main);
19          playVideoButton = (Button)findViewById(R.id.play_video);
20          playVideoButton.setOnClickListener(new OnClickListener() {
21              @Override
22              public void onClick(View v) {
```

```
23              Uri uri = Uri.parse(Environment.getExternalStorageDirectory().getPath() + "/
   test.mp4");
24              Intent intent = new Intent(Intent.ACTION_VIEW);
25              intent.setDataAndType(uri, "video/mp4");
26              startActivity(intent);
27          }
28      });
29  }
30
31  @Override
32  public boolean onCreateOptionsMenu(Menu menu) {
33      // Inflate the menu; this adds items to the action bar if it is present.
34      getMenuInflater().inflate(R.menu.main, menu);
35      return true;
36  }
37
38  @Override
39  public boolean onOptionsItemSelected(MenuItem item) {
40      // Handle action bar item clicks here. The action bar will
41      // automatically handle clicks on the Home/Up button, so long
42      // as you specify a parent activity in AndroidManifest.xml.
43      int id = item.getItemId();
44      if (id == R.id.action_settings) {
45          return true;
46      }
47      return super.onOptionsItemSelected(item);
48  }
49  }
```

第 14 行代码声明发送播放视频意图按钮，第 19 行代码表示找到 id 为 play_video 的按钮，第 24 行代码表示调用系统自带的播放器。

（4）部署 SelfDemo 工程，程序运行结果如图 9-2 所示。

图 9-2　程序运行结果

9.2.2　VideoView 播放视频

Android 中提供了一个 VideoView 组件，用于播放视频文件。要想使用 VideoView 组件播放视频，首先需要在布局文件中创建该组件，然后在 Activity 中获取该组件，并应用 setVideoPath()方法或 setVideoURI()方法加载要播放的视频，最后调用 VideoView 组件的 start()方法来播放视频。另外，VideoView 组件还提供了 stop()和 pause()方法来停止或暂停视频的播放。

在布局文件中创建 VideoView 组件的基本语法代码如下。

```
1  <VideoView
2      属性列表
3  </VideoView>
```

VideoView 组件支持的 XML 属性如表 9-3 所示。

表 9-3　VideoView 组件支持的 XML 属性

XML 属性	描　　述
android:id	用于设置组件的 ID
android:background	用于设置背景，可以设置背景图片，也可以设置背景颜色
android:layout_gravity	用于设置对齐方式
android：layout_width	用于设置宽度
android：layout_height	用于设置高度

Android 还提供了一个可以与 VideoView 组件结合使用的 MediaController 组件。MediaController 组件用于通过图形控制界面用于控制视频的播放。

下面通过实例介绍如何演示使用 VideoView 来播放视频文件。

(1) 创建一个新的 Android 工程，工程名为 VideoViewDemo，应用程序名为 VideoViewDemo，包名为 hlju.edu.cn，创建的 Activity 的名字为 MainActivity，最小 SDK 版本根据选择的目标 API 会自动添加。

(2) 修改 res 目录下 layout 文件夹中的 activity_main.xml 文件，设置线性布局，添加 1 个 VideoView 控件，对相应控件进行描述，并设置相关属性，代码如下：

```
1   <?xml version = "1.0" encoding = "utf - 8"?>
2   <LinearLayout xmlns:android = "http://schemas.android.com/apk/res/android"
3       android:orientation = "vertical"
4       android:layout_width = "fill_parent"
5       android:layout_height = "fill_parent"
6       >
7   <VideoView   android:id = "@ + id/video_view"
8           android:layout_width = "fill_parent"
9           android:layout_height = "wrap_content"
10          android:layout_gravity = "center"/>
11  </LinearLayout>
```

（3）修改 src 目录中 hlju.edu.cn 包下的 MainActivity.java 文件，代码如下：

```
1    package hlju.edu.cn;
2
3    import java.io.File;
4    import android.annotation.SuppressLint;
5    import android.app.Activity;
6    import android.media.MediaPlayer;
7    import android.media.MediaPlayer.OnCompletionListener;
8    import android.os.Bundle;
9    import android.widget.MediaController;
10   import android.widget.Toast;
11   import android.view.Menu;
12   import android.view.MenuItem;
13   import android.widget.VideoView;
14
15   public class MainActivity extends Activity {
16       private VideoView video;
17       @Override
18       public void onCreate(Bundle savedInstanceState) {
19           super.onCreate(savedInstanceState);
20           setContentView(R.layout.activity_main);
21           video = (VideoView) findViewById(R.id.video_view);
22           File file = new File("/storage/sdcard/test.mp4");
23           MediaController mc = new MediaController(MainActivity.this);
24           if(file.exists()){
25              video.setVideoPath(file.getAbsolutePath());
26              video.setMediaController(mc);
27              video.requestFocus();
28              try {
29                  video.start();
30                  } catch (Exception e) {
31                     e.printStackTrace();
32                  }
33              video.setOnCompletionListener(new OnCompletionListener() {
34
35           @Override
36           public void onCompletion(MediaPlayer mp) {
37              Toast.makeText(MainActivity.this, "视频播放完毕!", Toast.LENGTH_SHORT).show();
38           }
39         });
40           }else{
41         Toast.makeText(this, "要播放的视频文件不存在", Toast.LENGTH_SHORT).show();
42           }
43       }
44
45       @Override
46       public boolean onCreateOptionsMenu(Menu menu) {
```

```
47              // Inflate the menu; this adds items to the action bar if it is present.
48              getMenuInflater().inflate(R.menu.main, menu);
49              return true;
50          }
51
52          @Override
53          public boolean onOptionsItemSelected(MenuItem item) {
54              int id = item.getItemId();
55              if (id == R.id.action_settings) {
56                  return true;
57              }
58              return super.onOptionsItemSelected(item);
59          }
```

第16行代码用于声明 VideoView 对象,第21行代码用于获取 VideoView 组件,第22行代码用于获取 SD 卡上要播放的文件,第24行代码用于判断要播放的视频文件是否存在,第25行代码用于指定要播放的视频,第26行代码用于设置 VideoView 与 MediaController 相关联,第27行代码用于 VideoView 获得焦点,第29行代码用于开始播放视频,第31行代码用于输出异常信息,第33行代码用于 VideoView 添加完成事件监听器,第37行代码用于弹出消息提示框显示播放完毕,第41行代码用于弹出消息提示框提示文件不存在。

(4) 部署 VideoViewDemo 工程,程序运行结果如图 9-3 所示。

图 9-3　程序运行结果

9.2.3　MediaPlayer 结合 SurfaceView 播放视频

SurfaceView 是一个继承了 View,但是与一般的 View 有很大区别的类。这是由于 SurfaceView 的绘制方法与 View 的绘制方法不一样造成的。

SurfaceView 可以直接从内存或者 DMA 等硬件接口取得图像数据,是个非常重要的绘图容器。它的特性是可以在主线程之外的线程中向屏幕上绘图。这样可以避免画图任务繁重的时候造成主线程的阻塞,从而提高了程序的反应速度。在游戏开发中经常要用到 SurfaceView,游戏中的背景、任务、动画等尽量在画布 Canvas 中画出。

由于需要对 SurfaceView 进行监听,所以需要实现 SurfaceHolder.Callback 这个接口。这个接口需要实现 3 个方法。

- public void surfaceChanged(SurfaceHolder holder,int format,int width,int height):在 surface 的大小发生变化时调用。
- public void surfaceCreated(SurfaceHolder holder):在 surface 创建时调用,一般在这里调用画图的线程。
- public void surfaceDestoryed(SurfaceHolder holder):销毁时调用,一般在这里将画图的线程停止。

SurfaceHolder 可以被看成是 surface 的控制器,用于操作 surface。处理它在 Canvas 上画的效果、控制大小和像素等。SurfaceHolder 有几个比较常用的方法。

- public abstract void addCallBack(SurfaceHolder.CallBack):给 SurfaceView 当前的持有者添加一个回调对象。
- public abstract Canvas lockCanvas():锁定画布,一般在锁定后就可以通过其返回的画布对象 Canvas,在上面进行画图等操作。
- public abstract Canvas lockCanvas(Rect dirty):锁定画布的某个区域进行画图。相对部分内存要求比较高的游戏来说,可以不用重画 dirty 外的其他区域的像素,这样可以提高速度。
- public abstract unlockCanvasAndPost(Canvas canvas):结束锁定画图,并提交改变。

下面通过实例介绍如何利用 MediaPlayer 结合 SurfaceView 播放视频文件。

(1) 创建一个新的 Android 工程,工程名为 SurfaceViewDemo,应用程序名为 SurfaceViewDemo,包名为 hlju.edu.cn,创建的 Activity 的名字为 MainActivity,最小 SDK 版本根据选择的目标 API 会自动添加。

(2) 修改 res 目录下 layout 文件夹中的 activity_main.xml 文件,设置线性布局,添加 3 个 Button 控件和 1 个 SurfaceView 控件,对相应控件进行描述,并设置相关属性,代码如下:

```
1   <?xml version = "1.0" encoding = "utf - 8"?>
2   <LinearLayout xmlns:android = "http://schemas.android.com/apk/res/android"
3       android:layout_width = "fill_parent"
4       android:layout_height = "fill_parent"
5       android:orientation = "vertical" >
6       <SurfaceView
7           android:id = "@ + id/surfaceView"
8           android:layout_width = "fill_parent"
9           android:layout_height = "360px" />
10      <LinearLayout
```

```
11            android:layout_width = "fill_parent"
12            android:layout_height = "wrap_content"
13            android:gravity = "center_horizontal"
14            android:orientation = "horizontal" >
15        < Button
16            android:id = "@ + id/btnplay"
17            android:layout_width = "wrap_content"
18            android:layout_height = "wrap_content"
19        android:text = "播放"/>
20        < Button
21            android:id = "@ + id/btnpause"
22            android:layout_width = "wrap_content"
23            android:layout_height = "wrap_content"
24            android:text = "暂停"/>
25        < Button
26            android:id = "@ + id/btnstop"
27            android:layout_width = "wrap_content"
28            android:layout_height = "wrap_content"
29            android:text = "停止"/>
30      </LinearLayout >
31  </LinearLayout >
```

（3）修改 src 目录中 hlju.edu.cn 包下的 MainActivity.java 文件，代码如下：

```
1   package hlju.edu.cn;
2
3   import android.app.Activity;
4   import android.media.AudioManager;
5   import android.media.MediaPlayer;
6   import android.os.Bundle;
7   import android.os.Environment;
8   import android.util.Log;
9   import android.view.SurfaceHolder;
10  import android.view.SurfaceHolder.Callback;
11  import android.view.SurfaceView;
12  import android.view.View;
13  import android.view.View.OnClickListener;
14  import android.widget.Button;
15  import android.view.Menu;
16  import android.view.MenuItem;
17
18  public class MainActivity extends Activity   implements OnClickListener{
19      private static final String TAG = "MainActivity";
20      Button btnplay, btnstop, btnpause;
21      SurfaceView surfaceView;
22      MediaPlayer mediaPlayer;
23      int position;
24
```

```java
25      public void onCreate(Bundle savedInstanceState) {
26          super.onCreate(savedInstanceState);
27          setContentView(R.layout.activity_main);
28          btnplay = (Button)findViewById(R.id.btnplay);
29          btnpause = (Button)findViewById(R.id.btnpause);
30          btnstop = (Button)findViewById(R.id.btnstop);
31          btnplay.setOnClickListener(this);
32          btnpause.setOnClickListener(this);
33          btnstop.setOnClickListener(this);
34          mediaPlayer = new MediaPlayer();
35          surfaceView = (SurfaceView) this.findViewById(R.id.surfaceView);
36          surfaceView.getHolder().setType(SurfaceHolder.SURFACE_TYPE_PUSH_BUFFERS);
37          surfaceView.getHolder().addCallback(new Callback() {
38              @Override
39              public void surfaceDestroyed(SurfaceHolder holder) {
40                Log.d(TAG, "surfaceDestroyed method was invoked");
41              }
42              @Override
43              public void surfaceCreated(SurfaceHolder holder) {
44                  if (position > 0) {
45                      try {
46                          play();
47                          mediaPlayer.seekTo(position);
48                          position = 0;
49                      } catch (Exception e) {
50                        Log.d(TAG, "surfaceCreated occurs errors:" + e.getMessage());
51                      }
52                  }
53              }
54              @Override
55              public void surfaceChanged(SurfaceHolder holder, int format, int width,
56                      int height) {
57                Log.d(TAG, "surfaceChanged method was invoked");
58              }
59          });
60      }
61      @Override
62      public void onClick(View v) {
63          switch (v.getId()) {
64          case R.id.btnplay:
65            play();
66            break;
67          case R.id.btnpause:
68            if (mediaPlayer.isPlaying()) {
69                mediaPlayer.pause();
70            } else {
71                mediaPlayer.start();
72            }
73            break;
```

```java
74          case R.id.btnstop:
75              if (mediaPlayer.isPlaying()) {
76                  mediaPlayer.stop();
77              }
78              break;
79          default:
80              break;
81          }
82      }
83
84      @Override
85      protected void onPause() {
86          if (mediaPlayer.isPlaying()) {
87              position = mediaPlayer.getCurrentPosition() ;
88              mediaPlayer.stop();
89          }
90          super.onPause();
91      }
92
93      private void play() {
94          try {
95              mediaPlayer.reset();
96              mediaPlayer
97                  .setAudioStreamType(AudioManager.STREAM_MUSIC);
98              mediaPlayer.setDataSource(Environment.getExternalStorageDirectory() + "/test.mp4");
99              mediaPlayer.setDisplay(surfaceView.getHolder());
100             mediaPlayer.prepare();
101             mediaPlayer.start();
102         } catch (Exception e) {
103           Log.e(TAG, "play occurs errors:" + e.getMessage());
104         }
105     }
106
107     @Override
108     public boolean onCreateOptionsMenu(Menu menu) {
109          getMenuInflater().inflate(R.menu.main, menu);
110          return true;
111     }
112
113     @Override
114     public boolean onOptionsItemSelected(MenuItem item) {
115          int id = item.getItemId();
116         if (id == R.id.action_settings) {
117             return true;
118         }
119         return super.onOptionsItemSelected(item);
120     }
121 }
```

第 20 行代码表示声明播放、暂停和停止按钮,第 28 行表示代码找到 id 为 btnplay 的播放按钮,第 29 行代码表示找到 id 为 btnpause 的暂停按钮,第 30 行代码表示找到 id 为 btnstop 的停止按钮,第 31 行代码表示绑定单击事件监听器,第 36 行代码表示设置 SurfaceView 自己不管理的缓冲区,第 46 行代码表示开始播放,第 47 行代码表示直接从指定位置开始播放,第 64 行代码表示单击播放按钮,第 67 行代码表示单击暂停播放,第 74 行代码表示单击停止播放,第 86 行代码先判断是否正在播放,第 87 行代码表示如果正在播放就先保存这个播放位置,第 98 行代码表示设置需要播放的视频,第 99 行代码表示把视频画面输出到 SurfaceView,第 101 行代码表示视频播放。

(4)部署 SurfaceViewDemo 工程,程序运行结果如图 9-4 所示。

图 9-4　程序运行结果

9.3　音频录制

可以使用 Android SDK 提供的 MediaRecorder 类来完成音频的录制。

下面通过实例介绍如何利用 MediaRecorder 类来完成音频的录制。

(1)创建一个新的 Android 工程,工程名为 MediaRecorderVoiceDemo,应用程序名为 MediaRecorderVoiceDemo,包名为 hlju. edu. cn,创建的 Activity 的名字为 MainActivity,最小 SDK 版本根据选择的目标 API 会自动添加。

(2)修改 res 目录下 layout 文件夹中的 activity_main. xml 文件,设置线性布局,添加 4 个 Button 控件和 1 个 TextView 控件,对相应控件进行描述,并设置相关属性,代码如下:

```
1   <?xml version = "1.0" encoding = "utf - 8"?>
2   <LinearLayout xmlns:android = "http://schemas.android.com/apk/res/android"
3       android:layout_width = "fill_parent"
4       android:layout_height = "fill_parent"
```

```
5        android:orientation = "vertical" >
6          < TextView
7              android:id = "@ + id/state_info"
8              android:layout_width = "fill_parent"
9              android:layout_height = "wrap_content"
10         android:text = ""/>
11         < Button
12             android:id = "@ + id/btn_start"
13             android:layout_width = "fill_parent"
14             android:layout_height = "wrap_content"
15         android:text = "开始"/>
16          < Button
17             android:id = "@ + id/btn_stop"
18             android:layout_width = "fill_parent"
19             android:layout_height = "wrap_content"
20         android:text = "停止"/>
21         < Button
22             android:id = "@ + id/btn_play"
23             android:layout_width = "fill_parent"
24             android:layout_height = "wrap_content"
25         android:text = "播放"/>
26         < Button
27             android:id = "@ + id/btn_finish"
28             android:layout_width = "fill_parent"
29             android:layout_height = "wrap_content"
30         android:text = "结束"/>
31     </LinearLayout >
```

（3）由于需要录制音频,将需要录制好的音频文件保存在 SD Card 中,所以需要添加相关权限。需要在项目清单文件 AndroidManifest.xml 中添加相关的权限,代码如下:

```
1   < uses - permission android:name = "android.permission.WRITE_EXTERNAL_STORAGE"/>
2   < uses - permission android:name = "android.permission.MOUNT_UNMOUNT_FILESYSTEMS"/>
3   < uses - permission   android:name = "android.permission.RECORD_AUDIO"/>
```

（4）修改 src 目录中 hlju.edu.cn 包下的 MainActivity.java 文件,代码如下:

```
1    package hlju.edu.cn;
2    import java.io.File;
3    import android.app.Activity;
4    import android.content.ContentValues;
5    import android.media.MediaPlayer;
6    import android.media.MediaRecorder;
7    import android.net.Uri;
8    import android.os.Bundle;
9    import android.os.Environment;
10   import android.provider.MediaStore;
```

```java
11   import android.util.Log;
12   import android.view.Menu;
13   import android.view.MenuItem;
14   import android.view.View;
15   import android.view.View.OnClickListener;
16   import android.widget.Button;
17   import android.widget.TextView;
18
19   public class MainActivity extends Activity implements OnClickListener {
20       private static final String TAG = "MainActivity";
21       Button startButton, stopButton, playButton, finishButton;
22       TextView stateView;
23       private MediaRecorder recorder;
24       private MediaPlayer player;
25       private File audioFile;
26       private Uri fileUri;
27
28       @Override
29       public void onCreate(Bundle savedInstanceState) {
30           super.onCreate(savedInstanceState);
31           setContentView(R.layout.activity_main);
32           startButton = (Button) findViewById(R.id.btn_start);
33           stopButton = (Button) findViewById(R.id.btn_stop);
34           playButton = (Button) findViewById(R.id.btn_play);
35           finishButton = (Button) findViewById(R.id.btn_finish);
36           startButton.setOnClickListener(this);
37           stopButton.setOnClickListener(this);
38           playButton.setOnClickListener(this);
39           finishButton.setOnClickListener(this);
40           stateView = (TextView)findViewById(R.id.state_info);
41           stateView.setText("准备开始");
42       }
43
44       @Override
45       public void onClick(View v) {
46           int id = v.getId();
47           switch (id) {
48           case R.id.btn_start:
49                   recorder = new MediaRecorder();
50                   recorder.setAudioSource(MediaRecorder.AudioSource.MIC);
51                   recorder.setOutputFormat(MediaRecorder.OutputFormat.THREE_GPP);
52                   recorder.setAudioEncoder(MediaRecorder.AudioEncoder.AMR_NB);
53                   File fpath = new File(Environment.getExternalStorageDirectory().getAbsolutePath() + "/data/files/");
54                   fpath.mkdirs();
55                   try {
56                       audioFile = File.createTempFile("recording", ".3gp", fpath);
```

```java
57                     recorder.setOutputFile(audioFile.getAbsolutePath());
58                     recorder.prepare();
59                 } catch (Exception e) {
60                     Log.e(TAG, "Occurs errors:" + e.getMessage());
61                 }
62                 recorder.start();
63                 stateView.setText("正在录制");
64                 startButton.setEnabled(false);
65                 playButton.setEnabled(false);
66                 stopButton.setEnabled(true);
67                 break;
68             case R.id.btn_stop:
69                 recorder.stop();
70                 recorder.release();
71                 ContentValues values = new ContentValues();
72                 values.put(MediaStore.Audio.Media.TITLE, "this is my first record-audio");
73                 values.put(MediaStore.Audio.Media.DATE_ADDED, System.currentTimeMillis());
74                 values.put(MediaStore.Audio.Media.DATA, audioFile.getAbsolutePath());
75                 fileUri = this.getContentResolver().insert(MediaStore.Audio.Media.EXTERNAL_CONTENT_URI, values);
76                 player = new MediaPlayer();
77                 player.setOnCompletionListener(new MediaPlayer.OnCompletionListener() {
78
79                     @Override
80                     public void onCompletion(MediaPlayer arg0) {
81                         stateView.setText("准备录制");
82                         startButton.setEnabled(true);
83                         playButton.setEnabled(true);
84                         stopButton.setEnabled(false);
85                     }
86                 });
87                 try {
88                     player.setDataSource(audioFile.getAbsolutePath());
89                     player.prepare();
90                 } catch (Exception e) {
91                     Log.e(TAG, "Prepare for playing occurs errors:" + e.getMessage());
92                 }
93                 stateView.setText("准备播放");
94                 playButton.setEnabled(true);
95                 startButton.setEnabled(true);
96                 stopButton.setEnabled(false);
97                 break;
98             case R.id.btn_play:
99                 player.start();
100                stateView.setText("正在播放");
101                startButton.setEnabled(false);
102                stopButton.setEnabled(false);
```

```
103                playButton.setEnabled(false);
104                break;
105            case R.id.btn_finish:
106                break;
107            default:
108                finish();
109                break;
110        }
111    }
112    @Override
113    public boolean onCreateOptionsMenu(Menu menu) {
114        getMenuInflater().inflate(R.menu.main, menu);
115        return true;
116    }
117    @Override
118    public boolean onOptionsItemSelected(MenuItem item) {
119        int id = item.getItemId();
120        if (id == R.id.action_settings) {
121            return true;
122        }
123        return super.onOptionsItemSelected(item);
124    }
125 }
```

第21行声明开始录音、结束录音、播放录音以及结束Activity按钮。第22行显示录音状态文本。第32行代码表示找到id为btn_start的开始录音按钮。第33行代码表示找到id为btn_stop的停止录音按钮。第34行代码表示找到id为btn_play的播放录音按钮。第35行代码表示找到id为btn_finish的结束Activity按钮。第36行代码表示绑定单击事件监听器。第40行代码表示找到id为state_info的显示状态文本。第48行代码表示单击开始录音按钮,开始录制。第49行代码表示实例化一个MediaRecorder对象,然后进行相应的设置。第50行代码表示指定AudioSource为MIC(Microphone audio source)。第51行代码表示指定OutputFormat为3gp格式。THREE_GPP录制后文件是一个3gp文件,支持音频和视频录制;MPEG-4指定录制的文件为mpeg-4格式,可以保护Audio和Video;RAW_AMR录制原始文件,支持音频录制,同时要求音频编码为AMR_NB。第52行代码表示指定Audio编码方式。第53行代码表示指定录制后文件的存储路径。第54行代码表示创建文件夹。第56行代码表示创建临时文件。第62行代码表示开始录制。第68行代码表示结束录音按钮。第76行代码表示录制结束后,实例化一个MediaPlayer对象,然后准备播放。第87行代码表示准备播放。第98行代码表示单击播放录音按钮,开始播放录音。在录音结束的时候,已经实例化了MediaPlayer,做好了播放的准备,单击播放录音按钮即可播放。第105行代码表示单击结束按钮,结束Activity。

(5) 部署MediaRecorderVoiceDemo工程,程序运行结果如图9-5所示。

图 9-5　程序运行结果

9.4　视频录制

可以使用 Android SDK 提供的 MediaRecorder 类来完成视频的录制。

下面通过实例介绍如何利用 MediaRecorder 类来完成视频的录制。

（1）创建一个新的 Android 工程，工程名为 MediaRecorderViewDemo，应用程序名为 MediaRecorderViewDemo，包名为 hlju.edu.cn，创建的 Activity 的名字为 MainActivity，最小 SDK 版本根据选择的目标 API 会自动添加。

（2）修改 res 目录下 layout 文件夹中的 activity_main.xml 文件，设置线性布局，添加两个 Button 控件和 1 个 SurfaceView 控件，对相应控件进行描述，并设置相关属性，代码如下：

```
1   <LinearLayout xmlns:android = "http://schemas.android.com/apk/res/android"
2       android:layout_width = "fill_parent"
3       android:layout_height = "fill_parent"
4       android:orientation = "horizontal" >
5       <LinearLayout
6           android:layout_width = "fill_parent"
7           android:layout_height = "fill_parent"
8           android:layout_weight = "1" >
9           <SurfaceView
10              android:id = "@ + id/surfaceview"
11              android:layout_width = "fill_parent"
12              android:layout_height = "fill_parent" />
13      </LinearLayout>
14      <LinearLayout
15          android:layout_width = "fill_parent"
```

```
16          android:layout_height = "fill_parent"
17          android:layout_weight = "4"
18          android:gravity = "center"
19          android:orientation = "vertical" >
20          <Button
21              android:id = "@+id/start"
22              android:layout_width = "fill_parent"
23              android:layout_height = "wrap_content"
24              android:layout_weight = "1"
25              android:text = "Start" />
26          <Button
27              android:id = "@+id/stop"
28              android:layout_width = "fill_parent"
29              android:layout_height = "wrap_content"
30              android:layout_weight = "1"
31              android:text = "Stop" />
32      </LinearLayout>
33  </LinearLayout>
```

(3) 由于需要录制视频,将需要录制好的视频文件保存在 SD Card 当中,所以需要在项目清单文件 AndroidManifest.xml 中添加相关的权限,代码如下:

```
1  <uses-permission android:name = "android.permission.CAMERA" />
2  <uses-permission android:name = "android.permission.RECORD_AUDIO" />
3  <uses-permission android:name = "android.permission.WRITE_EXTERNAL_STORAGE" />
4  <uses-permission android:name = "android.permission.MOUNT_UNMOUNT_FILESYSTEMS"/>
```

(4) 修改 src 目录中 hlju.edu.cn 包下的 MainActivity.java 文件,代码如下:

```
1   package hlju.edu.cn;
2
3   import java.io.IOException;
4   import android.app.Activity;
5   import android.content.pm.ActivityInfo;
6   import android.graphics.PixelFormat;
7   import android.media.MediaRecorder;
8   import android.os.Bundle;
9   import android.view.Menu;
10  import android.view.MenuItem;
11  import android.view.SurfaceHolder;
12  import android.view.SurfaceView;
13  import android.view.View;
14  import android.view.View.OnClickListener;
15  import android.view.Window;
16  import android.view.WindowManager;
17  import android.widget.Button;
18
```

```java
19  public class MainActivity extends Activity implements SurfaceHolder.Callback {
20      private Button start;
21      private Button stop;
22      private MediaRecorder mediarecorder;
23      private SurfaceView surfaceview;
24      private SurfaceHolder surfaceHolder;
25
26      public void onCreate(Bundle savedInstanceState) {
27          super.onCreate(savedInstanceState);
28          requestWindowFeature(Window.FEATURE_NO_TITLE);
29          getWindow().setFlags(WindowManager.LayoutParams.FLAG_FULLSCREEN, WindowManager.LayoutParams.FLAG_FULLSCREEN);
30          setRequestedOrientation(ActivityInfo.SCREEN_ORIENTATION_LANDSCAPE);
31          getWindow().setFormat(PixelFormat.TRANSLUCENT);
32          setContentView(R.layout.activity_main);
33          init();
34      }
35      private void init() {
36          start = (Button) this.findViewById(R.id.start);
37          stop = (Button) this.findViewById(R.id.stop);
38          start.setOnClickListener(new TestVideoListener());
39          stop.setOnClickListener(new TestVideoListener());
40          surfaceview = (SurfaceView) this.findViewById(R.id.surfaceview);
41          SurfaceHolder holder = surfaceview.getHolder();
42          holder.addCallback(this);
43          holder.setType(SurfaceHolder.SURFACE_TYPE_PUSH_BUFFERS);
44      }
45      class TestVideoListener implements OnClickListener {
46          @Override
47          public void onClick(View v) {
48              if (v == start) {
49                  mediarecorder = new MediaRecorder();
50                  mediarecorder.setVideoSource(MediaRecorder.VideoSource.CAMERA);
51                  mediarecorder.setOutputFormat(MediaRecorder.OutputFormat.THREE_GPP);
52                  mediarecorder.setVideoEncoder(MediaRecorder.VideoEncoder.H264);
53                  mediarecorder.setVideoSize(176, 144);
54                  mediarecorder.setVideoFrameRate(20);
55                  mediarecorder.setPreviewDisplay(surfaceHolder.getSurface());
56                  mediarecorder.setOutputFile("/sdcard/1.3gp");
57                  try {
58                      mediarecorder.prepare();
59                      mediarecorder.start();
60                  } catch (IllegalStateException e) {
61                      e.printStackTrace();
62                  } catch (IOException e) {
63                      e.printStackTrace();
64                  }
65              }
66              if (v == stop) {
```

```java
67          if (mediarecorder != null) {
68              mediarecorder.stop();
69              mediarecorder.release();
70              mediarecorder = null;
71          }
72       }
73     }
74   }
75
76   @Override
77   public void surfaceChanged(SurfaceHolder holder, int format, int width,
78           int height) {
79       surfaceHolder = holder;
80   }
81
82   @Override
83   public void surfaceCreated(SurfaceHolder holder) {
84       surfaceHolder = holder;
85   }
86
87   @Override
88   public void surfaceDestroyed(SurfaceHolder holder) {
89       surfaceview = null;
90       surfaceHolder = null;
91       mediarecorder = null;
92   }
93
94   @Override
95   public boolean onCreateOptionsMenu(Menu menu) {
96       // Inflate the menu; this adds items to the action bar if it is present.
97       getMenuInflater().inflate(R.menu.main, menu);
98       return true;
99   }
100
101  @Override
102  public boolean onOptionsItemSelected(MenuItem item) {
103      // Handle action bar item clicks here. The action bar will
104      // automatically handle clicks on the Home/Up button, so long
105      // as you specify a parent activity in AndroidManifest.xml.
106      int id = item.getItemId();
107      if (id == R.id.action_settings) {
108          return true;
109      }
110      return super.onOptionsItemSelected(item);
111  }
112 }
```

第 20 行代码表示定义开始录制按钮。第 21 行代码表示定义停止录制按钮。第 22 行代码表示定义录制视频的类。第 23 行代码表示定义显示视频的控件。第 28 行代码表示去掉标题栏。第29行代码表示设置全屏。第30行代码表示设置横屏显示。第31行代码表示选择支持半透明模式，在有surfaceview的activity中使用。第41行代码表示取得holder。第42行代码表示holder加入回调接口。第43行代码表示设置setType。setType必须设置，否则出错。第49行代码表示创建mediarecorder对象。第50行代码表示设置录制视频源为Camera。第51行代码表示设置录制完成后视频的封装格式为3gp。第52行代码表示设置录制的视频编码为h264。第53行代码表示设置视频录制的分辨率。必须放在设置编码和格式的后面，否则报错。第54行代码表示设置录制的视频帧率。必须放在设置编码和格式的后面，否则报错。第56行代码表示设置视频文件输出的路径。第58行代码表示准备录制。第59行代码表示开始录制。第68行代码表示停止录制。第69行代码表示释放资源。第79行代码表示这个holder为开始在oncreat里面取得的holder，将它赋给surfaceHolder。第84行代码表示这个holder为开始在oncreat里面取得的holder，将它赋给surfaceHolder。第89行到第91行代码表示surfaceDestroyed的时候同时将对象设置为null。

（5）部署MediaRecorderViewDemo工程，程序运行结果如图9-6所示。

图9-6　程序运行结果

9.5　TTS的使用

TextToSpeech简称TTS,称为语音合成。是Android从版本1.6开始支持的新功能，能将所指定的文本转换成不同语言音频输出。TTS功能需要有TTS Engine的支持，下面首先来了解一下Android提供的TTS Engine。Android使用了叫做Pico支持多种语言的语音合成引擎，负责在后台分析输入的文本，把文本分解为它能识别的各个片段，再把合成的各个语音片段以听起来比较自然的方式连接在一起。TTS Engine依托于当前Android Platform所支持的几种主要的语言：English、French、German、Italian和Spanish共5种语言的语音输出。与此同时，对于个别的语言版本将取决于不同的时区。

下面通过实例介绍如何在 Android 应用程序当中使用 TTS 相关的 API。

(1) 创建一个新的 Android 工程,工程名为 TTSDemo,应用程序名为 TTSDemo,包名为 hlju.edu.cn,创建的 Activity 的名字为 MainActivity,最小 SDK 版本根据选择的目标 API 会自动添加。

(2) 修改 res 目录下 layout 文件夹中的 activity_main.xml 文件,设置线性布局,添加 1 个 Button 控件和 1 个 EditText 控件,对相应控件进行描述,并设置相关属性,代码如下:

```
1   <?xml version = "1.0" encoding = "utf - 8"?>
2   < LinearLayout xmlns:android = "http://schemas.android.com/apk/res/android"
3       android:layout_width = "fill_parent"
4       android:layout_height = "fill_parent"
5       android:orientation = "vertical" >
6       < EditText
7           android:id = "@ + id/inputText"
8           android:layout_width = "fill_parent"
9           android:layout_height = "wrap_content"
10          android:hint = "Input the text here!" >
11      </EditText >
12      < Button
13          android:id = "@ + id/speakBtn"
14          android:layout_width = "wrap_content"
15          android:layout_height = "wrap_content"
16          android:layout_gravity = "center_horizontal"
17          android:enabled = "false"
18          android:text = "Speak" >
19      </Button >
20  </LinearLayout >
```

(3) 修改 src 目录中 hlju.edu.cn 包下的 MainActivity.java 文件,代码如下:

```
1   package hlju.edu.cn;
2
3   import java.util.Locale;
4   import android.app.Activity;
5   import android.content.Intent;
6   import android.os.Bundle;
7   import android.view.Menu;
8   import android.view.MenuItem;
9   import android.speech.tts.TextToSpeech;
10  import android.speech.tts.TextToSpeech.OnInitListener;
11  import android.util.Log;
12  import android.view.View;
13  import android.view.View.OnClickListener;
14  import android.widget.Button;
15  import android.widget.EditText;
16
17  public class MainActivity extends Activity implements OnInitListener {
```

```java
18      private EditText inputText = null;
19      private Button speakBtn = null;
20      private static final int REQ_TTS_STATUS_CHECK = 0;
21      private static final String TAG = "TTS Demo";
22      private TextToSpeech mTts;
23
24      @Override
25      public void onCreate(Bundle savedInstanceState) {
26        super.onCreate(savedInstanceState);
27        setContentView(R.layout.activity_main);
28        Intent checkIntent = new Intent();
29        checkIntent.setAction(TextToSpeech.Engine.ACTION_CHECK_TTS_DATA);
30        startActivityForResult(checkIntent, REQ_TTS_STATUS_CHECK);
31        inputText = (EditText) findViewById(R.id.inputText);
32        speakBtn = (Button) findViewById(R.id.speakBtn);
33        inputText.setText("happy new year!");
34        speakBtn.setOnClickListener(new OnClickListener() {
35
36          public void onClick(View v) {
37            mTts.speak(inputText.getText().toString(), TextToSpeech.QUEUE_ADD, null);
38          }
39        });
40      }
41
42      @Override
43      public void onInit(int status) {
44        if (status == TextToSpeech.SUCCESS) {
45          int result = mTts.setLanguage(Locale.US);
46          if (result == TextToSpeech.LANG_MISSING_DATA || result == TextToSpeech.LANG_NOT_SUPPORTED)
47          {
48            Log.v(TAG, "Language is not available");
49            speakBtn.setEnabled(false);
50          } else {
51            mTts.speak("happy new year!", TextToSpeech.QUEUE_ADD, null);
52            speakBtn.setEnabled(true);
53          }
54        }
55      }
56
57      protected void onActivityResult(int requestCode, int resultCode, Intent data) {
58        if (requestCode == REQ_TTS_STATUS_CHECK) {
59          switch (resultCode) {
60          case TextToSpeech.Engine.CHECK_VOICE_DATA_PASS:
61          {
62            mTts = new TextToSpeech(this, this);
63            Log.v(TAG, "TTS Engine is installed!");
64          }
65            break;
```

```java
66          case TextToSpeech.Engine.CHECK_VOICE_DATA_BAD_DATA:
67          case TextToSpeech.Engine.CHECK_VOICE_DATA_MISSING_DATA:
68          case TextToSpeech.Engine.CHECK_VOICE_DATA_MISSING_VOLUME:
69          {
70            Log.v(TAG, "Need language stuff:" + resultCode);
71            Intent dataIntent = new Intent();
72            dataIntent.setAction(TextToSpeech.Engine.ACTION_INSTALL_TTS_DATA);
73            startActivity(dataIntent);
74          }
75            break;
76          case TextToSpeech.Engine.CHECK_VOICE_DATA_FAIL:
77          default:
78            Log.v(TAG, "Got a failure. TTS apparently not available");
79            break;
80          }
81      } else {
82      }
83    }
84
85    @Override
86    protected void onPause() {
87      super.onPause();
88      if (mTts != null)
89      {
90        mTts.stop();
91      }
92    }
93
94    @Override
95    protected void onDestroy() {
96      super.onDestroy();
97      mTts.shutdown();
98    }
99
100   @Override
101   public boolean onCreateOptionsMenu(Menu menu) {
102       // Inflate the menu; this adds items to the action bar if it is present.
103       getMenuInflater().inflate(R.menu.main, menu);
104       return true;
105   }
106
107   @Override
108   public boolean onOptionsItemSelected(MenuItem item) {
109       // Handle action bar item clicks here. The action bar will
110       // automatically handle clicks on the Home/Up button, so long
111       // as you specify a parent activity in AndroidManifest.xml.
112       int id = item.getItemId();
113       if (id == R.id.action_settings) {
114           return true;
```

```
115              }
116              return super.onOptionsItemSelected(item);
117          }
118  }
```

第 29 行代码表示检查 TTS 数据是否已经安装并且可用。第 37 行代码表示朗读输入框里的内容。第 43 行代码表示实现 TTS 初始化接口。第 44 行代码表示 TTS Engine 初始化完成。第 45 行代码表示设置发音语言。第 46 行代码表示判断语言是否可用。第 60 行代码表示这个返回结果表明 TTS Engine 可以用。第 66 行代码表示需要的语音数据已损坏。第 67 行代码表示缺少需要语言的语音数据。第 68 行代码表示缺少需要语言的发音数据。第 70 行代码表示这 3 种情况都表明数据有错,重新下载安装需要的数据。第 76 行代码表示检查失败。第 81 行代码表示其他 Intent 返回的结果。第 90 行代码表示 activity 暂停时也停止 TTS。第 97 行代码表示释放 TTS 的资源。

(4) 部署 TTSDemo 工程,程序运行结果如图 9-7 所示。

图 9-7　程序运行结果

习题

1. Android 提供的常见音频播放都有哪些。
2. Android 提供了哪几种方式来实现视频的播放。
3. Android 是否能够实现音频录制,如何实现。
4. Android 是否能够实现视频录制,如何实现。

第10章 Android 网络通信技术

Google 公司是以网络搜索引擎起家的,通过大胆的创意和不断的研发努力,目前已经成为网络世界的巨头。而出自于 Google 的 Android 平台,在进行网络编程方面,同样是非常优秀的。

本章主要学习内容:
- 掌握 Android 网络通信技术基础和 HTTP 通信应用;
- 掌握 WebKit 应用;
- 了解 Socket 通信。

10.1 Android 网络通信技术基础

10.1.1 无线网络技术

无线网络的产生为用户提供了很大的方便,通过无线网络,用户可以从任何地方接入网络。所谓无线网络,即采用无线传输媒介的网络。无线网络可以根据数据传输的距离分为无线局域网、无线个域网、低速率无线个域网、无线城域网和无线广域网。下面分别介绍。

无线局域网(WLAN):最常用的一种无线网络技术,可以使用户在本地创建无线连接。

无线个域网(WPAN):无线个域网技术使用户能够为个人操作空间(10 米以内的空间范围)设备(如手机、iPad、笔记本电脑等)创建临时无线通信。

低速率无线个域网(LR-WPAN):适用于工业监测、办公室和家庭自动化等。

无线城域网(WMAN):无线城域网技术使用户可以在城区的多个场所之间创建无线连接,使用无线电波或红外光波传送数据。

无线广域网(WWAN):无线广域网技术可以使用户通过远程公用网络或专用网络建立无线网络连接。2G、3G、4G 等都属于无线广域网。

10.1.2 Android 网络基础

Android 基于 Linux 内核,它包含一组优秀的联网功能。Android 平台有 3 种网络接口可以使用,分别是 java.net.*(标准 Java 接口)、org.apache(Apache 接口)和 android.net.*(Android 网络接口)。

java.net.*(标准 Java 接口)提供与联网有关的类,包括流和数据包套接字、Internet

协议、常见 HTTP 处理。例如创建 URL 及 URLConnection/HttpURLConnection 对象、设置连接参数、连接到服务器、向服务器写数据、从服务器读取数据等通信。下面是常见的使用 java.net 包的 HTTP 例子，代码如下：

```
1   try
2   {
3       URL url = new URL("http://www.google.com")
4       HttpURLConnection http = (HttpURLConnection)url.openConnection();
5       int nRC = http.getResponseCode();
6       if(nRC == HttpURLConnection.HTTP_OK)
7       {
8           InputStream is = http.getInputStream();
9           …
10      }
11  }
12  Catch (Exception e)
13  {
14  }
```

第 3 行代码表示定义一个连接地址。第 4 行代码表示打开连接。第 5 行代码表示获得连接的状态。第 8 行代码表示取得数据。第 9 行代码表示处理相关数据。

HTTP 协议可能是现状 Internet 上使用得最多、最重要的协议了。许多的 Java 应用程序需要通过 HTTP 协议来访问网络资源。虽然在 JDK 的 java.net 包中已经提供了访问 HTTP 协议的基本功能，但是对于大部分应用程序来说，JDK 库本身提供的功能远远不够。这时就需要 Android 提供的 Apache HttpClient 了。它是一个开源项目，功能更加完善，为客户端的 HTTP 编程提供高效、最新、功能丰富的工具包支持。Android 平台引入了 Apache HttpClient 的同时还提供了对它的一些封装和扩展，如设置默认的 HTTP 超时和缓存大小等。Android 使用的 HttpClient 主要包括创建 HttpClient、Get/Post、HttpRequest 等对象，设置连接参数，执行 HTTP 操作，处理服务器返回结果等功能。下面是一个使用 android.net.http.* 包的例子，代码如下：

```
1   try
2   {
3       HttpClient hc = new DefaultHttpClient();
4       HttpGet get = new HttpGet("http://www.google.com");
5       HttpResponse rp = hc.execute(get);
6       if(rp.getStatusLine().getStatusCode() == HttpStatus.sc_OK)
7       {
8           InputStream is = rp.getEntity().getContent();
9           …
10      }
11  }
12  catch(IOException e)
13  {
14  }
```

第 3 行代码表示创建 HttpClient,使用 DefaultHttpClient 表示默认属性。第 8 行代码表示取得数据。第 9 行代码表示处理相关数据。

android.net.＊包实际上是通过对 Apache 中 HttpClient 的封装来实现的一个 HTTP 编程接口,同时还提供了 HTTP 连接池管理,以提高并发请求情况下的处理效率,除此之外还有网络状态监视等接口、网络访问的 Socket、常用的 URI 类以及有关 Wi-Fi 相关的类等。

10.1.3 Android 中的蓝牙

蓝牙(Bluetooth)是目前广泛应用的无线通信协议,近距离无线通信的标准,主要针对短距离设备通信(10 米之内)。Android SDK 直到 2.0 以后的版本才开始支持蓝牙编程。蓝牙是一项无线电连接系统,它可以将不同的电子器材连接起来,常用于连接耳机、鼠标和移动通信设备等。

两个设备要想通过蓝牙协议进行数据通信的前提是两个设备上都要有蓝牙设备(蓝牙适配器)。例如:手机与蓝牙设备——手机开始扫描周围的蓝牙设备,扫描发现附件的蓝牙设备时,给它发出一个信号需要进行蓝牙配对并设置一个临时密钥,计算机收到请求信息,输入临时密钥,配对成功。

每个蓝牙设备都会有一个可见性的设置,如果将蓝牙设备设置为可见,那么其他近距离的蓝牙设备就可以搜索到这个蓝牙设备,如果将蓝牙设置为不可见,那么其他近距离的蓝牙设备就无法扫描到这个蓝牙设备。

有关蓝牙编程的核心类都位于 android.blue.tooth 包当中。android.bluetooth 包当中几个比较常用的核心类,如表 10-1 所示。

表 10-1　android.bluetooth 包中的类

名称	说明
BluetoothAdapter	代表本地的蓝牙适配器
BluetoothDevice	代表远程的蓝牙设备
BluetoothSocket	代表一个蓝牙 Socket 的接口
BluetoothServiceSocket	代表一个服务器 Socket,监听进入的连接请求

为了在应用程序当中使用蓝牙的相关功能 API,在项目清单文件当中至少需要声明两方面的权限:BLUETOOTH 权限和 BLUETOOTH_ADMIN 权限。只有在项目清单当中声明了 BLUETOOTH 权限,才能够实现蓝牙设备之间的互相通信,例如请求一个连接、接收一个连接以及数据的传输。只有在项目清单当中声明了 BLUETOOTH_ADMIN 这个权限,才能够发现近距离的蓝牙设备以及对蓝牙设备进行管理设置。要想使用 BLUETOOTH_ADMIN 这个权限,必须首先声明 BLUETOOTH 这个权限。

下面通过实例介绍如何使用蓝牙设备。

(1) 创建一个新的 Android 工程,工程名为 BluetoothDemo,应用程序名为 BluetoothDemo,包名为 hlju.edu.cn,创建的 Activity 的名字为 MainActivity,最小 SDK 版本根据选择的目标 API 会自动添加。

(2) 修改 res 目录下 layout 文件夹中的 activity_main.xml 文件,设置线性布局,添加 1 个 TextView 控件和 1 个 Button 控件,对相应控件进行描述,并设置相关属性,代码如下:

```
1   <LinearLayout xmlns:android = "http://schemas.android.com/apk/res/android"
2       xmlns:tools = "http://schemas.android.com/tools"
3       android:layout_width = "fill_parent"
4       android:layout_height = "fill_parent"
5       android:orientation = "vertical">
6       <TextView
7           android:id = "@ + id/show_devices"
8           android:layout_width = "wrap_content"
9           android:layout_height = "wrap_content"
10          android:textSize = "20sp"></TextView>
11      <Button
12          android:id = "@ + id/search_bt_devices"
13          android:layout_width = "fill_parent"
14          android:layout_height = "wrap_content"
15          android:text = "搜索周围蓝牙设备"/>
16  </LinearLayout>
```

(3) 在 AndroidManifest.xml 文件中添加相关权限,代码如下:

```
1   <?xml version = "1.0" encoding = "utf - 8"?>
2   <manifest xmlns:android = "http://schemas.android.com/apk/res/android"
3       package = "hlju.edu.cn"
4       android:versionCode = "1"
5       android:versionName = "1.0" >
6
7       <uses - permission android:name = "android.permission.BLUETOOTH"/>
8       <uses - permission android:name = "android.permission.BLUETOOTH_ADMIN"/>
9
10      <uses - sdk
11          android:minSdkVersion = "14"
12          android:targetSdkVersion = "21" />
13      <application
14          android:allowBackup = "true"
15          android:icon = "@drawable/ic_launcher"
16          android:label = "@string/app_name"
17          android:theme = "@style/AppTheme" >
18          <activity
19              android:name = ".MainActivity"
20              android:label = "@string/app_name" >
21              <intent - filter >
22                  <action android:name = "android.intent.action.MAIN" />
23                  <category android:name = "android.intent.category.LAUNCHER" />
24              </intent - filter >
25          </activity>
26      </application>
27  </manifest>
```

（4）在 src 目录中 hlju.edu.cn 包下创建 MainActivity.java 文件，代码如下：

```java
1   package hlju.edu.cn;
2
3   import android.app.Activity;
4   import android.bluetooth.BluetoothAdapter;
5   import android.bluetooth.BluetoothDevice;
6   import android.content.BroadcastReceiver;
7   import android.content.Context;
8   import android.content.Intent;
9   import android.content.IntentFilter;
10  import android.os.Bundle;
11  import android.util.Log;
12  import android.view.Menu;
13  import android.view.MenuItem;
14  import android.view.View;
15  import android.view.View.OnClickListener;
16  import android.widget.Button;
17  import android.widget.TextView;
18  import android.widget.Toast;
19
20  public class MainActivity extends Activity {
21      private static final String TAG = "MainActivity";
22      Button searchBTDevices;
23      TextView showDevices;
24      BluetoothReceiver bluetoothReceiver;
25      StringBuffer stringBuffer;
26      @Override
27      public void onCreate(Bundle savedInstanceState) {
28          super.onCreate(savedInstanceState);
29          setContentView(R.layout.activity_main);
30          showDevices = (TextView)findViewById(R.id.show_devices);
31          searchBTDevices = (Button)findViewById(R.id.search_bt_devices);
32          searchBTDevices.setOnClickListener(new SearchBtListener());
33          stringBuffer = new StringBuffer();
34      }
35      private class SearchBtListenner implements OnClickListener{
36      @Override
37      public void onClick(View v) {
38          BluetoothAdapter localAdapter = BluetoothAdapter.getDefaultAdapter();
39          if(localAdapter == null){
40              Toast.makeText(MainActivity.this, "设备不支持蓝牙", Toast.LENGTH_LONG).show();
41          }
42          if( ! localAdapter.isEnabled()){
43              Intent intent = new Intent(BluetoothAdapter.ACTION_REQUEST_ENABLE);
44          intent.putExtra(BluetoothAdapter.EXTRA_DISCOVERABLE_DURATION, 300);
45              MainActivity.this.startActivity(intent);
46          }
47          localAdapter.startDiscovery();
48      }
49      }
```

```
50        @Override
51        protected void onResume() {
52        super.onResume();
53        IntentFilter intentFilter = new IntentFilter ( BluetoothDevice.ACTION_FOUND);
54          bluetoothReceiver = new BluetoothReceiver ();
55          registerReceiver(bluetoothReceiver,intentFilter);
56        }
57        @Override
58        protected void onStop() {
59            super.onStop();
60            unregisterReceiver(bluetoothReceiver);
61        }
62         private class BluetoothReceiver extends BroadcastReceiver{
63            @Override
64            public void onReceive(Context context, Intent intent) {
65                BluetoothDevice device = intent.getParcelableExtra (BluetoothDevice.EXTRA_
    DEVICE);
66                Log.d(TAG, "new device address:" + device.getAddress());
67                stringBuffer.append(device.getAddress() + "\n");
68                showDevices.setText(stringBuffer);
69            }
70        }
71    }
```

第 22 行代码表示声明搜索临近蓝牙设备按钮。第 32 行代码表示为按钮绑定内部类形式的单击事件监听器。第 37 行代码表示处理单击事件回调。第 38 行代码表示获取运行当前程序设备的蓝牙适配器。第 40 行代码表示当设备不支持蓝牙,给出相关提示。第 42 行代码表示如果蓝牙设备未打开,打开蓝牙,并设置蓝牙可见性。第 44 行代码表示设置蓝牙可见性,最多 300 秒,当数值大于 300 时默认为 300。第 51 行代码表示用 onResume 注册发现相关蓝牙设备广播接收者。第 53 行代码表示设定广播接收的 filter。第 54 行代码表示创建蓝牙广播信息的 receiver。第 55 行代码表示注册广播接收器。

(5) 部署 BluetoothDemo 工程,程序运行结果,如图 10-1 所示。

图 10-1　程序运行结果

10.1.4 Android 中的 Wi-Fi

Wi-Fi(Wireless Fidelith)又称 802.11b 标准，它的最大优点是传输速度快，可以达到每秒 11Mbit/s。Wi-Fi 是无线网络通信技术的一个品牌，由 Wi-Fi 联盟(Wi-Fi Alliance)拥有，目的是改善基于 IEEE802.11 标准的无线网络产品之间的互通性。

Android 中提供了 android.net.wifi 包，供用户对 Wi-Fi 进行操作，其包含的类及说明如表 10-2 所示。

表 10-2 android.net.wifi 包含的类及说明

类	描述
ScanResult	描述已经检测出的接入点
WifiConfiguration	Wi-Fi 网络的配置，包括安全配置等
WifiConfiguration.AuthAlgorithm	公认的 IEEE802.11 认证算法
WifiConfiguration.GroupCipher	公认的密码
WifiConfiguration.KeyMgmt	公认的密钥管理方案
WifiConfiguration.PairwiseCipher	公认的 WPA 成对密码
WifiConfiguration.Protocol	公认的安全协议
WifiConfiguration.Status	网络配置可能的状态
WifiInfo	无线连接的描述，包括接入点、网络连接状态、隐藏的接入点、IP 地址、连接速度、MAC 地址、网络 ID 和信号强度等
WifiManager	提供管理 Wi-Fi 连接的大部分 API
WifiManager.MulticastLock	允许应用程序接收多播数据包 Wi-Fi
WifiManager.WifiLock	允许应用程序一直使用 Wi-Fi 无线网络
WpsInfo	表示 Wi-Fi 保护设置的类

下面给出一个使用 Wi-Fi 的实例。

(1) 创建一个新的 Android 工程，工程名为 WifiManagerDemo，应用程序名为 WifiManagerDemo，包名为 hlju.edu.cn，创建的 Activity 的名字为 MainActivity，最小 SDK 版本根据选择的目标 API 会自动添加。

(2) 修改 res 目录下 layout 文件夹中的 activity_main.xml 文件，设置线性布局，添加 1 个 TextView 控件和两个 Button 控件，对相应控件进行描述，并设置相关属性，代码如下：

```
1    <LinearLayout xmlns:android = "http://schemas.android.com/apk/res/android"
2        xmlns:tools = "http://schemas.android.com/tools"
3        android:layout_width = "fill_parent"
4        android:layout_height = "fill_parent"
5        android:orientation = "vertical">
6        <TextView
7            android:id = "@ + id/show_current_wifi"
8            android:layout_width = "wrap_content"
9            android:layout_height = "wrap_content"/>
10       <Button
11           android:id = "@ + id/get_current_wifi"
```

```
12        android:layout_width = "fill_parent"
13        android:layout_height = "wrap_content"
14        android:text = "获取当前 Wi-Fi 连接"/>
15    <Button
16        android:id = "@ + id/scan_wifi"
17        android:layout_width = "fill_parent"
18        android:layout_height = "wrap_content"
19        android:text = "扫描 Wi-Fi 热点" />
20 </LinearLayout>
```

(3) 在 AndroidManifest.xml 文件中添加相关权限,代码如下:

```
1 <uses-permission android:name = "android.permission.ACCESS_WIFI_STATE"/>
2 <uses-permission android:name = "android.permission.CHANGE_WIFI_STATE"/>
3 <uses-permission android:name = "android.permission.INTERNET"/>
```

(4) 修改 src 目录中 hlju.edu.cn 包下的 MainActivity.java 文件,代码如下:

```
1  package hlju.edu.cn;
2
3  import java.util.List;
4  import android.app.Activity;
5  import android.content.BroadcastReceiver;
6  import android.content.Context;
7  import android.content.Intent;
8  import android.content.IntentFilter;
9  import android.net.wifi.ScanResult;
10 import android.net.wifi.WifiInfo;
11 import android.net.wifi.WifiManager;
12 import android.os.Bundle;
13 import android.util.Log;
14 import android.view.Menu;
15 import android.view.MenuItem;
16 import android.view.View;
17 import android.view.View.OnClickListener;
18 import android.widget.Button;
19 import android.widget.TextView;
20
21 public class MainActivity extends Activity implements OnClickListener {
22     private static final String TAG = "MainActivity";
23     TextView showCurrentWIFI;
24     Button getCurrentWIFIBtn;
25     Button scanWIFIBtn;
26     WifiManager wifiManager = null;
27     @Override
28     public void onCreate(Bundle savedInstanceState) {
29         super.onCreate(savedInstanceState);
30         setContentView(R.layout.activity_main);
```

```java
31              showCurrentWIFI = (TextView) findViewById(R.id.show_current_wifi);
32              getCurrentWIFIBtn = (Button) findViewById(R.id.get_current_wifi);
33              getCurrentWIFIBtn.setOnClickListener(this);
34              scanWIFIBtn = (Button) findViewById(R.id.scan_wifi);
35              scanWIFIBtn.setOnClickListener(this);
36              wifiManager = (WifiManager) getSystemService(Context.WIFI_SERVICE);
37          }
38
39          @Override
40          public void onClick(View v) {
41              int viewId = v.getId();
42              switch (viewId) {
43                  case R.id.get_current_wifi:
44                      WifiInfo info = wifiManager.getConnectionInfo();
45                      String maxText = info.getMacAddress();
46                      String ipText = intToIp(info.getIpAddress());
47                      String status = "";
48                      if (wifiManager.getWifiState() == WifiManager.WIFI_STATE_ENABLED) {
49                          status = "WIFI_STATE_ENABLED";
50                      }
51                      String ssid = info.getSSID();
52                      int networkID = info.getNetworkId();
53                      int speed = info.getLinkSpeed();
54                      showCurrentWIFI.setText("mac: " + maxText + "\n\r" + "ip: " +
    ipText + "\n\r" + "wifi status :" + status + "\n\r" + "ssid :" + ssid + "\n\r" +
    "net work id :" + networkID + "\n\r" + "connection speed:" + speed + "\n\r");
55                      break;
56                  case R.id.scan_wifi:
57                      registerReceiver(new BroadcastReceiver() {
58
59                          @Override
60                          public void onReceive(Context context, Intent intent) {
61                              List<ScanResult> results = wifiManager.getScanResults();
62                              ScanResult bestSignal = null;
63                              for (ScanResult result : results) {
64                                  if (null == bestSignal || WifiManager.compareSignalLevel
    (bestSignal.level, result.level) < 0) {
65                                      bestSignal = result;
66                                  }
67                              }
68                              Log.d(TAG, results.size() + " networks found, " + bestSignal.
    SSID + " is the strongest");
69                          }
70                      }, new IntentFilter(WifiManager.SCAN_RESULTS_AVAILABLE_ACTION));
71                      wifiManager.startScan();
72                      break;
73              }
74          }
75
```

```
76        private String intToIp(int ip) {
77            return (ip & 0xFF) + "." + ((ip >> 8) & 0xFF) + "." + ((ip >> 16) & 0xFF) +
    "." + ((ip >> 24) & 0xFF);
78        }
79    }
```

第 23 行代码表示声明显示当前 Wi-Fi 状态信息文本。第 24 行代码表示获取当前 Wi-Fi 状态按钮。第 25 行代码表示扫描 Wi-Fi 热点按钮。第 26 行代码表示声明 Wi-Fi 管理器。第 36 行代码表示实例化 Wi-Fi 管理器。第 43 行代码表示获取当前 Wi-Fi 状态按钮。第 56 行代码表示扫描 Wi-Fi 热点按钮。

（5）部署 WifiManagerDemo 工程，程序运行结果，如图 10-2 所示。

图 10-2　程序运行结果

10.2　HTTP 通信

超文本传输协议（Hyper Text Transfer Protocol，HTTP）用于传送 WWW 方式的数据。HTTP 协议采用了请求/响应模型。客户端向服务器发送一个请求，请求头包含了请求的方法、URI、协议版本，以及包含请求修饰符、客户信息和内容类似于 MIME 的消息结构。服务器以一个状态行为作为响应，响应的内容包括消息协议的版本、成功或者错误编码，还包含服务器信息、实体元信息以及可能的实体内容。它是一个属于应用层面向对象的协议，由于其简洁和快速，它适用于分布式超媒体信息系统。

许多 HTTP 通信都是由一个用户代理初始化的，并且包括一个申请在源服务器上资源的请求，最简单的情况可能是在用户代理和服务器之间通过一个单独的连接来完成。在 Internet 上，HTTP 通信通常发生在 TCP/IP 连接之上，默认端口是 TCP 80，但其他的端口也是可用的。这并不预示着 HTTP 协议在 Internet 或其他网络的其他协议之上才能完成，

HTTP 只预示着一个可靠的传输。

随着智能手机和平板电脑等移动终端设备的迅速发展,现在的 Internet 已经不再是传统的有线互联网,还包括了移动互联网。同有线互联网一样,移动互联网也可以使用 HTTP 访问网络。在 Android 中,针对 HTTP 进行网络通信的方法主要有两种:一种是使用 HttpURLConnection 实现;另一种是使用 HttpClient 实现。

10.2.1　HttpURLConnection 接口

HttpURLConnection 位于 java.net 包中,用于发送 HTTP 请求和获取 HTTP 响应。由于该类是抽象类,不能直接实例化对象,所以需要使用 URL 的 openConnection()方法来获得。例如,要创建 http://www.google.com 网站对应的 HttpURLConnection 对象,代码如下:

```
1  URL url = new URL("http://www.google.com");
2  HttpURLConnection urlConn = (HttpURLConnection) url.openConnection();
```

通过 openConnection()方法创建的 HttpURLConnection 对象,并没有真正执行连接操作,只是创建了一个新的实例。在进行连接前,还可以设置一些属性,例如连接超时的时间和请求方式等。

创建了 HttpURLConnection 对象后,就可以使用该对象发送 HTTP 请求了。HTTP 请求通常分为 GET 请求和 POST 请求两种。

1. 发送 GET 请求

使用 Http 对象发送请求时,默认发送的是 GET 请求。因此,发送 GET 请求比较简单,只需要在指定连接地址时,先将要传递的参数通过"? 参数名=参数值"进行传递(多个参数间使用英文半角的逗号分隔。例如,要传递用户名和 E-mail 地址两个参数,可以使用"? user = hlj, email = hljxxkx@126.com"实现),然后获取流中的数据,最后关闭连接即可。

下面给出一个 GET 的实例。

(1) 创建一个新的 Android 工程,工程名为 URLConnectionGETDemo,应用程序名为 URLConnectionGETDemo,包名为 hlju.edu.cn,创建的 Activity 的名字为 MainActivity,最小 SDK 版本根据选择的目标 API 会自动添加。

(2) 修改 res 目录下 layout 文件夹中的 activity_main.xml 文件,设置线性布局,添加一个 EditText、一个 Button、一个 ScrollView 和一个 TextView 控件,对相应控件进行描述,并设置相关属性,代码如下:

```
1  <?xml version = "1.0" encoding = "utf - 8"?>
2  <LinearLayout xmlns:android = "http://schemas.android.com/apk/res/android"
3      android:layout_width = "fill_parent"
4      android:layout_height = "fill_parent"
```

```xml
5       android:gravity = "center_horizontal"
6       android:orientation = "vertical" >
7     < EditText
8         android:id = "@ + id/content"
9         android:layout_width = "match_parent"
10        android:layout_height = "wrap_content" />
11    < Button
12        android:id = "@ + id/button"
13        android:layout_width = "wrap_content"
14        android:layout_height = "wrap_content"
15        android:text = "@string/button" />
16    < ScrollView
17        android:id = "@ + id/scrollView1"
18        android:layout_width = "match_parent"
19        android:layout_height = "wrap_content"
20        android:layout_weight = "1" >
21      < LinearLayout
22          android:id = "@ + id/linearLayout1"
23          android:layout_width = "match_parent"
24          android:layout_height = "match_parent" >
25        < TextView
26            android:id = "@ + id/result"
27            android:layout_width = "match_parent"
28            android:layout_height = "wrap_content"
29            android:layout_weight = "1" />
30      </LinearLayout >
31    </ScrollView >
32  </LinearLayout >
```

(3) 修改 src 目录中 hlju.edu.cn 包下的 MainActivity.java 文件,代码如下:

```java
1   package hlju.edu.cn;
2
3   import java.io.BufferedReader;
4   import java.io.IOException;
5   import java.io.InputStreamReader;
6   import java.io.UnsupportedEncodingException;
7   import java.net.HttpURLConnection;
8   import java.net.MalformedURLException;
9   import java.net.URL;
10  import java.net.URLEncoder;
11  import android.app.Activity;
12  import android.os.Bundle;
13  import android.os.Handler;
14  import android.os.Message;
15  import android.util.Base64;
16  import android.view.Menu;
17  import android.view.MenuItem;
```

```java
18  import android.view.View;
19  import android.view.View.OnClickListener;
20  import android.widget.Button;
21  import android.widget.EditText;
22  import android.widget.TextView;
23  import android.widget.Toast;
24
25  public class MainActivity extends Activity {
26      private EditText content;
27      private Button button;
28      private Handler handler;
29      private String result = "";
30      private TextView resultTV;
31      @Override
32      protected void onCreate(Bundle savedInstanceState) {
33          super.onCreate(savedInstanceState);
34          setContentView(R.layout.activity_main);
35          content = (EditText) findViewById(R.id.content);
36          resultTV = (TextView) findViewById(R.id.result);
37          button = (Button) findViewById(R.id.button);
38          button.setOnClickListener(new OnClickListener() {
39              @Override
40              public void onClick(View v) {
41                  if ("".equals(content.getText().toString())) {
42                      Toast.makeText(MainActivity.this, "请输入要发表的内容!",
43                          Toast.LENGTH_SHORT).show();          //显示消息提示
44                      return;
45                  }
46                  new Thread(new Runnable() {
47                      public void run() {
48                          send();
49                          Message m = handler.obtainMessage();
50                          handler.sendMessage(m);
51                      }
52                  }).start();
53              }
54          });
55          handler = new Handler() {
56              @Override
57              public void handleMessage(Message msg) {
58                  if (result != null) {
59                      resultTV.setText(result);
60                      content.setText("");
61                  }
62                  super.handleMessage(msg);
63              }
64          };
65      }
66
```

```
67    public void send() {
68      String target = "";
69      target = "http://192.168.106.111:8080/blog/index.jsp?content = "
70          + base64(content.getText().toString().trim());
71      URL url;
72      try {
73        url = new URL(target);
74        HttpURLConnection urlConn = (HttpURLConnection) url
75            .openConnection();
76        InputStreamReader in = new InputStreamReader(
77            urlConn.getInputStream());
78        BufferedReader buffer = new BufferedReader(in);
79        String inputLine = null;
80        while ((inputLine = buffer.readLine()) != null) {
81          result += inputLine + "\n";
82        }
83        in.close();
84        urlConn.disconnect();
85      } catch (MalformedURLException e) {
86        e.printStackTrace();
87      } catch (IOException e) {
88        e.printStackTrace();
89      }
90    }
91    public String base64(String content){
92      try {
93        content = Base64.encodeToString(content.getBytes("utf-8"), Base64.DEFAULT);
94        content = URLEncoder.encode(content);
95      } catch (UnsupportedEncodingException e) {
96        e.printStackTrace();
97      }
98      return content;
99    }
100  }
```

第 26 行代码表示声明一个输入文本内容的编辑框对象。第 27 行代码表示声明一个发表按钮对象。第 28 行代码表示声明一个 Handler 对象。第 29 行代码表示声明一个代表显示内容的字符串。第 30 行代码表示声明一个显示结果的文本框对象。第 35 行代码表示获取输入文本内容的 EditText 组件。第 36 行代码表示获取显示结果的 TextView 组件。第 37 行代码表示获取"发表"按钮组。第 38 行代码表示为按钮添加单击事件监听器。第 46 行代码表示创建一个新线程，用于发送并读取微博信息。第 48 行代码表示发送文本内容到 Web 服务器。第 49 行代码表示获取一个 Message。第 50 行代码表示发送消息。第 52 行代码表示开启线程。第 55 行代码表示创建一个 Handler 对象。第 59 行代码表示显示获得的结果。第 60 行代码表示清空文本框。第 69 行代码表示要访问的 URL 地址。第 74 行代码表示创建一个 HTTP 连接。第 76 行代码表示获得读取的内容。第 78 行代码表示获取输入流对象。第 80 行代码表示通过循环逐行读取输入流中的内容。第 83 行代码表示关闭字符输入流对象。第84行代码表示断开连接。第91行代码表示对字符串进行Base64编码。

（4）在AndroidManifest.xml文件中指定访问的权限，代码如下：

```
<uses-permission android:name="android.permission.INTERNET"/>
```

（5）部署URLConnectionGETDemo工程，程序运行结果，如图10-3所示。输入要发表的内容，然后单击"发表"按钮即可发表一条信息，如图10-4所示。

图10-3　程序运行结果

图10-4　发表信息

2. 发送POST请求

由于采用GET方式发送请求只适合发送大小在1024个字节以内的数据，所以在要发送的数据比较大时，就需要使用POST方式来发送该请求。在Android中，使用HttpURLConnection类在发送请求时，默认采用的是GET请求，如果要发送POST请求，需要通过其setRequestMethod()方法进行指定。例如，创建一个HTTP连接，并为该连接指定请求的发送方式为POST，代码如下：

```
1  HttpURLConnection urlConn = (HttpURLConnection) url.openConnection();
2  urlConn.setRequestMethod("POST");
```

第2行指定请求方式为POST。

发送POST请求时要比发送GET请求复杂一些，它经常需要通过HttpURLConnection类及其父类URLConnection提供的方法设置相关内容，发送POST请求时常用的方法如表10-3所示。

表 10-3　发送 POST 请求时常用的方法

方　　法	描　　述
setDoInput(Boolean new Value)	用于设置是否向连接中写入数据,如果参数值为 true,表示写入数据;否则不写入数据
setDoOutput(Boolean new Value)	用于设置是否从连接中读取数据,如果参数值为 true,表示读取数据;否则不读取数据
setUseCaches(Boolean new Value)	用于设置是否缓存数据,如果参数值为 true,表示缓存数据;否则表示禁用缓存
setInstanceFollowRedirects(Boolean followRedirects)	用于设置是否应该自动执行 HTTP 重定向,如果参数值为 true,表示自动执行;否则不自动执行
setRequestProperty(String field, Sting new Value)	用于设置一般请求属性,例如,要设置内容类型为表单数据,可以进行设置:setRequestProperty("Content-Type","application/x-www-form-urlencoded")

下面通过实例介绍如何使用不需要与组件交互的本地服务。

(1) 创建一个新的 Android 工程,工程名为 URLConnectionPOSTDemo,应用程序名为 URLConnectionPOSTDemo,包名为 hlju.edu.cn,创建的 Activity 的名字为 MainActivity,最小 SDK 版本根据选择的目标 API 会自动添加。

(2) 修改 res 目录下 layout 文件夹中的 activity_main.xml 文件,设置线性布局,添加两个 EditText、一个 Button、一个 ScrollView 和一个 TextView 控件,对相应控件进行描述,并设置相关属性,代码如下:

```
1   <?xml version = "1.0" encoding = "utf-8"?>
2   <LinearLayout xmlns:android = "http://schemas.android.com/apk/res/android"
3       android:orientation = "vertical"
4       android:gravity = "center_horizontal"
5       android:layout_width = "fill_parent"
6       android:layout_height = "fill_parent">
7       <EditText android:id = "@ + id/nickname"
8           android:hint = "@string/nickname"
9           android:layout_width = "match_parent"
10          android:layout_height = "wrap_content"/>
11      <EditText android:id = "@ + id/content"
12          android:layout_height = "wrap_content"
13          android:layout_width = "match_parent"
14          android:inputType = "textMultiLine"/>
15      <Button android:id = "@ + id/button"
16          android:layout_width = "wrap_content"
17          android:layout_height = "wrap_content"
18          android:text = "@string/button"/>
19      <ScrollView
20          android:id = "@ + id/scrollView1"
21          android:layout_width = "match_parent"
22          android:layout_height = "wrap_content"
23          android:layout_weight = "1" >
```

```
24          <LinearLayout
25              android:id = "@ + id/linearLayout1"
26              android:layout_width = "match_parent"
27              android:layout_height = "match_parent" >
28          <TextView
29              android:id = "@ + id/result"
30              android:layout_width = "match_parent"
31              android:layout_height = "wrap_content"
32              android:layout_weight = "1" />
33          </LinearLayout>
34      </ScrollView>
35  </LinearLayout>
```

(3) 修改 src 目录中 hlju.edu.cn 包下的 MainActivity.java 文件,代码如下:

```
1   package hlju.edu.cn;
2   import java.io.BufferedReader;
3   import java.io.DataOutputStream;
4   import java.io.IOException;
5   import java.io.InputStreamReader;
6   import java.net.HttpURLConnection;
7   import java.net.MalformedURLException;
8   import java.net.URL;
9   import java.net.URLEncoder;
10  import android.app.Activity;
11  import android.os.Bundle;
12  import android.os.Handler;
13  import android.os.Message;
14  import android.view.Menu;
15  import android.view.MenuItem;
16  import android.view.View;
17  import android.view.View.OnClickListener;
18  import android.widget.Button;
19  import android.widget.EditText;
20  import android.widget.TextView;
21  import android.widget.Toast;
22  public class MainActivity extends Activity {
23      private EditText nickname;
24      private EditText content;
25      private Button button;
26      private Handler handler;
27      private String result = "";
28      private TextView resultTV;
29      @Override
30      protected void onCreate(Bundle savedInstanceState) {
31          super.onCreate(savedInstanceState);
32          setContentView(R.layout.activity_main);
33          content = (EditText) findViewById(R.id.content);
```

```java
34      resultTV = (TextView) findViewById(R.id.result);
35      nickname = (EditText) findViewById(R.id.nickname);
36      button = (Button) findViewById(R.id.button);
37      button.setOnClickListener(new OnClickListener() {
38        @Override
39        public void onClick(View v) {
40          if ("".equals(nickname.getText().toString())
41              || "".equals(content.getText().toString())) {
42          Toast.makeText(MainActivity.this, "请将内容输入完整!",
43              Toast.LENGTH_SHORT).show();
44            return;
45          }
46          new Thread(new Runnable() {
47            public void run() {
48                send();
49                Message m = handler.obtainMessage();
50                handler.sendMessage(m);
51            }
52          }).start();
53        }
54      });
55      handler = new Handler() {
56        @Override
57        public void handleMessage(Message msg) {
58          if (result != null) {
59            resultTV.setText(result);
60            content.setText("");
61            nickname.setText("");
62          }
63          super.handleMessage(msg);
64        }
65      };
66    }
67    public void send() {
68      String target = "http://192.168.106.111:8080/blog/dealPost.jsp";
69      URL url;
70      try {
71        url = new URL(target);
72        HttpURLConnection urlConn = (HttpURLConnection) url.openConnection();
73        urlConn.setRequestMethod("POST");
74        urlConn.setDoInput(true);
75        urlConn.setDoOutput(true);
76        urlConn.setUseCaches(false);
77        urlConn.setInstanceFollowRedirects(true);
78        urlConn.setRequestProperty("Content-Type",
79            "application/x-www-form-urlencoded");
80        DataOutputStream out = new DataOutputStream( urlConn.getOutputStream());
81        String param = "nickname="
82            + URLEncoder.encode(nickname.getText().toString(), "utf-8")
```

```
83                + "&content = "
84                + URLEncoder.encode(content.getText().toString(), "utf - 8");
85         out.writeBytes(param);
86         out.flush();
87         out.close();
88         if (urlConn.getResponseCode() == HttpURLConnection.HTTP_OK) {
89            InputStreamReader in = new InputStreamReader(
90                urlConn.getInputStream());
91            BufferedReader buffer = new BufferedReader(in);
92            String inputLine = null;
93            while ((inputLine = buffer.readLine()) != null) {
94              result += inputLine + "\n";
95            }
96            in.close();
97         }
98         urlConn.disconnect();
99       } catch (MalformedURLException e) {
100        e.printStackTrace();
101      } catch (IOException e) {
102        e.printStackTrace();
103      }
104    }
105  }
```

第 23 行代码表示声明了一个输入昵称的编辑框对象。第 24 行代码表示声明了一个输入文本内容的编辑框对象。第 25 行代码表示声明了一个发表按钮对象。第 26 行代码表示声明一个 Handler 对象。第 27 行代码表示声明一个代表显示内容的字符串。第 28 行代码表示声明一个显示结果的文本框对象。第 33 行代码表示获取输入文本内容的 EditText 组件。第 34 行代码表示获取显示结果的 TextView 组件。第 35 行代码表示获取输入昵称的 EditText 组件。第 36 行代码表示获取"发表"按钮组件。第 37 行代码表示为按钮添加单击事件监听器。第 46 行代码表示创建一个新线程，用于从网络上获取文件。第 49 行代码表示获取一个 Message。第 50 行代码表示发送消息。第 52 行代码表示开启线程。第 59 行代码表示显示获得的结果。第 60 行代码表示清空内容编辑框。第 61 行代码表示清空昵称编辑框。第 68 行代码表示要提交的目标地址。第 72 行代码表示创建一个 HTTP 连接。第 73 行代码表示指定使用 POST 请求方式。第 74 行代码表示向连接中写入数据。第 75 行代码表示从连接中读取数据。第 76 行代码表示禁止缓存。第 77 行代码表示自动执行 HTTP 重定向。第 79 行代码表示设置内容类型。第 80 行代码表示获取输出流。第 84 行代码表示连接要提交的数据。第 85 行代码表示将要传递的数据写入数据输出流。第 86 行代码表示输出缓存。第 87 行代码表示关闭数据输出流。第 88 行代码表示判断是否响应成功。第 90 行代码表示获得读取的内容。第 91 行代码表示获取输入流对象。第 96 行代码表示关闭字符输入流。第 98 行代码表示断开连接。

(4) 在 AndroidManifest.xml 文件中指定访问的权限，代码如下：

```
< uses - permission android:name = "android.permission.INTERNET" />
```

（5）部署 URLConnectionPOSTDemo 工程，程序运行结果，如图 10-5 所示。输入"昵称"和发表的内容，单击"发表"按钮即可发表一条信息，如图 10-6 所示。

图 10-5　程序运行结果

图 10-6　发表信息

10.2.2　HttpClient 接口

在通常情况下，用 java.net 包中的 HttpURLConnection 来访问网络，如果只需要到某个简单页面提交请求并获取服务器的响应，完全可以使用该技术来实现。不过，对于比较复杂的联网操作，使用 HttpURLConnection 就不一定能满足要求，这时，可以使用 Apache 组织提供的 HttpClient 来实现。Android 中已经成功地集成了 HttpClient，所以可以直接在 Android 中使用 HttpClient 来访问网络。

HttpClient 实际上是对 Java 提供的访问网络的方法进行了封装。HttpURLConnection 类中的输入、输出流操作在这个 HttpClient 中被统一封装成了 HttpGet 和 HttpResponse 类，这样就降低了操作的烦琐性。其中，HttpGet 类代表发送 GET 请求，HttpPost 类代表发送 POST 请求，HttpResponse 类代表处理响应的对象。

同使用 HttpURLConnection 类一样，使用 HttpClient 发送 HTTP 请求也可以分为 GET 请求和 POST 请求两种。

1. 发送 GET 请求

同 HttpURLConnection 类一样，使用 HttpClient 发送 GET 请求的方法也比较简单，大致可以分为以下 5 个步骤。

（1）创建 HttpClient 对象。

（2）创建 HttpGet 对象。

（3）如果需要发送请求参数，可以直接将要发送的参数连接到 URL 地址中，也可以调用 HttpGet 的 setParams()方法来添加请求参数。

（4）调用 HttpClient 对象的 execute()方法发送请求。执行该方法将返回一个 HttpResponse 对象。

（5）调用 HttpResponse 的 getEntity()方法，可获得包含服务器响应内容的 HttpEntity 对象，通过该对象可以获取服务器的响应内容。

下面给出一个使用 HttpClient 发送 GET 请求的实例。

（1）创建一个新的 Android 工程，工程名为 ClientGETDemo，应用程序名为 ClientGETDemo，包名为 hlju.edu.cn，创建的 Activity 的名字为 MainActivity，最小 SDK 版本根据选择的目标 API 会自动添加。

（2）修改 res 目录下 layout 文件夹中的 activity_main.xml 文件，设置线性布局，添加一个 Button 和一个 TextView 个控件，对相应控件进行描述，并设置相关属性，代码如下：

```
1   <?xml version = "1.0" encoding = "utf - 8"?>
2   <LinearLayout xmlns:android = "http://schemas.android.com/apk/res/android"
3       android:layout_width = "fill_parent"
4       android:layout_height = "fill_parent"
5       android:gravity = "center_horizontal"
6       android:orientation = "vertical" >
7       <Button
8           android:id = "@ + id/button"
9           android:layout_width = "wrap_content"
10          android:layout_height = "wrap_content"
11          android:text = "@string/button" />
12      <TextView
13          android:id = "@ + id/result"
14          android:layout_width = "match_parent"
15          android:layout_height = "wrap_content" />
16  </LinearLayout>
```

（3）在 AndroidManifest.xml 文件中指定访问的权限，代码如下：

```
<uses - permission android:name = "android.permission.INTERNET"/>
```

（4）修改 src 目录中 hlju.edu.cn 包下的 MainActivity.java 文件，代码如下：

```
1   package hlju.edu.cn;
2   import java.io.IOException;
3   import org.apache.http.HttpResponse;
4   import org.apache.http.HttpStatus;
5   import org.apache.http.client.ClientProtocolException;
6   import org.apache.http.client.HttpClient;
7   import org.apache.http.client.methods.HttpGet;
8   import org.apache.http.impl.client.DefaultHttpClient;
9   import org.apache.http.util.EntityUtils;
```

```java
10  import android.app.Activity;
11  import android.os.Bundle;
12  import android.view.Menu;
13  import android.view.MenuItem;
14  import android.os.Handler;
15  import android.os.Message;
16  import android.view.View;
17  import android.view.View.OnClickListener;
18  import android.widget.Button;
19  import android.widget.TextView;
20  public class MainActivity extends Activity {
21      private Button button;
22      private Handler handler;
23      private String result = "";
24      private TextView resultTV;
25  
26      @Override
27      protected void onCreate(Bundle savedInstanceState) {
28          super.onCreate(savedInstanceState);
29          setContentView(R.layout.activity_main);
30          resultTV = (TextView) findViewById(R.id.result);
31          button = (Button) findViewById(R.id.button);
32          button.setOnClickListener(new OnClickListener() {
33              @Override
34              public void onClick(View v) {
35                  new Thread(new Runnable() {
36                      public void run() {
37                          send();
38                          Message m = handler.obtainMessage();
39                          handler.sendMessage(m);
40                      }
41                  }).start();
42              }
43          });
44          handler = new Handler() {
45              @Override
46              public void handleMessage(Message msg) {
47                  if (result != null) {
48                      resultTV.setText(result);
49                  }
50                  super.handleMessage(msg);
51              }
52          };
53      }
54      public void send() {
55          String target = "http://192.168.106.111:8080/blog/deal_httpclient.jsp?param=get";
56          HttpClient httpclient = new DefaultHttpClient();
57          HttpGet httpRequest = new HttpGet(target);
58          HttpResponse httpResponse;
```

```
59              try {
60                  httpResponse = httpclient.execute(httpRequest);
61                  if (httpResponse.getStatusLine().getStatusCode() == HttpStatus.SC_OK){
62          result = EntityUtils.toString(httpResponse.getEntity());
63                  }else{
64                      result = "请求失败!";
65                  }
66              } catch (ClientProtocolException e) {
67                  e.printStackTrace();
68              } catch (IOException e) {
69                  e.printStackTrace();
70              }
71          }
72      }
```

第 21 行代码表示声明一个发表按钮对象。第 22 行代码表示声明一个 Handler 对象。第 23 行代码表示声明一个代表显示结果的字符串。第 24 行代码表示声明一个显示结果的文本框对象。第 30 行代码表示获取显示结果的 TextView 组件。第 31 行代码表示获取"发表"按钮组件。第 32 行代码表示为按钮添加单击事件监听器。第 35 行代码表示创建一个新线程,用于发送并获取 GET 请求。第 38 行代码表示获取一个 Message。第 39 行代码表示发送消息。第 41 行代码表示开启线程。第 48 行代码表示显示获得的结果。第 55 行代码表示要提交的目标地址。第 56 行代码表示创建 HttpClient 对象。第 57 行代码表示创建 HttpGet 连接对象。第 60 行代码表示执行 HttpClient 请求。第 62 行代码表示获取返回的字符串。

(5) 部署 ClientGETDemo 工程,程序运行结果,如图 10-7 所示。单击"发送 GET 请求"按钮,如果请求发送成功,如图 10-8 所示。

图 10-7　程序运行结果　　　　　　　　图 10-8　请求发送成功

2. 发送 POST 请求

同 HttpURLConnection 类发送请求一样，HttpClient 对于复杂的请求数据也需要使用 POST 方法发送。使用 HttpClient 发送 POST 请求大致可以分为以下 5 个步骤。

(1) 创建 HttpClient 对象。

(2) 创建 HttpPost 对象。

(3) 如果需要发送请求参数，可以调用 HttpPost 的 setParams()方法来添加请求参数。也可以调用 setEntity()方法来设置请求参数。

(4) 调用 HttpClient 对象的 execute()方法发送请求。执行该方法将返回一个 HttpResponse 对象。

(5) 调用 HttpResponse 的 getEntity()方法，可获得包含服务器响应内容的 HttpEntity 对象，通过该对象可以获取服务器的响应内容。

下面给出一个使用 HttpClient 发送 POST 请求的实例。

(1) 创建一个新的 Android 工程，工程名为 ClientPOSTDemo，应用程序名为 ClientPOSTDemo，包名为 hlju.edu.cn，创建的 Activity 的名字为 MainActivity，最小 SDK 版本根据选择的目标 API 会自动添加。

(2) 修改 res 目录下 layout 文件夹中的 activity_main.xml 文件，设置线性布局，添加两个 EditText、一个 Button、一个 ScrollView 和一个 TextView 控件，对相应控件进行描述，并设置相关属性，代码如下：

```
1   <?xml version = "1.0" encoding = "utf - 8"?>
2   < LinearLayout xmlns:android = "http://schemas.android.com/apk/res/android"
3       android:orientation = "vertical"
4       android:gravity = "center_horizontal"
5       android:layout_width = "fill_parent"
6       android:layout_height = "fill_parent">
7       < EditText
8           android:id = "@ + id/nickname"
9           android:hint = "@string/nickname"
10          android:layout_width = "match_parent"
11          android:layout_height = "wrap_content" />
12      < EditText
13          android:id = "@ + id/content"
14          android:layout_height = "wrap_content"
15          android:layout_width = "match_parent"
16          android:inputType = "textMultiLine"/>
17      < Button
18          android:id = "@ + id/button"
19          android:layout_width = "wrap_content"
20          android:layout_height = "wrap_content"
21          android:text = "@string/button" />
22      < ScrollView
23          android:id = "@ + id/scrollView1"
24          android:layout_width = "match_parent"
```

```
25        android:layout_height = "wrap_content"
26        android:layout_weight = "1" >
27    < LinearLayout
28        android:id = "@ + id/linearLayout1"
29        android:layout_width = "match_parent"
30        android:layout_height = "match_parent" >
31    < TextView android:id = "@ + id/result"
32        android:layout_width = "match_parent"
33        android:layout_height = "wrap_content"
34        android:layout_weight = "1" />
35    </LinearLayout>
36  </ScrollView>
37 </LinearLayout>
```

（3）在 AndroidManifest.xml 文件中指定访问的权限，代码如下：

```
1  < uses - permission android:name = "android.permission.INTERNET"/>
```

（4）修改 src 目录中 hlju.edu.cn 包下的 MainActivity.java 文件，代码如下：

```
1   package hlju.edu.cn;
2   import java.io.IOException;
3   import java.io.UnsupportedEncodingException;
4   import java.util.ArrayList;
5   import java.util.List;
6   import org.apache.http.HttpResponse;
7   import org.apache.http.HttpStatus;
8   import org.apache.http.NameValuePair;
9   import org.apache.http.client.ClientProtocolException;
10  import org.apache.http.client.HttpClient;
11  import org.apache.http.client.entity.UrlEncodedFormEntity;
12  import org.apache.http.client.methods.HttpPost;
13  import org.apache.http.impl.client.DefaultHttpClient;
14  import org.apache.http.message.BasicNameValuePair;
15  import org.apache.http.util.EntityUtils;
16  import android.app.Activity;
17  import android.os.Bundle;
18  import android.os.Handler;
19  import android.os.Message;
20  import android.view.Menu;
21  import android.view.MenuItem;
22  import android.view.View;
23  import android.view.View.OnClickListener;
24  import android.widget.Button;
25  import android.widget.EditText;
26  import android.widget.TextView;
27  import android.widget.Toast;
```

```java
28  public class MainActivity extends Activity {
29      private EditText nickname;
30      private EditText content;
31      private Button button;
32      private Handler handler;
33      private String result = "";
34      private TextView resultTV;
35
36      @Override
37      protected void onCreate(Bundle savedInstanceState) {
38          super.onCreate(savedInstanceState);
39          setContentView(R.layout.activity_main);
40          content = (EditText) findViewById(R.id.content);
41          resultTV = (TextView) findViewById(R.id.result);
42          nickname = (EditText) findViewById(R.id.nickname);
43          button = (Button) findViewById(R.id.button);
44          button.setOnClickListener(new OnClickListener() {
45              @Override
46              public void onClick(View v) {
47                  if ("".equals(nickname.getText().toString())
48                          || "".equals(content.getText().toString())) {
49                      Toast.makeText(MainActivity.this, "请将内容输入完整!",
50                              Toast.LENGTH_SHORT).show();
51                      return;
52                  }
53                  new Thread(new Runnable() {
54                      public void run() {
55                          send();
56                          Message m = handler.obtainMessage();
57                          handler.sendMessage(m);
58                      }
59                  }).start();
60              }
61          });
62          handler = new Handler() {
63              @Override
64              public void handleMessage(Message msg) {
65                  if (result != null) {
66                      resultTV.setText(result);
67                      content.setText("");
68                      nickname.setText("");
69                  }
70                  super.handleMessage(msg);
71              }
72          };
73      }
74      public void send() {
```

```
75          String target = "http://127.0.0.1:8080/blog/deal_httpclient.jsp";
76          HttpClient httpclient = new DefaultHttpClient();
77          HttpPost httpRequest = new HttpPost(target);
78          List<NameValuePair> params = new ArrayList<NameValuePair>();
79          params.add(new BasicNameValuePair("param", "post"));
80          params.add(new BasicNameValuePair("nickname", nickname.getText().toString()));
81          params.add(new BasicNameValuePair("content", content.getText().toString()));
82          try {
83              httpRequest.setEntity(new UrlEncodedFormEntity(params, "utf-8"));
84              HttpResponse httpResponse = httpclient.execute(httpRequest);
85              if (httpResponse.getStatusLine().getStatusCode() == HttpStatus.SC_OK){
86                  result += EntityUtils.toString(httpResponse.getEntity());
87              }else{
88                  result = "请求失败!";
89              }
90          } catch (UnsupportedEncodingException e1) {
91              e1.printStackTrace();
92          } catch (ClientProtocolException e) {
93              e.printStackTrace();
94          } catch (IOException e) {
95              e.printStackTrace();
96          }
97      }
98  }
```

第 29 行代码表示声明一个输入昵称的编辑框对象。第 30 行代码表示声明一个输入文本内容的编辑框对象。第 31 行代码表示声明一个发表按钮对象。第 32 行代码表示声明一个 Handler 对象。第 33 行代码表示声明一个代表显示内容的字符串。第 34 行代码表示声明一个显示结果的文本框对象。第 40 行代码表示获取输入文本内容的 EditText 组件。第 41 行代码表示获取显示结果的 TextView 组件。第 42 行代码表示获取输入昵称的 EditText 组件。第 43 行代码表示获取"发表"按钮组件。第 44 行代码表示为按钮添加单击事件监听器。第 53 行代码表示创建一个新线程,用于从网络上获取文件。第 56 行代码表示获取一个 Message。第 57 行代码表示发送消息。第 59 行代码表示开启线程。第 66 行代码表示显示获得的结果。第 67 行代码表示清空内容编辑框。第 68 行代码表示清空昵称编辑框。第 75 行代码表示要提交的目标地址。第 76 行代码表示创建 HttpClient 对象。第 77 行代码表示创建 HttpPost 对象。第 78 行代码表示将要传递的参数保存到 List 集合中。第 79 行代码表示标记参数。第 80 行代码表示添加昵称。第 81 行代码表示添加内容。第 83 行代码表示设置编码方式。第 84 行代码表示执行 HttpClient 请求。第 85 行代码表示判断请求是否成功。第 86 行代码表示获取返回的字符串。

(5) 部署 ClientPOSTDemo 工程,程序运行结果,如图 10-9 所示。输入相关内容,单击"发表"按钮,如图 10-10 所示。

图 10-9　程序运行界面　　　　　　　　　图 10-10　发表信息

10.3　WebKit 应用

Android 浏览器的内核是 WebKit 引擎，Webkit 是一个开源浏览器网页排版引擎。

10.3.1　WebKit 概述

Android 提供了内置的浏览器，该浏览器使用了开源的 WebKit 引擎。WebKit 不仅能够搜索网址、查看电子邮件，而且包含播放视频节目、触摸屏以及上网等功能。在 Android 中使用内置的浏览器需要通过 WebView 组件来实现。

10.3.2　WebView 浏览网页

WebView 组件是专门用于浏览网页的，它的使用方法与其他组件一样，既可以在 XML 布局文件中使用 <WebView> 标记添加，又可以在 Java 文件中通过 new 关键字创建。推荐采用第一种方法，也就是通过 <WebView> 标记在 XML 布局文件中添加 WebView 组件。在 XML 布局文件中添加一个 WebView 组件，代码如下：

```
1    < WebView
2      android: id = " @ + id/ WebView1 "
3      android: layout_width = "match_parent"
4      android: layout_height = "match_parent"/>
```

添加 WebView 组件后，就可以应用该组件提供的方法来执行浏览器操作。WebView 组件提供的常用方法，如表 10-4 所示。

表 10-4　WebView 组件提供的常用方法

方　　法	描　　述
loadUrl(String url)	用于加载指定 URL 对应的网页
loadData(String data，String mimeType，String ecoding)	用于将指定的字符串数据加载到浏览器中
loadDataWithBaseURL(String baseUrl，String data，String mimeType，String encoding，String historyUrl)	用于基于 URL 加载指定的数据
capturePicture()	用于创建当前屏幕的快照
goBack()	执行后退操作，相当于浏览器上的后退按钮的功能
goForward()	执行前进操作，相当于浏览器上的前进按钮的功能
stopLoading()	用于停止加载当前页面
reload()	用于刷新当前页面

下面给出一个使用 WebView 组件浏览的实例。

（1）创建一个新的 Android 工程，工程名为 WebViewloadUrlDemo，应用程序名为 WebViewloadUrlDemo，包名为 hlju.edu.cn，创建的 Activity 的名字为 MainActivity，最小 SDK 版本根据选择的目标 API 会自动添加。

（2）修改 res 目录下 layout 文件夹中的 activity_main.xml 文件，设置线性布局，添加 1 个 WebView 控件，对相应控件进行描述，并设置相关属性，代码如下：

```
1   <?xml version = "1.0" encoding = "utf - 8"?>
2   <LinearLayout xmlns:android = "http://schemas.android.com/apk/res/android"
3       android:layout_width = "fill_parent"
4       android:layout_height = "fill_parent"
5       android:orientation = "vertical" >
6       <WebView
7           android:id = "@ + id/webView1"
8           android:layout_width = "match_parent"
9           android:layout_height = "match_parent" />
10  </LinearLayout>
```

（3）在 AndroidManifest.xml 文件中指定访问的权限，代码如下：

```
<uses - permission android:name = "android.permission.INTERNET"/>
```

（4）修改 src 目录中 hlju.edu.cn 包下的 MainActivity.java 文件，代码如下：

```
1   public class MainActivity extends Activity {
2       @Override
3       protected void onCreate(Bundle savedInstanceState) {
4           super.onCreate(savedInstanceState);
5           setContentView(R.layout.activity_main);
6           WebView webview = (WebView)findViewById(R.id.webView1);
7           webview.loadUrl("http://192.168.106.111:8080/bbs/");
8       } }
```

第 6 行代码表示获取布局管理器中添加的 WebView 组件。第 7 行代码表示指定要加载的网页，对应的地址和端口按照实际相应改变。

(5) 部署 WebViewloadUrlDemo 工程，程序运行结果，如图 10-11 所示。

图 10-11　程序运行结果

10.3.3　WebView 加载 HTML 代码

在进行 Android 开发时，对于一些游戏的帮助信息，使用 HTML 代码进行显示比较实用，这样不仅可以让界面更加美观，而且可以让开发更加简单和快捷。WebView 组件提供了 loadData( ) 方法和 loadDataWithBaseURL( ) 方法来加载 HTML 代码。但是，使用 loadData( ) 方法加载带中文的 HTML 内容时，会产生乱码，使用 loadDataWithBaseURL( ) 方法就不会出现中文乱码的情况。loadDataWithBaseURL( ) 方法的基本语法格式如下：

```
loadDataWithBaseURL(String baseUrl, String data, String mimeType, String encoding, String historyUrl)
```

loadDataWithBaseURL( ) 方法的各参数说明，如表 10-5 所示。

表 10-5　loadDataWithBaseURL( ) 方法的参数说明

参　数	描　述
baseUrl	用于指定当前页使用的基本 URL，如果为 null，则使用默认的 about:blank，也就是空白页
data	用于指定要显示的字符串数据
mimeType	用于指定要显示内容的 MIME 类型，如果为 null，默认使用 text/html
encoding	用于指定数据的编码方式
historyUrl	用于指定当前页的历史 URL，也就是进入该页前显示页的 URL，如果为 null，则使用默认的 about:blank

下面给出一个使用 WebView 组件加载 HTML 的实例。

（1）创建一个新的 Android 工程，工程名为 WebViewHTMLDemo，应用程序名为 WebViewHTMLDemo，包名为 hlju.edu.cn，创建的 Activity 的名字为 MainActivity，最小 SDK 版本根据选择的目标 API 会自动添加。

（2）修改 res 目录下 layout 文件夹中的 activity_main.xml 文件，设置线性布局，添加一个 WebView 控件，对相应控件进行描述，并设置相关属性，代码如下：

```xml
1  <?xml version = "1.0" encoding = "utf - 8"?>
2  <LinearLayout xmlns:android = "http://schemas.android.com/apk/res/android"
3      android:layout_width = "fill_parent"
4      android:layout_height = "fill_parent"
5      android:orientation = "vertical" >
6      <WebView
7          android:id = "@ + id/webView1"
8          android:layout_width = "match_parent"
9          android:layout_height = "match_parent" />
10 </LinearLayout>
```

（3）修改 src 目录中 hlju.edu.cn 包下的 MainActivity.java 文件，代码如下：

```java
1  public class MainActivity extends Activity {
2      @Override
3      protected void onCreate(Bundle savedInstanceState) {
4          super.onCreate(savedInstanceState);
5          setContentView(R.layout.activity_main);
6      WebView webview = (WebView)findViewById(R.id.webView1);
7      StringBuilder sb = new StringBuilder();
8      sb.append("<div>四大名著：</div>");
9      sb.append("<ul>");
10     sb.append("<li>《水浒传》,施耐庵(1296—1370?),元末明初。</li>");
11     sb.append("<li>《西游记》,吴承恩(1510?—1582?),明代。</li>");
12     sb.append("<li>《三国演义》,罗贯中(1330?～1440?),元末明初。</li>");
13     sb.append("<li>《红楼梦》,曹雪芹(1715—1763),清代。</li>");
14     sb.append("</ul>");
15     webview.loadDataWithBaseURL(null, sb.toString(), "text/html", "utf - 8", null);
16     }
17 }
```

第 6 行代码表示获取布局管理器中添加的 WebView 组件。第 7 行代码表示创建一个字符串构建器，将要显示的 HTML 内容放置在该构建器中。第 15 行代码表示加载数据。

（4）部署 WebViewHTMLDemo 工程，程序运行结果，如图 10-12 所示。

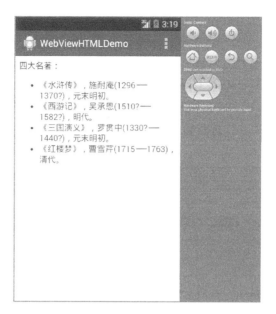

图 10-12　程序运行结果

10.3.4　WebView 与 JavaScript

在默认的情况下，WebView 组件是不支持 JavaScript 的，但是在运行某些不得不使用 JavaScript 代码的网站时，还需要让它支持 JavaScript。让 WebView 组件支持 JavaScript 需要两个步骤。

（1）使用 WebView 组件的 WebSetting 对象提供的 setJavaScriptEnabled()方法让 JavaScript 可用。例如，存在一个名称为"webview"的 WebView 组件，要设置在该组件中允许使用 JavaScript，代码如下：

```
WebView.getSetting().setJavaScriptEnabled(true);
```

（2）经过以上设置后，网页中的大部分 JavaScript 代码均可用，但是，对通过 window.alert()方法弹出的对话框并不可用。要想显示弹出的对话框。需要使用 WebView 组件的 setWebChromeClient()方法来处理 JavaScript 的对话框。

```
webview.setWebChromeClient(new WebChromeClient());
```

这样设置后，在使用 WebView 显示带弹出 JavaScript 对话框的网页时，网页中弹出的对话框将不会被屏蔽掉。

下面通过给出一个实例，本例在 Android 中制作一个包含前进、后退和支持 JavaScript 的网页浏览器。

（1）创建一个新的 Android 工程，工程名为 WebViewJavaScriptDemo，应用程序名为 WebViewJavaScriptDemo，包名为 hlju.edu.cn，创建的 Activity 的名字为 MainActivity，最

小 SDK 版本根据选择的目标 API 会自动添加。

（2）修改 res 目录下 layout 文件夹中的 activity_main.xml 文件，设置线性布局，添加三个 Button、一个 EditText 和一个 WebView 控件，对相应控件进行描述，并设置相关属性，代码如下：

```xml
1  <?xml version = "1.0" encoding = "utf-8"?>
2  <LinearLayout xmlns:android = "http://schemas.android.com/apk/res/android"
3      android:orientation = "vertical"
4      android:layout_width = "fill_parent"
5      android:layout_height = "fill_parent"  >
6    <LinearLayout
7        android:orientation = "horizontal"
8        android:layout_width = "fill_parent"
9        android:layout_height = "wrap_content"  >
10     <Button
11         android:id = "@+id/forward"
12         android:layout_width = "wrap_content"
13         android:layout_height = "wrap_content"
14         android:text = "前进" />
15     <Button
16         android:id = "@+id/back"
17         android:layout_width = "wrap_content"
18         android:layout_height = "wrap_content"
19         android:text = "后退" />
20     <EditText
21       android:layout_weight = "1"
22       android:id = "@+id/editText_url"
23       android:layout_height = "wrap_content"
24       android:layout_width = "wrap_content"
25       android:text = "http://192.168.106.111:8080/bbs/"
26       android:lines = "1" />
27     <Button
28         android:id = "@+id/button_go"
29         android:layout_width = "wrap_content"
30         android:layout_height = "wrap_content"
31         android:text = "@string/go" />
32   </LinearLayout>
33   <WebView android:id = "@+id/webView1"
34       android:layout_width = "fill_parent"
35       android:layout_height = "0dip"
36       android:focusable = "false"
37       android:layout_weight = "1.0" />
38 </LinearLayout>
```

（3）在 AndroidManifest.xml 文件中指定访问的权限，代码如下：

```xml
<uses-permission android:name = "android.permission.INTERNET"/>
```

(4) 修改 src 目录中 hlju.edu.cn 包下的 MainActivity.java 文件，代码如下：

```
1   package hlju.edu.cn;
2   import android.app.Activity;
3   import android.app.AlertDialog;
4   import android.content.DialogInterface;
5   import android.os.Bundle;
6   import android.util.Log;
7   import android.view.Menu;
8   import android.view.MenuItem;
9   import android.view.KeyEvent;
10  import android.view.View;
11  import android.view.View.OnClickListener;
12  import android.view.View.OnKeyListener;
13  import android.webkit.WebChromeClient;
14  import android.webkit.WebView;
15  import android.webkit.WebViewClient;
16  import android.widget.Button;
17  import android.widget.EditText;
18  import android.widget.Toast;
19  public class MainActivity extends Activity {
20      private WebView webView;
21      private EditText urlText;
22      private Button goButton;
23      @Override
24      protected void onCreate(Bundle savedInstanceState) {
25          super.onCreate(savedInstanceState);
26          setContentView(R.layout.activity_main);
27          urlText = (EditText)findViewById(R.id.editText_url);
28          goButton = (Button)findViewById(R.id.button_go);
29          webView = (WebView)findViewById(R.id.webView1);
30          webView.getSettings().setJavaScriptEnabled(true);
31          webView.setWebChromeClient(new WebChromeClient());
32          webView.setWebViewClient(new WebViewClient());
33          Button forward = (Button)findViewById(R.id.forward);
34          forward.setOnClickListener(new OnClickListener() {
35              @Override
36              public void onClick(View v) {
37                  webView.goForward();
38              }
39          });
40          Button back = (Button)findViewById(R.id.back);
41          back.setOnClickListener(new OnClickListener() {
42            @Override
43            public void onClick(View v) {
44              webView.goBack();
45            }
46          });
47          urlText.setOnKeyListener(new OnKeyListener() {
```

```
48        @Override
49        public boolean onKey(View v, int keyCode, KeyEvent event) {
50          if(keyCode == KeyEvent.KEYCODE_ENTER){
51            if(!"".equals(urlText.getText().toString())){
52                openBrowser();
53                return true;
54            }else{
55              showDialog();
56            }
57          }
58          return false;
59        }
60      });
61      goButton.setOnClickListener(new OnClickListener() {
62        @Override
63        public void onClick(View v) {
64          if(!"".equals(urlText.getText().toString())){
65              openBrowser();
66          }else{
67              showDialog();
68          }
69        }
70      });
71    }
72    private void openBrowser(){
73      webView.loadUrl(urlText.getText().toString());
74      Toast.makeText(this, "正在加载:" + urlText.getText().toString(), Toast.LENGTH_SHORT).show();
75    }
76    private void showDialog(){
77      new AlertDialog.Builder(MainActivity.this)
78      .setTitle("网页浏览器")
79      .setMessage("请输入要访问的网址")
80      .setPositiveButton("确定",new DialogInterface.OnClickListener(){
81          public void onClick(DialogInterface dialog,int which){
82            Log.d("WebWiew","单击确定按钮");
83          }
84      }).show();
85    } }
```

第 20 行代码表示声明 WebView 组件的对象。第 21 行代码表示声明作为地址栏的 EditText 对象。第 22 行代码表示声明 GO 按钮对象。第 27 行代码表示获取布局管理器中添加的地址栏。第 28 行代码表示获取布局管理器中添加的 GO 按钮。第 29 行代码表示获取 WebView 组件。第 30 行代码表示设置 JavaScript 可用。第 31 行代码表示处理 JavaScript 对话框。第 32 行代码表示处理各种通知和请求事件,如果不使用该句代码,将使用内置浏览器访问网页。第 33 行代码表示获取布局管理器中添加的"前进"按钮。第 37 行代码表示前进。第 40 行代码表示获取布局管理器中添加的"后退"按钮。第 44 行代码表示后退。第 47 行代码表示为地址栏添加键盘键被按下的事件监听器。第 50 行代码表示如果为回车键。第 52 行代码表示打开浏览器。第 55 行代码表示弹出提示对话框。第 61 行代码表示为 GO 按钮添加单击事件监听器。第 67 行代码表示弹出提示对话框。第 72 行代码表示用于打开网页的方法。第 76 行代码表示用于显示对话框的方法。

（5）部署 WebViewJavaScriptDemo 工程，程序运行结果，如图 10-13 所示。输入网址，单击按钮 GO 加载网址，如图 10-14 所示。如果加载成功，则会显示网页信息，如图 10-15 所示。

图 10-13　程序运行结果

图 10-14　加载网址

图 10-15　显示网页

10.4　Socket 通信

Socket 通常也称作"套接字"，用于描述 IP 地址和端口，是一个通信链的句柄。应用程序通常通过"套接字"向网络发出请求或者回复请求。它是通信的基石，是支持 TCP/IP 网络通信的基本操作单元。它是网络通信过程中端点的抽象表示，包含进行网络通信必需的 5 种信息：连接使用的协议、本地主机的地址、本机进程的协议端口、远程主机的 IP 地址和远程进程的协议端口。

10.4.1　Socket 传输模式

Socket 有两种主要的操作模式：面向连接的和无连接的。面向连接的 Socket 操作就像一部电话，必须建立一个连接和一个呼叫。所有的事情到达时的顺序与它们出发时的顺序是一样的。无连接的 Socket 操作就像是一个邮件投递，没什么保证，多个邮件到达时的顺序可能与出发时的顺序不一样。根据应用程序的需要决定 Socket 传输模式的选择。如果可靠性更高的话，用面向连接的操作会好一些。例如文件服务器需要数据的正确性和有序性，如果一些数据丢失了其时效性导致系统的有效性降低。但同时确保数据的正确性和有序性需要额外的操作,这会带来内存消耗,额外的费用将会降低系统的回应速率。

无连接的操作使用数据报协议。一个数据报是一个独立的单元。它包含了这次投递的所有信息。它包含了目的地址和要发送的内容,这个模式下的 Socket 不需要连接一个目的 Socket,它只是简单地投出数据报。无连接的操作是快递和高效的,但是数据安全性不佳。

面向连接的操作使用 TCP 协议。该模式下的 Socket 必须在发送数据之前与目的地的 Socket 取得连接。一旦连接建立了,Socket 就可以使用一个流接口进行打开、读、写和关闭操作。所有发送的信息都会在另一端以同样的顺序被接收。面向连接的操作比无连接操作的效率要低,但是数据的安全性要高。

10.4.2 Socket 编程原理

1. Socket 构造

Java 在包 java.net 中提供了两个类 Socket 和 ServerSocket,分别用于表示双向连接的客户端和服务端。其构造方法如代码所示:

```
1   Socket(InetAddress address, int port);
2   Socket(InetAddress address, int port, oolean stream);
3   Socket(String host, int port);
4   Socket(String host, int port, oolean stream);
5   Socket(SocketImpl impl);
6   Socket(String host, int port, InetAddress address, int localPort);
7   Socket(InerAddress address, int port, InetAddress localAddr, int localPort);
8   Socket(int port);
9   Socket(int port, int backlog);
10  Socket(int port, int backlog,InetAddress bindAddr);
```

第 1 行的参数 address 是双向连接中另一方的 IP 地址,参数 port 是双向连接中另一方的端口号,后面几行中的 address 和 port 都表示同一个意思。第 2 行的参数 stream 指明 Socket 是流 Socket 还是数据报 Socket,后面几行中的 stream 都表示同一个意思。第 3 行的参数 host 是双向连接中另一方的主机名,后面几行中的 host 都表示同一个意思。第 5 行的参数 impl 是 Socket 的父类,既可以用于创建 ServerSocket,又可以用于创建 Socket。第 7 行的参数 localAddr 是本地机器的地址。第 10 行的参数 bindAddr 是本地机器的地址。

在选择端口时必须小心。每个端口提供一种特定的服务,只有给出正确的端口,才能获得相应的服务。0~1023 的端口号为系统所保留,例如 HTTP 服务的端口号为 80,Telnet 服务的端口号为 21 等。所以在选择端口号时,最后选择一个大于 1023 的数,防止发生冲突。在创建 Socket 时,如果发生错误,将产生 IOException,在程序中必须对其进行处理。

2. 客户端 Socket

要想使用 Socket 来与一个服务器通信,就必须先在客户端创建一个 Socket,并指出需要连接的服务器的 IP 地址和端口,这也是使用 Socket 通信的第一步,代码如下:

```
1  try
2  {
3      socket = new Socket("127.0.0.1",12345);
4  }
5  catch(Exception e){}
```

3. ServerSocket

下面给出一个 Server 的典型工作模式，代码如下：

```
1  try
2  {
3      ServerSocket serverSocket = new ServerSocket(12345);
4      Socket client = serverSocket.accept();
5  }
6  catch(Exception e){}
```

上面的程序创建了一个 ServerSocket 在端口 12345 监听客户请求，在这里 Server 只能接收一个请求，接收后 Server 就退出了。实际的应用中总是让它不停地循环接收，一旦有客户请求，Server 总是会创建一个服务线程来服务新来的客户，而自己继续监听。程序中 accept() 是一个阻塞方法，表示该方法在被调用后等待客户的请求，直到有一个客户启动并请求连接到相同的端口，然后 accept() 返回一个对应于客户的 Socket。这时，客户端和服务端都建立了用于通信的 Socket，接下来就是由各个 Socket 分别打开各自的输入、输出流。

4. 输入流和输出流

Socket 提供了方法 getInputStream() 和 getOutputStream() 来得到对应的输入流和输出流以进行读和写的操作，这两个方法分别返回 InputStream 和 OutputStream 类对象。为了便于读数据和写数据，可以在返回的输入、输出流对象上建立过滤流，如 DataInputStream、DataOutputStream 或 PrintStream 类对象，对于文本方式流对象，可以采用 InputStreamReader 和 OutputStreamWriter、PrintWriter 等处理。代码如下：

```
1  PrintStream os = new PrintStream( new BufferedOutputStream (socket.getOutputStream()));
2  DataInputStream is = new DataInputStream(socket. getInputStream());
3  PrintWriter out = new PrintWriter(socket.getOutputStream(), true);
4  BufferedReader in = new BufferedReader(new InputStreamReader(socket. getInputStream())));
```

5. 关闭 Socket

每一个 Socket 存在时都将占用一定的资源，在 Socket 对象使用完毕时，要使其关闭，关闭 Socket 可以调用 Socket 的 close() 方法。在关闭 Socket 之前，应将与 Socket 相关的所有输入流、输出流全部关闭，以释放所有的资源。而且要注意关闭的顺序，与 Socket 相关的所有的输入、输出应先关闭，然后再关闭 Socket。尽管 Java 有自动回收机制，网络资源最终

会被释放,但是为了有效利用资源,建议按照合理的顺序主动释放资源。代码如下:

```
1   os.close();
2   is.close();
3   socket.close();
```

可以利用 Java 标准 API 来开发网络应用,实现一个简单的服务器和客户端通信,客户端发送数据且接收服务器发回的数据并显示。

首先是服务器端的建立,创建服务器的步骤如下。

- 指定端口实例化一个 ServerSocket。
- 调用 ServerSocket 的 accept() 以在等待连接期间造成阻塞。
- 获取位于该底层 Socket 的流以进行读、写操作。
- 将数据封装成流。
- 对 Socket 进行读、写。
- 关闭打开的流。

然后是客户端的实现,实现客户端的步骤如下。

- 通过 IP 地址和端口实例化 Socket,请求连接服务器。
- 获取 Socket 上的流以进行读、写。
- 将流包装进 BufferedReader、PrintWriter 的实例。
- 对 Socket 进行读、写。
- 关闭打开的流。

利用 Socket 可以使得客户端定时向服务器发送连接请求,服务器在收到该请求后对客户端进行回复,表明知道客户端"在线"。若服务器长时间无法收到客户端的请求,则认为客户端"下线";若客户端长时间无法收到服务器的回复,则认为网络已经断开。很多情况下,服务器会主动向客户端发送数据,保持客户端与服务器数据的实时与同步。这样不仅可以保持客户端程序的在线状态,同时也是在"询问"服务器是否有新的数据,如果有就将数据传给客户端。

习题

1. 无线网络技术都有哪些。
2. 什么是蓝牙,在 Android 中是否支持蓝牙编程。
3. 什么是 Wi-Fi,在 Android 中是否支持 Wi-Fi 编程。
4. 简述 HTTP 通信。
5. 简述 Socket 的操作模式。

第11章 图形和图像

Android 处理图形的能力非常强大。图像与动画处理技术在 Android 中非常重要,特别是在开发益智类游戏或者 2D 游戏时,都离不开图形与动画处理技术的支持。

本章主要学习内容:
- 掌握图片浏览器的应用和访问图片;
- 掌握 2D 绘图;
- 掌握图像特效的应用;
- 了解内存优化。

11.1 图片浏览器

实现图片浏览器的功能,可以通过单独使用 Gallery 控件的方式,也可以使用 Gallery 控件和 ImageSwitcher 控件组合的方式。

11.1.1 Gallery

Android 的 Gallery 控件是一个水平的列表选择框,一般用于展示一组图片。Gallery 控件的水平列表可以让用户以滑动的方式切换列表项,所以用户体验也比较好。

下面通过实例介绍如何使用 Gallery 控件来实现图片浏览器的功能。

(1) 创建一个新的 Android 工程,工程名为 GalleryDemo,应用程序名为 GalleryDemo,包名为 hlju. edu. cn,创建的 Activity 的名字为 MainActivity,最小 SDK 版本根据选择的目标 API 会自动添加。

(2) 修改 res 目录下 layout 文件夹中的 activity_main. xml 文件,设置线性布局,添加 1 个 Gallery 控件,对相应控件进行描述,并设置相关属性,代码如下:

```
1   <?xml version = "1.0" encoding = "utf - 8"?>
2   < LinearLayout xmlns:android = "http://schemas.android.com/apk/res/android"
3       android:layout_width = "fill_parent"
4       android:layout_height = "fill_parent"
5       android:orientation = "vertical" >
6       < Gallery
7           android:id = "@ + id/gallery"
```

```
8           android:layout_width = "fill_parent"
9           android:layout_height = "fill_parent" >
10      </Gallery>
11  </LinearLayout>
```

(3) 修改 src 目录中 hlju.edu.cn 包下的 MainActivity.java 文件,代码如下:

```
1   package hlju.edu.cn;
2   import java.lang.reflect.Field;
3   import java.util.ArrayList;
4   import android.app.Activity;
5   import android.content.Context;
6   import android.graphics.Bitmap;
7   import android.graphics.BitmapFactory;
8   import android.os.Bundle;
9   import android.view.Menu;
10  import android.view.MenuItem;
11  import android.util.Log;
12  import android.view.View;
13  import android.view.ViewGroup;
14  import android.widget.AdapterView;
15  import android.widget.AdapterView.OnItemClickListener;
16  import android.widget.BaseAdapter;
17  import android.widget.Gallery;
18  import android.widget.ImageView;
19  import android.widget.ListView;
20  public class MainActivity extends Activity {
21      private static final String TAG = "MainActivity";
22      private Gallery mGallery;
23          @Override
24      public void onCreate(Bundle savedInstanceState) {
25          super.onCreate(savedInstanceState);
26          setContentView(R.layout.activity_main);
27          mGallery = (Gallery)findViewById(R.id.gallery);
28          try {
29              mGallery.setAdapter(new ImageAdapter(this));
30          } catch (Exception e) {
31              Log.e(TAG, "set adpater occurs errors:" + e.toString());
32          }
33          mGallery.setOnItemClickListener(new OnItemClickListener() {
34              public void onItemClick(AdapterView parent, View v, int position, long id) {
35                  MainActivity.this.setTitle(String.valueOf(position));
36              }
37          });
38      }
39      private class ImageAdapter extends BaseAdapter{
40          private Context mContext;
41          private ArrayList < Integer > imgList = new ArrayList < Integer >();
```

```java
42          private ArrayList<Object> imgSizes = new ArrayList<Object>();
43              public ImageAdapter(Context c) throws IllegalArgumentException,
    IllegalAccessException{
44             mContext = c;
45          Field[] fields = R.drawable.class.getDeclaredFields();
46          for (Field field : fields)
47          {
48              if (!"ic_action_search".equals(field.getName()) && ! "ic_launcher".equals
    (field.getName()))
49              {
50                  int index = field.getInt(R.drawable.class);
51                  imgList.add(index);
52                  int size[] = new int[2];
53                  Bitmap bmImg = BitmapFactory.decodeResource (getResources(),index);
54                  size[0] = bmImg.getWidth();
55                  size[1] = bmImg.getHeight();
56                  imgSizes.add(size);
57              }
58          }
59          }
60          @Override
61          public int getCount() {
62              return imgList.size();
63          }
64
65          @Override
66          public Object getItem(int position) {
67              return position;
68          }
69
70          @Override
71          public long getItemId(int position) {
72              return position;
73          }
74
75          @Override
76          public View getView(int position, View convertView, ViewGroup parent) {
77              ImageView imageView = null;
78              if(convertView == null){
79                 imageView = new ImageView(mContext);
80                 imageView.setImageResource(imgList.get(position). intValue());
81                  imageView.setScaleType(ImageView.ScaleType.FIT_XY);
82                  int size[] = new int[2];
83                  size = (int[]) imgSizes.get(position);
84                  imageView.setLayoutParams(new Gallery.LayoutParams (size[0], size[1]));
85              }else{
86                 imageView = (ImageView)convertView;
87              }
88              return imageView;
```

```
89              }
90          };
91
92      @Override
93      public boolean onCreateOptionsMenu(Menu menu) {
94          // Inflate the menu; this adds items to the action bar if it is present.
95          getMenuInflater().inflate(R.menu.main, menu);
96          return true;
97      }
98
99      @Override
100      public boolean onOptionsItemSelected(MenuItem item) {
101         // Handle action bar item clicks here. The action bar will
102         // automatically handle clicks on the Home/Up button, so long
103         // as you specify a parent activity in AndroidManifest.xml.
104         int id = item.getItemId();
105         if (id == R.id.action_settings) {
106            return true;
107         }
108         return super.onOptionsItemSelected(item);}}
```

第 22 行代码表示声明 Gallery 控件。第 27 行代码表示找到 id 为 gallery 的控件。第 34 行代码表示用匿名内部类形式绑定 Gallery 控件的各选项单击事件监听器。第 34 行代码表示各选项单击时回调。第 35 行代码表示设置应用程序标题。第 46 行代码用反射机制来获取资源中的图片 ID 和尺寸。第 48 行代码表示除了 ic_action_search 和 ic_launcher 之外的图片。第 52 行代码表示保存图片 ID。第 53 行代码表示保存图片大小。第 81 行代码表示从 imgList 取得图片 ID。第 84 行代码表示从 imgSizes 取得图片大小。

（4）部署 GalleryDemo 工程，程序运行结果如图 11-1 所示。

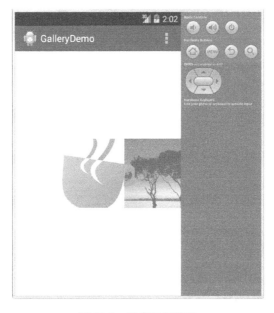

图 11-1　程序运行结果

11.1.2　ImageSwither

ImageSwither控件继承于android.widget.FrameLayout父类。ImageSwither控件的功能与ImageView控件的功能类似，都可以显示单张图片。ImageSwither控件在切换图片显示的时候可以添加相应的动画效果。

下面通过实例介绍如何使用Gallery控件结合ImageSwither控件来实现图片浏览器的功能。

（1）创建一个新的Android工程，工程名为ImageSwitherDemo，应用程序名为ImageSwitherDemo，包名为hlju.edu.cn，创建的Activity的名字为MainActivity，最小SDK版本根据选择的目标API会自动添加。

（2）修改res目录下layout文件夹中的activity_main.xml文件，设置线性布局，添加1个ImageSwitcher控件和1个Gallery控件，对相应控件进行描述，并设置相关属性，代码如下：

```
1   <?xml version = "1.0" encoding = "utf-8"?>
2   <RelativeLayout xmlns:android = "http://schemas.android.com/apk/res/android"
3       android:layout_width = "match_parent"
4       android:layout_height = "match_parent">
5       <ImageSwitcher android:id = "@+id/switcher"
6           android:layout_width = "match_parent"
7           android:layout_height = "match_parent"/>
8       <Gallery android:id = "@+id/gallery"
9           android:background = "#55000000"
10          android:layout_width = "match_parent"
11          android:layout_alignParentBottom = "true"
12          android:layout_alignParentLeft = "true"
13          android:gravity = "center_vertical"
14          android:spacing = "16dp" android:layout_height = "100dp"/>
15  </RelativeLayout>
```

（3）修改src目录中hlju.edu.cn包下的MainActivity.java文件，代码如下：

```
1   package hlju.edu.cn;
2
3   import java.lang.reflect.Field;
4   import java.util.ArrayList;
5   import android.app.Activity;
6   import android.content.Context;
7   import android.os.Bundle;
8   import android.view.Menu;
9   import android.view.MenuItem;
10  import android.util.Log;
11  import android.view.MotionEvent;
12  import android.view.View;
13  import android.view.View.OnTouchListener;
```

```java
14  import android.view.ViewGroup;
15  import android.view.animation.AnimationUtils;
16  import android.widget.AdapterView;
17  import android.widget.AdapterView.OnItemSelectedListener;
18  import android.widget.BaseAdapter;
19  import android.widget.Gallery;
20  import android.widget.Gallery.LayoutParams;
21  import android.widget.ImageSwitcher;
22  import android.widget.ImageView;
23  import android.widget.ViewSwitcher.ViewFactory;
24
25      public class MainActivity extends Activity implements ViewFactory {
26          private static final String TAG = "MainActivity";
27          private ImageSwitcher is;
28          private Gallery gallery;
29          private int downX, upX;
30          private ArrayList<Integer> imgList = new ArrayList<Integer>();
31
32          @Override
33          protected void onCreate(Bundle savedInstanceState) {
34              super.onCreate(savedInstanceState);
35              setContentView(R.layout.activity_main);
36          Field[] fields = R.drawable.class.getDeclaredFields();
37              for (Field field : fields) {
38                  if (!"ic_action_search".equals(field.getName()) && !"ic_launcher".equals(field.getName()))
39                  {
40                      int index = 0;
41                      try {
42                          index = field.getInt(R.drawable.class);
43                      } catch (Exception e) {
44                          Log.e(TAG, "occurs errors:" + e.getMessage());
45                      }
46                      imgList.add(index);
47                  }
48              }
49
50              is = (ImageSwitcher) findViewById(R.id.switcher);
51              is.setFactory(this);
52              is.setInAnimation(AnimationUtils.loadAnimation(this, android.R.anim.fade_in));
53              is.setOutAnimation(AnimationUtils.loadAnimation(this, android.R.anim.fade_out));
54              is.setOnTouchListener(new OnTouchListener() {
55                  @Override
56                  public boolean onTouch(View v, MotionEvent event) {
57                      if (event.getAction() == MotionEvent.ACTION_DOWN) {
58                          downX = (int) event.getX();
59                          return true;
60                      } else if (event.getAction() == MotionEvent.ACTION_UP) {
```

```java
61                          upX = (int) event.getX();
62                          int index = 0;
63                          if (upX - downX > 100)
64                          {
65                              if (gallery.getSelectedItemPosition() == 0) {
66                                  index = gallery.getCount() - 1;
67                              } else {
68                                  index = gallery.getSelectedItemPosition() - 1;
69                              }
70                          } else if (downX - upX > 100)
71                          {
72                              if (gallery.getSelectedItemPosition() == (gallery.getCount() - 1)) {
73                                  index = 0;
74                              } else {
75                                  index = gallery.getSelectedItemPosition() + 1;
76                              }
77                          }
78                          gallery.setSelection(index, true);
79                          return true;
80                      }
81                  return false;
82              }
83          });
84
85          gallery = (Gallery) findViewById(R.id.gallery);
86          gallery.setAdapter(new ImageAdapter(this));
87          gallery.setOnItemSelectedListener(new OnItemSelectedListener(){
88              @Override
89              public void onItemSelected(AdapterView<?> arg0, View arg1, int position, long arg3) {
90                  is.setImageResource(imgList.get(position));
91              }
92
93              @Override
94              public void onNothingSelected(AdapterView<?> arg0) {
95              }
96          });
97      }
98
99      @Override
100     public View makeView() {
101         ImageView i = new ImageView(this);
102         i.setBackgroundColor(0xFF000000);
103         i.setScaleType(ImageView.ScaleType.CENTER);
104         i.setLayoutParams(new ImageSwitcher.LayoutParams(
105             LayoutParams.FILL_PARENT, LayoutParams.FILL_PARENT));
106         return i;
107     }
```

```
108
109        public class ImageAdapter extends BaseAdapter {
110            private Context mContext;
111
112            public ImageAdapter(Context c) {
113                mContext = c;
114            }
115            public int getCount() {
116                return imgList.size();
117            }
118            public Object getItem(int position) {
119                return position;
120            }
121
122            public long getItemId(int position) {
123                return position;
124            }
125
126            public View getView(int position, View convertView, ViewGroup parent) {
127                ImageView imageView = null;
128                if (convertView == null) {
129                    imageView = new ImageView(mContext);
130                    imageView.setImageResource(imgList.get(position));
131                    imageView.setAdjustViewBounds(true);
132                    imageView.setLayoutParams(new Gallery.LayoutParams(LayoutParams.WRAP_CONTENT, LayoutParams.WRAP_CONTENT));
133                } else {
134                    imageView = (ImageView) convertView;
135                }
136                return imageView;
137            }
138        }
139    @Override
140    public boolean onCreateOptionsMenu(Menu menu) {
141        getMenuInflater().inflate(R.menu.main, menu);
142        return true;
143    }
144
145    @Override
146    public boolean onOptionsItemSelected(MenuItem item) {
147        int id = item.getItemId();
148        if (id == R.id.action_settings) {
149            return true;
150        }
151        return super.onOptionsItemSelected(item);
152    }
153 }
```

第 27 行代码表示声明图像切换控件。第 28 行代码表示声明 Gallery 控件。第 30 行代码表示图像 I。第 36 行代码表示用反射机制来获取资源中的图片 ID。第 38 行代码表示除了 ic_action_search 和 ic_launcher 之外的图片。第 46 行代码表示保存图片 ID。第 50 行代码表示设置 ImageSwitcher 控件。第 51 行代码表示显示 makeView() 返回的 ImageView 控件。第 54 行代码表示为 ImageSwitcher 控件添加触摸事件监听器。第 56 行代码表示在 ImageSwitcher 控件上滑动可以切换图片。第 58 行代码表示取得按下时的坐标。第 61 行代码表示取得松开时的坐标。第 63 行代码表示从左拖到右,即看前一张。第 65 行代码表示如果是第一,则去到尾部。第 70 行代码表示从右拖到左,即看后一张。第 72 行代码表示如果是最后,则去到第一。第 79 行代码表示改变 gallery 图片所选,自动触发 Gallery 控件的 setOnItemSelectedListener 的回调 onItemSelected 方法。第 85 行代码表示设置 gallery 控件。第 86 行代码表示绑定 Gallery 数据源。第 87 行代码表示绑定 Gallery 选项选中监听器。第 100 行代码表示设置 ImgaeSwitcher。第 103 行代码表示居中。第 104 行代码表示自适应图片大小。

(4)部署 ImageSwitherDemo 工程,程序运行结果如图 11-2 所示。

图 11-2　程序运行结果

11.2　访问图片

Android 对于静态图形的处理,也就是不经常变化的图片,如 Icon 和 Logo 等,一般是通过各种 Drawable 类来进行处理的。

11.2.1　Drawable

Android 访问图片的时候,使用 Drawable 类及其子类 BitmapDrawable、LayerDrawable 和 ShapeDrawable 等类处理。下面分别给出使用 Java 代码的方式和 XML 文件引用的方式来使用 Drawable 对象访问图片。

下面通过实例介绍如何使用 Java 代码的方式来实现图片浏览器的功能。

(1) 创建一个新的 Android 工程,工程名为 DrawableJavaDemo,应用程序名为 DrawableJavaDemo,包名为 hlju.edu.cn,创建的 Activity 的名字为 MainActivity,最小 SDK 版本根据选择的目标 API 会自动添加。

(2) 修改 src 目录中 hlju.edu.cn 包下的 MainActivity.java 文件,代码如下:

```java
package hlju.edu.cn;
import android.app.Activity;
import android.os.Bundle;
import android.view.Menu;
import android.view.MenuItem;
import android.view.ViewGroup.LayoutParams;
import android.widget.Gallery;
import android.widget.ImageView;
import android.widget.LinearLayout;

public class MainActivity extends Activity {
    @Override
    public void onCreate(Bundle savedInstanceState) {
        super.onCreate(savedInstanceState);
        LinearLayout mLinearLayout = new LinearLayout(this);
        ImageView mImageView = new ImageView(this);
        mImageView.setImageResource(R.drawable.images);
        mImageView.setLayoutParams(new Gallery.LayoutParams(LayoutParams.WRAP_CONTENT,
LayoutParams.WRAP_CONTENT));
        mLinearLayout.addView(mImageView);
        setContentView(mLinearLayout);
    }
    @Override
    public boolean onCreateOptionsMenu(Menu menu) {
        getMenuInflater().inflate(R.menu.main, menu);
        return true;
    }
    @Override
    public boolean onOptionsItemSelected(MenuItem item) {
        int id = item.getItemId();
        if (id == R.id.action_settings) {
            return true;
        }
        return super.onOptionsItemSelected(item);
    }
}
```

(3) 部署 DrawableJavaDemo 工程,程序运行结果如图 11-3 所示。

下面通过实例介绍如何使用 XML 文件引用的方式来实现图片浏览器的功能。

(1) 创建一个新的 Android 工程,工程名为 DrawableXMLDemo,应用程序名为 DrawableXMLDemo,包名为 hlju.edu.cn,创建的 Activity 的名字为 MainActivity,最小 SDK 版本根据选择的目标 API 会自动添加。

(2) 修改 res 目录下 layout 文件夹中的 activity_main.xml 文件,设置线性布局,添加 1 个 ImageView 控件,对相应控件进行描述,并设置相关属性,代码如下:

图 11-3　程序运行结果

```
1   <LinearLayout xmlns:android = "http://schemas.android.com/apk/res/android"
2       xmlns:tools = "http://schemas.android.com/tools"
3       android:layout_width = "fill_parent"
4       android:layout_height = "fill_parent"
5       android:orientation = "vertical">
6       <ImageView
7           android:layout_width = "wrap_content"
8           android:layout_height = "wrap_content"
9           android:src = "@drawable/images" />
10  </LinearLayout>
```

（3）部署 DrawableXMLDemo 工程，程序运行结果如图 11-4 所示。

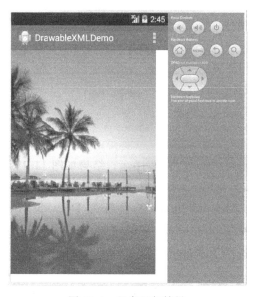

图 11-4　程序运行结果

11.2.2 Bitmap 和 BitmapFactory

Bitmap 是 Android 系统中图像处理的最重要类之一。用它可以获取图像文件信息，进行图像裁剪、旋转、缩放等操作，并可以指定格式保存图像文件。Bitmap 位于 android.graphics 包中。由于 Bitmap 类的构造函数是私有的，所以在类的外面并不能对其实例化。

利用 BitmapFactory 可以从一个指定文件中，调用 decodeFile()方法返回 Bitmap 对象；也可以调用 decodeResource()从工程的资源文件中返回 Bitmap 对象。Bitmap 类常用方法如表 11-1 所示。

表 11-1 Bitmap 类常用方法

方 法	描 述
recycle()	回收位图占用的内存空间
isRecycle()	判断位图内存是否已释放
getWidth()	获取位图的宽度
getHeight()	获取位图的高度
isMutable()	判断图片是否可修改
getScaledWidth(Canvas canvas)	获取指定密度转换后的图像宽度
getScaledHeight(Canvas canvas)	获取指定密度转换后的图像高度
compress(CompressFormat format, int quality, OutputStream stream)	按指定的图片格式以及画质，将图片转换为输出流

下面通过实例介绍如何使用 Bitmap 类以及 BitmapFactory 类来实现图片浏览器的功能。

（1）创建一个新的 Android 工程，工程名为 BitmapFactoryDemo，应用程序名为 BitmapFactoryDemo，包名为 hlju.edu.cn，创建的 Activity 的名字为 MainActivity，最小 SDK 版本根据选择的目标 API 会自动添加。

（2）修改 src 目录中 hlju.edu.cn 包下的 MainActivity.java 文件，代码如下：

```
1    package hlju.edu.cn;
2
3    import android.app.Activity;
4    import android.graphics.Bitmap;
5    import android.graphics.BitmapFactory;
6    import android.os.Bundle;
7    import android.view.Menu;
8    import android.view.MenuItem;
9    import android.widget.ImageView;
10
11   public class MainActivity extends Activity {
12
13       @Override
14       public void onCreate(Bundle savedInstanceState) {
15           super.onCreate(savedInstanceState);
```

```
16          setContentView(R.layout.activity_main);
17          Bitmap bm = BitmapFactory.decodeResource(getResources(), R.drawable.images);
18          ImageView imageView = new ImageView(this);
19          imageView.setImageBitmap(bm);
20          this.setContentView(imageView);
21      }
22
23      @Override
24      public boolean onCreateOptionsMenu(Menu menu) {
25          // Inflate the menu; this adds items to the action bar if it is present.
26          getMenuInflater().inflate(R.menu.main, menu);
27          return true;
28      }
29
30      @Override
31      public boolean onOptionsItemSelected(MenuItem item) {
32          // Handle action bar item clicks here. The action bar will
33          // automatically handle clicks on the Home/Up button, so long
34          // as you specify a parent activity in AndroidManifest.xml.
35          int id = item.getItemId();
36          if (id == R.id.action_settings) {
37              return true;
38          }
39          return super.onOptionsItemSelected(item);
40      }
41  }
```

（3）部署 BitmapFactoryDemo 工程，程序运行结果如图 11-5 所示。

图 11-5　程序运行结果

11.3 内存优化

由于移动设备内存的大小限制，使得在开发 Android 应用程序的时候，内存的优化始终是贯彻于整个开发过程的一条线索。

11.3.1 Drawable 与 Bitmap 占用内存比较

下面通过实例比较 Drawable 与 Bitmap 占用内存的大小。

（1）创建一个新的 Android 工程，工程名为 MemoryDrawableDemo，应用程序名为 MemoryDrawableDemo，包名为 hlju.edu.cn，创建的 Activity 的名字为 MainActivity，最小 SDK 版本根据选择的目标 API 会自动添加。

（2）修改 src 目录中 hlju.edu.cn 包下的 MainActivity.java 文件，代码如下：

```java
package hlju.edu.cn;

import android.app.Activity;
import android.graphics.drawable.BitmapDrawable;
import android.graphics.drawable.Drawable;
import android.os.Bundle;
import android.util.Log;
import android.view.Menu;
import android.view.MenuItem;

public class MainActivity extends Activity {
    private static final String TAG = "MainActivity";
    int number = 1000;
    Drawable[] array ;

    @Override
    public void onCreate(Bundle savedInstanceState) {
        super.onCreate(savedInstanceState);
        setContentView(R.layout.activity_main);
        array = new BitmapDrawable[number];
        for(int i = 0; i < number; i++)
        {
            Log.d(TAG, "测试第" + (i + 1) + "张图片");
            array[i] = getResources().getDrawable(R.drawable.images);
        }
    }

    @Override
    public boolean onCreateOptionsMenu(Menu menu) {
        // Inflate the menu; this adds items to the action bar if it is present.
        getMenuInflater().inflate(R.menu.main, menu);
        return true;
```

```
33          }
34
35          @Override
36          public boolean onOptionsItemSelected(MenuItem item) {
37              // Handle action bar item clicks here. The action bar will
38              // automatically handle clicks on the Home/Up button, so long
39              // as you specify a parent activity in AndroidManifest.xml.
40              int id = item.getItemId();
41              if (id == R.id.action_settings) {
42                  return true;
43              }
44              return super.onOptionsItemSelected(item);
45          }
46      }
```

（3）部署 MemoryDrawableDemo 工程，程序运行结果如图 11-6 所示。

图 11-6　程序运行结果

（4）创建一个新的 Android 工程，工程名为 MemoryBitmapDemo，应用程序名为 MemoryBitmapDemo，包名为 hlju.edu.cn，创建的 Activity 的名字为 MainActivity，最小 SDK 版本根据选择的目标 API 会自动添加。

（5）修改 src 目录中 hlju.edu.cn 包下的 MainActivity.java 文件，代码如下：

```
1   package hlju.edu.cn;
2
3   import android.app.Activity;
4   import android.graphics.Bitmap;
5   import android.graphics.BitmapFactory;
6   import android.os.Bundle;
7   import android.util.Log;
8   import android.view.Menu;
9   import android.view.MenuItem;
10
11  public class MainActivity extends Activity {
12      private static final String TAG = "MainActivity";
13      int number = 1000;
```

```
14    Bitmap[ ] array ;
15
16    @Override
17    public void onCreate(Bundle savedInstanceState) {
18        super.onCreate(savedInstanceState);
19        setContentView(R.layout.activity_main);
20        array = new Bitmap[number];
21        for(int i = 0; i < number; i++)
22        {
23            Log.d(TAG, "测试第" + (i + 1) + "张图片");
24            array[i] = BitmapFactory.decodeResource(getResources(), R.drawable.images);
25        }
26    }
27
28    @Override
29    @Override
30    public boolean onCreateOptionsMenu(Menu menu) {
31        // Inflate the menu; this adds items to the action bar if it is present.
32        getMenuInflater().inflate(R.menu.main, menu);
33        return true;
34    }
35
36    @Override
37    public boolean onOptionsItemSelected(MenuItem item) {
38        // Handle action bar item clicks here. The action bar will
39        // automatically handle clicks on the Home/Up button, so long
40        // as you specify a parent activity in AndroidManifest.xml.
41        int id = item.getItemId();
42        if (id == R.id.action_settings) {
43            return true;
44        }
45        return super.onOptionsItemSelected(item);
46    }
47 }
```

（6）部署 MemoryBitmapDemo 工程，程序运行结果如图 11-7 所示。

图 11-7　程序运行结果

通过以上两个程序的运行结果可以看出使用 Drawable 对象保存图片时,占用更小的内存空间。而使用 Bitmap 对象保存图片时,则会占用很大的空间,很容易出现 OOM(Out of memory)。

11.3.2 防止内存溢出

通过使用 Bitmap 的情况可以看出经常会发生内存溢出的错误,特别是在处理比较大的图片的时候,发生 OOM 错误的概率会更高。使用 Bitmap 对象来保存图片,或者对图片进行处理的时候,要做到避免内存溢出问题的发生。

下面通过实例介绍如何在 Bitmap 的时候,防止内存溢出的发生。

(1) 创建一个新的 Android 工程,工程名为 AvoidMemoryOverflowDemo,应用程序名为 AvoidMemoryOverflowDemo,包名为 hlju.edu.cn,创建的 Activity 的名字为 MainActivity,最小 SDK 版本根据选择的目标 API 会自动添加。

(2) 修改 res 目录下 layout 文件夹中的 activity_main.xml 文件,设置线性布局,添加 1 个 Button 控件和 1 个 ImageView 控件,对相应控件进行描述,并设置相关属性,代码如下:

```
1   <LinearLayout xmlns:android = "http://schemas.android.com/apk/res/android"
2       xmlns:tools = "http://schemas.android.com/tools"
3       android:layout_width = "fill_parent"
4       android:layout_height = "fill_parent"
5       android:orientation = "vertical" >
6       <Button
7           android:id = "@ + id/launch_camera"
8           android:layout_width = "fill_parent"
9           android:layout_height = "wrap_content"
10          android:text = "启动 Camera"></Button>
11      <ImageView
12          android:id = "@ + id/show_image"
13          android:layout_width = "wrap_content"
14          android:layout_height = "wrap_content"></ImageView>
15  </LinearLayout>
```

(3) 由于需要调用系统 Camera,需要在 SD Card 创建文件以及写入数据,所以需要添加相关权限。需要在项目清单文件 AndroidManifest.xml 中添加相关的权限,代码如下:

```
1   <uses - permission  android:name = "android.permission.CAMERA"/>
2   <uses - permission android:name = "android.permission.MOUNT_UNMOUNT_FILESYSTEMS"/>
3   <uses - permission android:name = "android.permission.WRITE_EXTERNAL_STORAGE"/>
```

(4) 由于调用系统 Camera 拍摄的照片会自动旋转 90°,所以需要对拍摄后的照片进行旋转处理。同时需要防止处理过大图片引起的内存溢出问题。在 src 目录中 hlju.edu.cn 包下新建 ImageUtil.java 文件,代码如下:

```java
1   package hlju.edu.cn;
2   import java.io.File;
3   import java.io.FileInputStream;
4   import java.io.FileNotFoundException;
5   import java.io.IOException;
6   import android.graphics.Bitmap;
7   import android.graphics.BitmapFactory;
8   import android.graphics.Matrix;
9   import android.util.Log;
10  public final class ImageUtil {
11      private static final String TAG = "ImageUtil";
12      public static Bitmap getBitmap(String url) {
13          Log.d(TAG, "Get Image Url:" + url);
14          Bitmap bm = null;
15          File file = new File(url);
16          FileInputStream fs = null;
17          try {
18              fs = new FileInputStream(file);
19          } catch (FileNotFoundException e) {
20              Log.d(TAG, "download image occurs errors:" + e);
21          }
22          BitmapFactory.Options bfOptions = new BitmapFactory.Options();
23          bfOptions.inDither = false;
24          bfOptions.inPurgeable = true;
25          bfOptions.inInputShareable = true;
26          bfOptions.inTempStorage = new byte[32 * 1024];
27          try {
28              if (fs != null)
29                  bm = BitmapFactory.decodeFileDescriptor(fs.getFD(), null, bfOptions);
30          } catch (IOException e) {
31              e.printStackTrace();
32          } finally {
33              if (fs != null) {
34                  try {
35                      fs.close();
36                  } catch (IOException e) {
37                      e.printStackTrace();
38                  }
39              }
40          }
41          return bm;
42      }
43      public static Bitmap getRotatedBitmap(String url) {
44          Bitmap bm = getBitmap(url);
45          Matrix matrix = new Matrix();
46          matrix.postRotate(90);
47          bm = Bitmap.createBitmap(bm, 0, 0, bm.getWidth(), bm.getHeight(), matrix, true);
48          return bm;
49      }
```

```
50      public static Bitmap rotateMatrixBitmap(String fromUrl){
51          Bitmap bm = getRotatedBitmap(fromUrl);
52          return bm;
53      }
54  }
```

第 28 行代码表示解决 java. lang. OutOfMemoryError：bitmap size exceeds VM budget 异常。

（5）由于对拍摄后的照片进行处理比较耗时。可以在 src 目录中 hlju. edu. cn 包下新建 CameraAsyncTask. java 文件，代码如下：

```
1   package hlju.edu.cn;
2   import android.content.Context;
3   import android.graphics.Bitmap;
4   import android.os.AsyncTask;
5   import android.widget.ImageView;
6   import android.widget.Toast;
7   public class CameraAsyncTask extends AsyncTask<Void, Void, Bitmap>
    {   private Context context;
8       private String url;
9       private ImageView imageView;
10      public CameraAsyncTask(Context context, String url, ImageView imageView){
11          this.context = context;
12          this.url = url;
13          this.imageView = imageView;
14      }
15      @Override
16      protected void onPreExecute() {
17          super.onPreExecute();
18          Toast.makeText(context, "图片正在进行压缩,请您耐心等待", Toast.LENGTH_
    SHORT).show();
19      }
20      @Override
21      protected Bitmap doInBackground(Void…params) {
22          return ImageUtil.rotateMatrixBitmap(url);       }
23      @Override
24      protected void onPostExecute(Bitmap result) {
25          super.onPostExecute(result);
26          imageView.setImageBitmap(result);
27      } }
```

（6）修改 src 目录中 hlju. edu. cn 包下的 MainActivity. java 文件，代码如下：

```
1   package hlju.edu.cn;
2   import java.io.File;
3   import java.io.IOException;
4   import java.text.SimpleDateFormat;
```

```java
5    import java.util.Date;
6    import android.app.Activity;
7    import android.content.Intent;
8    import android.graphics.Bitmap;
9    import android.net.Uri;
10   import android.os.AsyncTask;
11   import android.os.Bundle;
12   import android.os.Environment;
13   import android.provider.MediaStore;
14   import android.util.Log;
15   import android.view.Menu;
16   import android.view.MenuItem;
17   import android.view.View;
18   import android.view.View.OnClickListener;
19   import android.widget.Button;
20   import android.widget.ImageView;
21   import android.widget.Toast;
22
23   public class MainActivity extends Activity {
24       private static final String TAG = "MainActivity";
25       public static final int  OPEN_CAMERA_REQUEST_CODE = 0x0001;
26       public static String CAMERA_CAPTURE_SAVE_PATH = "";
27       Button launchCameraButton;
28       ImageView showImage;
29
30       @Override
31       public void onCreate(Bundle savedInstanceState) {
32           super.onCreate(savedInstanceState);
33           setContentView(R.layout.activity_main);
34           launchCameraButton = (Button)findViewById(R.id.launch_camera);
35           showImage = (ImageView)findViewById(R.id.show_image);
36           launchCameraButton.setOnClickListener(new OnClickListener() {
37               @Override
38               public void onClick(View v) {
39                   open();
40               }
41           });
42       }
43       @Override
44       protected void onActivityResult(int requestCode, int resultCode, Intent data) {
45             super.onActivityResult(requestCode, resultCode, data);
46             if (requestCode == OPEN_CAMERA_REQUEST_CODE && resultCode == RESULT_OK) {
47                 AsyncTask<Void, Void, Bitmap> cameraAsyncTask = new CameraAsyncTask(this, CAMERA_CAPTURE_SAVE_PATH, showImage);
48                 cameraAsyncTask.execute((Void[])null);
49             }
50       }
51       private void open(){
52           if (Environment.MEDIA_MOUNTED.equals(Environment.getExternalStorageState())) {
```

```java
53              File dir = new File(Environment.getExternalStorageDirectory() + "/DCIM/Camera");
54              File destFile = new File(dir, generatePhotoFileName());
55              if (!dir.exists()){
56                  dir.mkdirs();
57              }
58              if (!destFile.exists()){
59                  try {
60                      destFile.createNewFile();
61                  } catch (IOException e) {
62                      Log.e(TAG,"创建文件抛出异常:" + e);
63                  }
64              }
65              CAMERA_CAPTURE_SAVE_PATH = destFile.getPath();
66
67              Intent cameraIntent = new Intent();
68              cameraIntent.putExtra(MediaStore.EXTRA_OUTPUT, Uri.fromFile(destFile));
69              cameraIntent.setAction(MediaStore.ACTION_IMAGE_CAPTURE);
70              startActivityForResult(cameraIntent, OPEN_CAMERA_REQUEST_CODE);
71          } else {
72              Toast.makeText(this, "没有找到 SD Card", Toast.LENGTH_LONG).show();
73          }
74      }
75
76      private String generatePhotoFileName(){
77          Date date = new Date(System.currentTimeMillis());
78          SimpleDateFormat dateFormat = new SimpleDateFormat("'IMG'_yyyyMMdd_HHmmss");
79          return dateFormat.format(date) + ".JPG";
80      }
81      @Override
82      public boolean onCreateOptionsMenu(Menu menu) {
83          getMenuInflater().inflate(R.menu.main, menu);
84          return true;
85      }
86      @Override
87      public boolean onOptionsItemSelected(MenuItem item) {
88          // Handle action bar item clicks here. The action bar will
89          // automatically handle clicks on the Home/Up button, so long
90          // as you specify a parent activity in AndroidManifest.xml.
91          int id = item.getItemId();
92          if (id == R.id.action_settings) {
93              return true;
94          }
95          return super.onOptionsItemSelected(item);
96      }
97  }
```

（7）部署 AvoidMemoryOverflowDemo 工程，程序运行结果如图 11-8 所示。

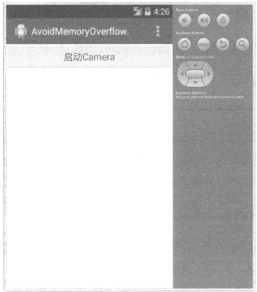

图 11-8　程序运行结果

11.4　2D 绘图

2D 图形的接口实际上是 Android 图形系统的基础，GUI 的各种可见元素也是基于 2D 图形接口构建的，各种控件实际上是基于图形 API 绘制出来的。Android 系统提供的 UI 控件通过继承 android.view.View 类，并实现其中的 onDraw() 函数来实现绘制的工作，绘制的工作主要是由 android.graphics 包来实现的。android.graphics 包中的内容是 Android 系统的 2D 图形 API。

11.4.1　View 类

任何自定义的控件都需要继承 android.view.View 类，通过重写 android.view.View 父类的 onDraw() 函数来完成绘制的工作。自定义控件的 onDraw() 方法不能被外部类直接调用，想要刷新自定义控件的界面，根据所在线程的不同，可以分为两种情况：UI 主线程中直接调用自定义控件的 invalidate() 方法；子线程当中直接调用自定义控件的 postInvalidate() 方法。

11.4.2　SurfaceView 类

SurfaceView 类继承 View 类，它通过一个新的线程来更新界面。因此，SurfaceView 类更适合需要快速加载 UI，或渲染代码阻塞 UI 主线程的时间过长的情形。SurfaceView 封装了一个 Surface 对象，而不是 Canvas 对象，这一点对于那些资源敏感的操作特别有用。SurfaceView 一般通过使用 SurfaceHolder 类来控制 Canvas 在其 Surface 上的操作，

SurfaceHolder 类的实例可以通过 SurfaceHolder 对象的 getHolder()方法来获得。

11.4.3　Paint 类

Paint 类代表画笔,用于描述图形的颜色和风格,如线宽、颜色、透明度和填充效果等信息。使用 Paint 类时,需要先创建该类的对象,这可以通过该类提供的构造方法来实现。通常情况下,只需要使用 Paint()方法来创建一个使用默认设置的 Paint 对象。具体代码如下:

```
Paint pait = new Paint();
```

创建 Paint 类的对象后,还可以通过该对象提供的方法来对画笔的默认设置进行改变。用于改变画笔设置的常用方法如表 11-2 所示。

表 11-2　Paint 类的常用方法

方　　法	描　　述
setARGB(int a,int g,int b)	用于设置颜色,各参数值均为 0~255 之间的整数,分别用于表示透明度、红色、绿色和蓝色值
setColor(int color)	用于设置颜色,参数 color 可以通过 Color 类提供的颜色常量指定,也可以通过 Color.rgb(int red,int green,int blue)方法指定
setAlpha(int a)	用于设置透明度,值为 0~255 之间的整数
setAntiAlias(boolean aa)	用于指定是否使用抗锯齿功能,如果使用会使绘图速度变慢
setDither(boolean dither)	用于指定是否使用图像抖动处理,如果使用会使图像颜色更加平滑和饱满,且更加清晰
setPathEffect(PathEffect effect)	用于设置绘制路径时的路径效果
setShader(Shader shader)	用于设置渐变,可以使用 LinearGradient、RadialGradient 或者 SweepGradient
setShadowLayer(float radius,float dx,float dy,int color)	用于设置阴影,参数 radius 为阴影的角度,dx 和 dy 为阴影在 x 轴和 y 轴上的距离,color 为阴影的颜色。如果参数 radius 的值为 0,那么将没有阴影
setStrokeCap(Paint.Cap cap)	用于当画笔的填充样式为 STROKE 或 FILL_AND_STROKE 时,设置笔刷的图形样式,参数值可以是 Cap.BUTT、Cap.BOUND 或 Cap.SQUARE,主要体现在线的端点上
setStrokeJoin(Paint.Join join)	用于设置画笔转弯处的连接风格,参数值为 Join.BEVEL、Join.MITER 或 Join.ROUND
setStrokeWidth(float width)	用于设置笔触的宽度
setStyle(Paint.Style style)	用于设置填充风格,参数值为 Style.FILL、Style.FILL_AND_STROKE 或 Style.STROKE
setTextAlign(Paint.Align align)	用于设置绘制文本时文字的对齐方式,参数值为 Align.CENTER、Align.LEFT 或 Align.RIGHT
setTextSize(float textSize)	用于设置绘制文本时文字的大小
setFakeBoldText(boolean fakeBoldText)	用于设置是否为粗体文字
setXfermode(Xfermode xfermode)	用于设置图形重叠时的处理方式,例如合并、取交集或并集,经常用于制作橡皮的擦除效果

下面通过实例介绍如何利用 Paint 进行绘制。

（1）创建一个新的 Android 工程，工程名为 PaintDemo，应用程序名为 PaintDemo，包名为 hlju.edu.cn，创建的 Activity 的名字为 MainActivity，最小 SDK 版本根据选择的目标 API 会自动添加。

（2）修改 src 目录中 hlju.edu.cn 包下的 MainActivity.java 文件，代码如下：

```
1   package hlju.edu.cn;
2   import android.app.Activity;
3   import android.content.Context;
4   import android.graphics.Canvas;
5   import android.graphics.Color;
6   import android.graphics.LinearGradient;
7   import android.graphics.Paint;
8   import android.graphics.RadialGradient;
9   import android.graphics.Shader;
10  import android.graphics.SweepGradient;
11  import android.os.Bundle;
12  import android.view.Menu;
13  import android.view.MenuItem;
14  import android.view.View;
15  import android.widget.FrameLayout;
16  public class MainActivity extends Activity {
17      @Override
18      protected void onCreate(Bundle savedInstanceState) {
19          super.onCreate(savedInstanceState);
20          setContentView(R.layout.activity_main);
21          FrameLayout ll = (FrameLayout)findViewById(R.id.frameLayout1);
22          ll.addView(new MyView(this));
23      }
24      public class MyView extends View{
25          public MyView(Context context) {
26              super(context);
27          }
28          @Override
29          protected void onDraw(Canvas canvas) {
30              Paint paint = new Paint();
31              Shader shader = new LinearGradient(0, 0, 50, 50, Color.RED, Color.GREEN, Shader.TileMode.MIRROR);
32              paint.setShader(shader);
33              canvas.drawRect(10, 70, 100, 150, paint);
34              shader = new RadialGradient(160, 110, 50, Color.RED, Color.GREEN, Shader.TileMode.MIRROR);
35              paint.setShader(shader);
36              canvas.drawRect(115,70,205,150, paint);
37              shader = new SweepGradient(265,110,new int[]{Color.RED,Color.GREEN,Color.BLUE},null);
38              paint.setShader(shader);
39              canvas.drawRect(220, 70, 310, 150, paint);
```

```
40              super.onDraw(canvas);
41          }
42      }
43      @Override
44      public boolean onCreateOptionsMenu(Menu menu) {
45          getMenuInflater().inflate(R.menu.main, menu);
46          return true;
47      }
48      @Override
49      public boolean onOptionsItemSelected(MenuItem item) {
50          int id = item.getItemId();
51          if (id == R.id.action_settings) {
52              return true;
53          }
54          return super.onOptionsItemSelected(item);
55      }
56  }
```

(3) 部署 PaintDemo 工程，程序运行结果如图 11-9 所示。

图 11-9　程序运行结果

11.4.4　Canvas 类

Canvas 类代表画布，通过该类提供的方法，可以绘制各种图形。在通常情况下，要在 Android 中绘图，首先需要创建一个继承 View 类的视图，并且在该类中重写它的 onDraw（Canvas canvas）方法，然后在显示绘图的 Activity 中添加该视图。

下面通过实例介绍如何利用 Canvas 进行绘制。

（1）创建一个新的 Android 工程，工程名为 CanvasDemo，应用程序名为 CanvasDemo，

包名为 hlju.edu.cn,创建的 Activity 的名字为 MainActivity,最小 SDK 版本根据选择的目标 API 会自动添加。

(2) 修改 res 目录下 layout 文件夹中的 activity_main.xml 文件,添加 1 个帧布局管理器,并设置相关属性,代码如下:

```
1   <?xml version = "1.0" encoding = "utf-8"?>
2   <FrameLayout xmlns:android = "http://schemas.android.com/apk/res/android"
3       android:layout_width = "fill_parent"
4       android:layout_height = "fill_parent"
5       android:orientation = "vertical" >
6       <hlju.edu.cn.DrawView
7           android:id = "@ + id/drawView1"
8           android:layout_width = "wrap_content"
9           android:layout_height = "wrap_content" />
10  </FrameLayout>
```

(3) 在 src 目录中 hlju.edu.cn 包下新建 DrawView.java 文件,添加构造方法和重写 onDraw(Canvas canvas)方法,代码如下:

```
1   package hlju.edu.cn;
2   import android.content.Context;
3   import android.graphics.Canvas;
4   import android.graphics.Color;
5   import android.graphics.Paint;
6   import android.util.AttributeSet;
7   import android.view.View;
8   public class DrawView extends View {
9       public DrawView(Context context, AttributeSet attrs) {
10          super(context, attrs);
11      }
12      @Override
13      protected void onDraw(Canvas canvas) {
14          Paint paint = new Paint();
15          paint.setColor(Color.RED);
16          paint.setShadowLayer(2, 20, 20, Color.rgb(180, 180, 180));
17          canvas.drawRect(40, 80, 400, 250, paint);
18          super.onDraw(canvas);
19      }
20  }
```

(4) 修改 src 目录中 hlju.edu.cn 包下的 MainActivity.java 文件,代码如下:

```
1   package hlju.edu.cn;
2   import android.app.Activity;
3   import android.os.Bundle;
4   import android.view.Menu;
5   import android.view.MenuItem;
```

```
 6    public class MainActivity extends Activity {
 7        @Override
 8        protected void onCreate(Bundle savedInstanceState) {
 9            super.onCreate(savedInstanceState);
10            setContentView(R.layout.activity_main);
11        }
12        @Override
13        public boolean onCreateOptionsMenu(Menu menu) {
14            getMenuInflater().inflate(R.menu.main, menu);
15            return true;
16        }
17        @Override
18        public boolean onOptionsItemSelected(MenuItem item) {
19            int id = item.getItemId();
20            if (id == R.id.action_settings) {
21                return true;
22            }
23            return super.onOptionsItemSelected(item);
24        }
25    }
```

（5）部署 CanvasDemo 工程，程序运行结果如图 11-10 所示。

图 11-10　程序运行结果

11.4.5　绘制几何图形

比较常见的几何图形包括点、线、弧、圆形和矩形等。在 Android 中，Canvas 类提供了丰富的绘制几何图形的方法，通过这些方法可以绘制出各种几何图形。常用的绘制几何图形的方法如表 11-3 所示。

表 11-3　Canvas 类提供的绘制几何图形的方法

方　　法	描　　述
drawArc(RectF oval,float startAngle,float sweepAngle,boolean useCenter,Paint paint)	绘制弧
drawCircle(float cx,float cy,float radius,Paint paint)	绘制圆形
drawLine(float startX,float startY,float stopX,float stopY,Paint paint)	绘制一条线
drawLines(float[] pts,Paint paint)	绘制多条线
drawOval(RectF oval,Paint paint)	绘制椭圆
drawPoint(float x,float y,Paint paint)	绘制一个点
drawPoints(float[] pts,Paint paint)	绘制多个点
drawRect(float left,float top,float right,float bottom,Paint paint)	绘制矩形
drawRoundRect(RectF rect,float rx,float ry,Paint paint)	绘制圆角矩形

下面通过实例介绍如何绘制几何图形。

(1) 创建一个新的 Android 工程,工程名为 DrawFigureDemo,应用程序名为 DrawFigureDemo,包名为 hlju. edu. cn,创建的 Activity 的名字为 MainActivity,最小 SDK 版本根据选择的目标 API 会自动添加。

(2) 修改 res 目录下 layout 文件夹中的 activity_main. xml 文件,添加 1 个帧布局管理器,并设置相关属性,代码如下:

```
1  <?xml version = "1.0" encoding = "utf - 8"?>
2  < FrameLayout xmlns:android = "http://schemas.android.com/apk/res/android"
3      android:id = "@ + id/frameLayout1"
4      android:layout_width = "fill_parent"
5      android:layout_height = "fill_parent"
6      android:orientation = "vertical" >
7  </FrameLayout >
```

(3) 修改 src 目录中 hlju. edu. cn 包下的 MainActivity. java 文件,代码如下:

```
1   package hlju.edu.cn;
2   import android.app.Activity;
3   import android.content.Context;
4   import android.graphics.Canvas;
5   import android.graphics.Color;
6   import android.graphics.Paint;
7   import android.graphics.Paint.Style;
8   import android.graphics.RectF;
9   import android.os.Bundle;
10  import android.view.Menu;
11  import android.view.MenuItem;
12  import android.view.View;
13  import android.widget.FrameLayout;
14  public class MainActivity extends Activity {
15      @Override
```

```java
16      protected void onCreate(Bundle savedInstanceState) {
17          super.onCreate(savedInstanceState);
18          setContentView(R.layout.activity_main);
19          FrameLayout ll = (FrameLayout)findViewById(R.id.frameLayout1);
20          ll.addView(new MyView(this));
21      }
22      public class MyView extends View{
23          public MyView(Context context) {
24              super(context);
25          }
26          @Override
27          protected void onDraw(Canvas canvas) {
28              canvas.drawColor(Color.WHITE);
29              Paint paint = new Paint();
30              paint.setAntiAlias(true);
31              paint.setStrokeWidth(3);
32              paint.setStyle(Style.STROKE);
33              paint.setColor(Color.BLUE);
34              canvas.drawCircle(200, 50, 30, paint);
35              paint.setColor(Color.YELLOW);
36              canvas.drawCircle(250, 50, 30, paint);
37              paint.setColor(Color.BLACK);
38              canvas.drawCircle(300, 50, 30, paint);
39              paint.setColor(Color.GREEN);
40              canvas.drawCircle(225, 90, 30, paint);
41              paint.setColor(Color.RED);
42              canvas.drawCircle(275, 90, 30, paint);
43              RectF rectf = new RectF(250, 180, 470, 260);
44              canvas.drawOval(rectf,paint);
45              RectF rectf1 = new RectF(40, 200, 180, 250);
46              canvas.drawRoundRect(rectf1, 6, 6, paint);
47              paint.setStyle(Style.FILL);
48              super.onDraw(canvas);
49          }
50      }
51      @Override
52      public boolean onCreateOptionsMenu(Menu menu) {
53          getMenuInflater().inflate(R.menu.main, menu);
54          return true;
55      }
56      @Override
57      public boolean onOptionsItemSelected(MenuItem item) {
58          int id = item.getItemId();
59          if (id == R.id.action_settings) {
60              return true;
61          }
62          return super.onOptionsItemSelected(item);
63      }
64  }
```

第 19 行代码表示获取布局文件中添加的帧布局管理器。第 20 行代码表示将自定义的 MyView 视图添加到帧布局管理器中。第 29 行代码表示创建采用默认设置的画笔。第 30 行代码表示使用抗锯齿功能。第 31 行代码表示设置笔触的宽度。第 32 行代码表示设置填充样式为描边。第 34 行代码表示绘制蓝色的圆形。第 36 行代码表示绘制黄色的圆形。第 38 行代码表示绘制黑色的圆形。第 40 行代码表示绘制绿色的圆形。第 42 行代码表示绘制红色的圆形。第 43 行代码表示绘制椭圆。第 45 行代码表示绘制圆角矩形。

（4）部署 DrawFigureDemo 工程，程序运行结果如图 11-11 所示。

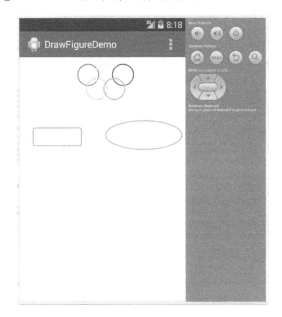

图 11-11　程序运行结果

11.4.6　绘制文本

在 Android 中，虽然可以通过 TextView 或是图片显示文本，但是在开发游戏时，特别是开发角色类游戏时，会包含很多文字，使用 TextView 和图片显示文本不太合适，这时就需要通过绘制文本的方式来实现。Canvas 类提供了一系列绘制文本的方法，下面分别进行介绍。

drawText()方法用于在画布的指定位置绘制文字。该方法的语法格式如下：

```
drawText(String text,float x,float y,Paint paint)
```

在语法中，参数 text 用于指定要绘制的文字，x 用于指定文字起始位置的 x 轴坐标，y 用于指定文字起始位置的 y 轴坐标，paint 用于指定使用的画笔。

drawPostText()方法也用于画布上绘制文字，与 drawText()方法不同的是，使用该字符串时，需要为每个字符指定一个位置。该方法的语法格式如下：

```
drawPostText(String text,float[] pos,Paint paint)
```

在语法中,参数 text 用于指定要绘制的文字,pos 用于指定每一个字符的位置,paint 用于指定要使用的画笔。

下面通过实例介绍如何绘制几何图形。

(1) 创建一个新的 Android 工程,工程名为 DrawTextDemo,应用程序名为 DrawTextDemo,包名为 hlju.edu.cn,创建的 Activity 的名字为 MainActivity,最小 SDK 版本根据选择的目标 API 会自动添加。

(2) 修改 res 目录下 layout 文件夹中的 activity_main.xml 文件,添加 1 个帧布局管理器,并设置相关属性,代码如下:

```
1  <?xml version = "1.0" encoding = "utf - 8"?>
2  <FrameLayout xmlns:android = "http://schemas.android.com/apk/res/android"
3      android:id = "@ + id/frameLayout1"
4      android:layout_width = "fill_parent"
5      android:layout_height = "fill_parent"
6      android:background = "@drawable/background"
7      android:orientation = "vertical" >
8  </FrameLayout>
```

(3) 修改 src 目录中 hlju.edu.cn 包下的 MainActivity.java 文件,代码如下:

```
1   package hlju.edu.cn;
2   import android.app.Activity;
3   import android.content.Context;
4   import android.graphics.Canvas;
5   import android.graphics.Paint;
6   import android.graphics.Paint.Align;
7   import android.os.Bundle;
8   import android.view.Menu;
9   import android.view.MenuItem;
10  import android.view.View;
11  import android.widget.FrameLayout;
12  public class MainActivity extends Activity {
13      @Override
14      protected void onCreate(Bundle savedInstanceState) {
15          super.onCreate(savedInstanceState);
16          setContentView(R.layout.activity_main);
17      FrameLayout ll = (FrameLayout)findViewById(R.id.frameLayout1);
18          ll.addView(new MyView(this));
19      }
20      public class MyView extends View{
21        public MyView(Context context) {
22            super(context);
23        }
24        @Override
25        protected void onDraw(Canvas canvas) {
26          Paint paintText = new Paint();
27          paintText.setColor(0xFFFF6600);
```

```
28           paintText.setTextAlign(Align.LEFT);
29           paintText.setTextSize(16);
30           paintText.setAntiAlias(true);
31           canvas.drawText("不,我不想去!", 230,45, paintText);
32      float[] pos = new float[]{100,145, 125,145, 150,145, 175,145,
33      75,170, 100,170, 125,170, 150,170, 175,170, 200,170, 220,170};
34           canvas.drawPosText("你想和我一起去探险吗?", pos, paintText);
35           super.onDraw(canvas);
36        }
37     }
38     @Override
39     public boolean onCreateOptionsMenu(Menu menu) {
40        getMenuInflater().inflate(R.menu.main, menu);
41        return true;
42     }
43     @Override
44     public boolean onOptionsItemSelected(MenuItem item) {
45        int id = item.getItemId();
46        if (id == R.id.action_settings) {
47           return true;
48        }
49        return super.onOptionsItemSelected(item);
50     }
51  }
```

（4）部署 DrawTextDemo 工程，程序运行结果如图 11-12 所示。

图 11-12　程序运行结果

11.4.7 绘制路径

在 Android 中提供了绘制路径的功能。绘制一条路径可以分别为创建路径和绘制定义好的路径两部分。

创建路径可以使用 android.graphics.Path 类来实现。Path 类包含一组矢量绘图方法，例如绘制圆、矩形、弧和线条等。常用的绘图方法如表 11-4 所示。

表 11-4 Path 类的常用方法

方法	描述
addArc(RectF oval,float startAngle,float sweetAngle)	添加弧形路径
addCircle(float x,float y,float radius,Path.Direction dir)	添加圆形路径
addOval(RectF oval,Path.Direction dir)	添加椭圆形路径
addRect(RectF rect,Path.Direction dir)	添加矩形路径
addRoundRect(RectF rect, float rx, float ry, Path.Direction dir)	添加圆角矩形路径
moveTo(float x,float y)	设置开始绘制直线的起始点
lineTo(float x,float y)	在 moveTo()方法设置的起始点与该方法指定的结束点之间画一条，如果在调用该方法之前没有使用 moveTo()方法设置起始点，那么将从(0,0)点开始绘制直线
quadTo(float x1,float y1,float x2,float y2)	用于根据指定的参数绘制一条线段轨迹
close()	闭合路径

使用 Canvas 类提供的 drawPath()方法可以将定义好的路径绘制在画布上。

下面通过实例介绍如何绘制路径。

(1) 创建一个新的 Android 工程，工程名为 DrawPathDemo，应用程序名为 DrawPathDemo，包名为 hlju.edu.cn，创建的 Activity 的名字为 MainActivity，最小 SDK 版本根据选择的目标 API 会自动添加。

(2) 修改 src 目录中 hlju.edu.cn 包下的 MainActivity.java 文件，代码如下：

```
1    package hlju.edu.cn;
2    import android.app.Activity;
3    import android.content.Context;
4    import android.graphics.Canvas;
5    import android.graphics.Paint;
6    import android.graphics.Path;
7    import android.graphics.Paint.Style;
8    import android.os.Bundle;
9    import android.view.Menu;
10   import android.view.MenuItem;
11   import android.view.View;
12   import android.widget.FrameLayout;
13   public class MainActivity extends Activity {
```

```
14      @Override
15      protected void onCreate(Bundle savedInstanceState) {
16          super.onCreate(savedInstanceState);
17          setContentView(R.layout.activity_main);
18          FrameLayout ll = (FrameLayout)findViewById(R.id.linearLayout1);
19          ll.addView(new MyView(this));
20      }
21      public class MyView extends View{
22          public MyView(Context context) {
23              super(context);
24          }
25          @Override
26          protected void onDraw(Canvas canvas) {
27              Paint paint = new Paint();
28              paint.setAntiAlias(true);
29              paint.setColor(0xFFFF6600);
30              paint.setTextSize(18);
31              paint.setStyle(Style.STROKE);
32              Path pathCircle = new Path();
33              pathCircle.addCircle(70, 70, 40, Path.Direction.CCW);
34              canvas.drawPath(pathCircle, paint);
35              Path pathLine = new Path();
36              pathLine.moveTo(150, 100);
37              pathLine.lineTo(200, 45);
38              pathLine.lineTo(250, 100);
39              pathLine.lineTo(300, 80);
40              canvas.drawPath(pathLine, paint);
41              Path pathTr = new Path();
42              pathTr.moveTo(70,300);
43              pathTr.lineTo(120, 270);
44              pathTr.lineTo(170, 300);
45              pathTr.close();
46              canvas.drawPath(pathTr, paint);
47              String str = "宝剑锋自磨砺出,梅花香自苦寒来。";
48              Path path = new Path();
49              path.addCircle(200, 200, 48, Path.Direction.CW);
50              paint.setStyle(Style.FILL);
51              canvas.drawTextOnPath(str, path,0, -18, paint);
52              super.onDraw(canvas);
53          }
54      }
55      @Override
56      public boolean onCreateOptionsMenu(Menu menu) {
57          getMenuInflater().inflate(R.menu.main, menu);
58          return true;
59      }
60      @Override
61      public boolean onOptionsItemSelected(MenuItem item) {
62          int id = item.getItemId();
```

```
63          if (id == R.id.action_settings) {
64            return true;
65          }
66          return super.onOptionsItemSelected(item);
67       }
68   }
```

第 27 行代码表示创建一个画笔。第 32 行代码表示绘制圆形路径。第 35 行代码表示绘制折线路径。第 41 行代码表示绘制三角形路径。第 47 行代码表示绘制绕路径的环形文字。

(3) 部署 DrawPathDemo 工程,程序运行结果如图 11-13 所示。

图 11-13　程序运行结果

11.5　为图像添加特效

在 Android 中,不仅可以绘制图形,还可以为图形添加特效。

11.5.1　旋转图像实例

使用 Android 提供的 android.graphics.Matrix 类的 setRotate()、postRotate() 和 preRotate() 方法,可以对图像进行旋转。这 3 种方法除了方法名不同外,其他语法格式均相同,下面以 setRotate() 方法为例来介绍其语法格式。setRotate() 方法有以下两种语法格式。

```
setRotate(float degrees)
```

使用该语法格式可以控制 Matrix 进行旋转，float 类型的参数用于指定旋转的角度。另一种语法格式如下。

```
setRotate(float degrees,float px,float py)
```

使用该语法格式可以控制 Matrix 以参数 px 和 py 为轴心进行旋转，float 类型的参数用于指定旋转的角度。

创建 Matrix 的对象，对其进行旋转后，还需要应用该 Matrix 对图像或组件进行控制。在 Canvas 类中提供了一个 drawBitmap(Bitmap bitmap, Matrix matrix, Paint paint)方法，可以在绘制图像的同时应用 Matrix 上的变化。例如，将一个图像旋转 45°后，再绘制到画布上，可以使用下面的代码。

```
1  Paint paint = new Paint();
2  Bitmap bitmap = BitmapFactory.decodeResource(MainActivity.this.getResource(),R.drawable.
   rabbit);
3  Matrix matrix = new Matrix();
4  matrix.setRotate(45);
5  canvas.drawBitmap(bitmap,matrix,paint);
```

下面通过实例介绍如何实现旋转图像。

（1）创建一个新的 Android 工程，工程名为 MatrixRotateDemo，应用程序名为 MatrixRotateDemo，包名为 hlju.edu.cn，创建的 Activity 的名字为 MainActivity，最小 SDK 版本根据选择的目标 API 会自动添加。

（2）修改 src 目录中 hlju.edu.cn 包下的 MainActivity.java 文件，代码如下：

```
1   package hlju.edu.cn;
2   import android.app.Activity;
3   import android.content.Context;
4   import android.graphics.Bitmap;
5   import android.graphics.BitmapFactory;
6   import android.graphics.Canvas;
7   import android.graphics.Matrix;
8   import android.graphics.Paint;
9   import android.os.Bundle;
10  import android.view.Menu;
11  import android.view.MenuItem;
12  import android.view.View;
13  import android.widget.FrameLayout;
14  public class MainActivity extends Activity {
15      @Override
16      protected void onCreate(Bundle savedInstanceState) {
17          super.onCreate(savedInstanceState);
18          setContentView(R.layout.activity_main);
19          FrameLayout ll = (FrameLayout)findViewById(R.id.frameLayout1);
20          ll.addView(new MyView(this));
```

```java
21      }
22      public class MyView extends View{
23          public MyView(Context context) {
24              super(context);
25          }
26          @Override
27          protected void onDraw(Canvas canvas) {
28              Paint paint = new Paint();
29              paint.setAntiAlias(true);
30              Bitmap bitmap_bg = BitmapFactory.decodeResource(MainActivity.this.getResources(), R.drawable.background);
31              canvas.drawBitmap(bitmap_bg, 0, 0, paint);
32              Bitmap bitmap_rabbit = BitmapFactory.decodeResource (MainActivity.this.getResources(), R.drawable.rabbit);
33              canvas.drawBitmap(bitmap_rabbit, 0, 0, paint);
34              Matrix matrix = new Matrix();
35              matrix.setRotate(30);
36              canvas.drawBitmap(bitmap_rabbit, matrix, paint);
37              Matrix m = new Matrix();
38              m.setRotate(90,150,150);
39              canvas.drawBitmap(bitmap_rabbit, m, paint);
40              super.onDraw(canvas);
41          }
42      }
43      @Override
44      public boolean onCreateOptionsMenu(Menu menu) {
45          getMenuInflater().inflate(R.menu.main, menu);
46          return true;
47      }
48      @Override
49      public boolean onOptionsItemSelected(MenuItem item) {
50          int id = item.getItemId();
51          if (id == R.id.action_settings) {
52              return true;
53          }
54          return super.onOptionsItemSelected(item);
55      }
56  }
```

第 19 行代码表示获取布局文件中的帧布局管理器。第 20 行代码表示将自定义视图添加到帧布局管理器中。第 28 行代码表示定义一个画笔。第 31 行代码表示绘制背景图像。第 33 行代码表示绘制原图。第 35 行代码表示以(0,0)点为轴心转换 30°。第 36 行代码表示绘制图像并应用 matrix 的变换。第 38 行代码表示以(150,150)点为轴心转换 90°。第 39 行代码表示绘制图像并应用 matrix 的变换。

（3）部署 MatrixRotateDemo 工程,程序运行结果如图 11-14 所示。

图 11-14　程序运行结果

11.5.2　缩放图像实例

使用 Android 提供的 android.graphics.Matrix 类的 setScale()、postScale()和 preScale()方法，可以对图像进行缩放。这3种方法除了方法名不同外，其他语法格式均相同，下面以 setScale()方法为例来介绍其语法格式。setScale()方法有以下两种语法格式。

```
setScale(float sx,float sy)
```

使用该语法格式可以控制 Matrix 进行缩放，参数 sx 和 sy 用于指定 x 轴和 y 轴的缩放比例。

另一种语法格式如下。

```
setScale(float sx,float sy,float px,float py)
```

使用该语法格式可以控制 Matrix 以参数 px 和 py 为轴心进行缩放，参数 sx 和 sy 用于指定 x 轴和 y 轴的缩放比例。

创建 Matrix 的对象，对其进行缩放后，还需要应用该 Matrix 对图像或组件进行控制。同旋转图像一样，也可以应用 Canvas 类中提供的 drawBitmap(Bitmap bitmap,Matrix matrix,Paint paint)方法，在绘制图像的同时应用 Matrix 上的变化。

下面通过实例介绍如何实现缩放图像。

（1）创建一个新的 Android 工程，工程名为 MatrixZoomDemo，应用程序名为 MatrixZoomDemo，包名为 hlju.edu.cn，创建的 Activity 的名字为 MainActivity，最小 SDK 版本根据选择的目标 API 会自动添加。

（2）修改 src 目录中 hlju.edu.cn 包下的 MainActivity.java 文件，代码如下：

```
1   package hlju.edu.cn;
2   import android.app.Activity;
3   import android.content.Context;
4   import android.graphics.Bitmap;
5   import android.graphics.BitmapFactory;
6   import android.graphics.Canvas;
7   import android.graphics.Matrix;
8   import android.graphics.Paint;
9   import android.os.Bundle;
10  import android.view.Menu;
11  import android.view.MenuItem;
12  import android.view.View;
13  import android.widget.FrameLayout;
14  public class MainActivity extends Activity {
15      @Override
16      protected void onCreate(Bundle savedInstanceState) {
17          super.onCreate(savedInstanceState);
18          setContentView(R.layout.activity_main);
19          FrameLayout ll = (FrameLayout)findViewById(R.id.frameLayout1);
20          ll.addView(new MyView(this));
21      }
22      public class MyView extends View{
23          public MyView(Context context) {
24              super(context);
25          }
26          @Override
27          protected void onDraw(Canvas canvas) {
28              Paint paint = new Paint();
29              paint.setAntiAlias(true);
30              Bitmap bitmap_bg = BitmapFactory.decodeResource (MainActivity.this.get-
    Resources(), R.drawable.background);
31              canvas.drawBitmap(bitmap_bg, 0, 0, paint);
32              Bitmap bitmap_rabbit = BitmapFactory.decodeResource (MainActivity.this.
    getResources(), R.drawable.rabbit);
33              Matrix matrix = new Matrix();
34              matrix.setScale(3f, 3f);
35              canvas.drawBitmap(bitmap_rabbit, matrix, paint);
36              Matrix m = new Matrix();
37              m.setScale(0.8f,0.8f,400,400);
38              canvas.drawBitmap(bitmap_rabbit, m, paint);
39              canvas.drawBitmap(bitmap_rabbit, 0, 0, paint);
40              super.onDraw(canvas);
41          }
42      }
43      @Override
44      public boolean onCreateOptionsMenu(Menu menu) {
45          getMenuInflater().inflate(R.menu.main, menu);
```

```
46              return true;
47          }
48          @Override
49          public boolean onOptionsItemSelected(MenuItem item) {
50              int id = item.getItemId();
51              if (id == R.id.action_settings) {
52                  return true;
53              }
54              return super.onOptionsItemSelected(item);
55          }
56      }
```

第19行代码表示获取布局文件中的帧布局管理器。第20行代码表示将自定义视图添加到帧布局管理器中。第28行代码表示定义一个画笔。第31行代码表示绘制背景。第34行代码表示以$(0,0)$点为轴心将图像在x轴和y轴均缩放300％。第35行代码表示绘制图像并应用matrix的变换。第37行代码表示以$(400,400)$点为轴心将图像在x轴和y轴均缩放80％。第38行代码表示绘制图像并应用matrix的变换。第39行代码表示绘制原图。

（3）部署MatrixZoomDemo工程，程序运行结果如图11-15所示。

图11-15　程序运行结果

11.5.3　倾斜图像实例

使用Android提供的android.graphics.Matrix类的setSkew()、postSkew()和preSkew()方法，可以对图像进行倾斜。这3种方法除了方法名不同外，其他语法格式均相同，下面以setSkew()方法为例来介绍其语法格式。setSkew()方法有以下两种语法格式。

```
setSkew(float kx,float ky)
```

使用该语法格式可以控制 Matrix 进行倾斜,参数 kx 和 ky 用于指定 x 轴和 y 轴的倾斜量。

另一种语法格式如下。

```
setSkew(float kx,float ky,float px,float py)
```

使用该语法格式可以控制 Matrix 以参数 px 和 py 为轴心进行倾斜,参数 kx 和 ky 用于指定 x 轴和 y 轴的倾斜量。

创建 Matrix 的对象,对其进行倾斜后,还需要应用该 Matrix 对图像或组件进行控制。同旋转图像一样,也可应用 Canvas 类中提供的 drawBitmap(Bitmap bitmap, Matrix matrix, Paint paint)方法,在绘制图像的同时应用 Matrix 上的变化。

下面通过实例介绍如何实现倾斜图像。

(1)创建一个新的 Android 工程,工程名为 MatrixSkewDemo,应用程序名为 MatrixSkewDemo,包名为 hlju.edu.cn,创建的 Activity 的名字为 MainActivity,最小 SDK 版本根据选择的目标 API 会自动添加。

(2)修改 src 目录中 hlju.edu.cn 包下的 MainActivity.java 文件,代码如下:

```
1   package hlju.edu.cn;
2   import android.app.Activity;
3   import android.content.Context;
4   import android.graphics.Bitmap;
5   import android.graphics.BitmapFactory;
6   import android.graphics.Canvas;
7   import android.graphics.Matrix;
8   import android.graphics.Paint;
9   import android.os.Bundle;
10  import android.view.Menu;
11  import android.view.MenuItem;
12  import android.view.View;
13  import android.widget.FrameLayout;
14  public class MainActivity extends Activity {
15      @Override
16      protected void onCreate(Bundle savedInstanceState) {
17          super.onCreate(savedInstanceState);
18          setContentView(R.layout.activity_main);
19          FrameLayout ll = (FrameLayout)findViewById(R.id.frameLayout1);
20          ll.addView(new MyView(this));
21      }
22      public class MyView extends View{
23          public MyView(Context context) {
24              super(context);
25          }
```

```
26          @Override
27          protected void onDraw(Canvas canvas) {
28              Paint paint = new Paint();
29              paint.setAntiAlias(true);
30              Bitmap bitmap_bg = BitmapFactory.decodeResource (MainActivity.this.get-
    Resources(), R.drawable.background);
31              canvas.drawBitmap(bitmap_bg, 0, 0, paint);
32              Bitmap bitmap_rabbit = BitmapFactory.decodeResource (MainActivity.this.
    getResources(), R.drawable.rabbit);
33              Matrix matrix = new Matrix();
34              matrix.setSkew(2f, 1f);
35              canvas.drawBitmap(bitmap_rabbit, matrix, paint);
36              Matrix m = new Matrix();
37              m.setSkew(-0.5f, 0f, 78, 69);
38              canvas.drawBitmap(bitmap_rabbit, m, paint);
39              canvas.drawBitmap(bitmap_rabbit, 0, 0, paint);
40              super.onDraw(canvas);
41          }
42      }
43      @Override
44      public boolean onCreateOptionsMenu(Menu menu) {
45          getMenuInflater().inflate(R.menu.main, menu);
46          return true;
47      }
48      @Override
49      public boolean onOptionsItemSelected(MenuItem item) {
50          int id = item.getItemId();
51          if (id == R.id.action_settings) {
52              return true;
53          }
54          return super.onOptionsItemSelected(item);
55      }
56  }
```

第 19 行代码表示获取布局文件中的帧布局管理器。第 20 行代码表示将自定义视图添加到帧布局管理器中。第 28 行代码表示定义一个画笔。第 31 行代码表示绘制背景。第 34 行代码表示以 (0,0) 点为轴心将图像在 x 轴上倾斜 2，在 y 轴上倾斜 1。第 35 行代码表示绘制图像并应用 matrix 的变换。第 37 行代码表示以 (78,69) 点为轴心将图像在 x 轴上倾斜 −0.5。第 38 行代码表示绘制图像并应用 matrix 的变换。第 39 行代码表示绘制原图。

（3）部署 MatrixSkewDemo 工程，程序运行结果如图 11-16 所示。

11.5.4 平移图像实例

使用 Android 提供的 android.graphics.Matrix 类的 setTranslate()、postTranslate() 和 preTranslate() 方法，可以对图像进行平移。这 3 种方法除了方法名不同外，其他语法格式均相同，下面以 setTranslate() 方法为例来介绍其语法格式。setTranslate() 方法的语法格式如下。

图 11-16　程序运行结果

```
setTranslate(float dx,float dy)
```

在该语法中,参数 dx 和 dy 用于指定将 Matrix 移动到的位置的 x 和 y 坐标。

创建 Matrix 的对象,对其进行平移后,还需要应用该 Matrix 对图像或组件进行控制。同旋转图像一样,也可应用 Canvas 类中提供的 drawBitmap(Bitmap bitmap,Matrix matrix,Paint paint)方法,在绘制图像的同时应用 Matrix 上的变化。

下面通过实例介绍如何实现将图像旋转后再平移。

(1) 创建一个新的 Android 工程,工程名为 MatrixTranslateDemo,应用程序名为 MatrixTranslateDemo,包名为 hlju.edu.cn,创建的 Activity 的名字为 MainActivity,最小 SDK 版本根据选择的目标 API 会自动添加。

(2) 修改 src 目录中 hlju.edu.cn 包下的 MainActivity.java 文件,代码如下:

```
1    package hlju.edu.cn;
2    import android.app.Activity;
3    import android.content.Context;
4    import android.graphics.Bitmap;
5    import android.graphics.BitmapFactory;
6    import android.graphics.Canvas;
7    import android.graphics.Matrix;
8    import android.graphics.Paint;
9    import android.os.Bundle;
10   import android.view.Menu;
11   import android.view.MenuItem;
12   import android.view.View;
13   import android.widget.FrameLayout;
```

```
14  public class MainActivity extends Activity {
15      @Override
16      protected void onCreate(Bundle savedInstanceState) {
17          super.onCreate(savedInstanceState);
18          setContentView(R.layout.activity_main);
19          FrameLayout ll = (FrameLayout)findViewById(R.id.frameLayout1);
20          ll.addView(new MyView(this));
21      }
22      public class MyView extends View{
23          public MyView(Context context) {
24              super(context);
25          }
26          @Override
27          protected void onDraw(Canvas canvas) {
28              Paint paint = new Paint();
29              paint.setAntiAlias(true);
30              Bitmap bitmap_bg = BitmapFactory.decodeResource (MainActivity.this.get-
    Resources(), R.drawable.background);
31              canvas.drawBitmap(bitmap_bg, 0, 0, paint);
32              Bitmap bitmap_rabbit = BitmapFactory.decodeResource (MainActivity.this.
    getResources(), R.drawable.rabbit);
33              canvas.drawBitmap(bitmap_rabbit, 0, 0, paint);
34              Matrix matrix = new Matrix();
35              matrix.setRotate(30);
36              matrix.postTranslate(200,100);
37              canvas.drawBitmap(bitmap_rabbit, matrix, paint);
38              super.onDraw(canvas);
39          }
40      }
41      @Override
42      public boolean onCreateOptionsMenu(Menu menu) {
43          getMenuInflater().inflate(R.menu.main, menu);
44          return true;
45      }
46      @Override
47      public boolean onOptionsItemSelected(MenuItem item) {
48          int id = item.getItemId();
49          if (id == R.id.action_settings) {
50              return true;
51          }
52          return super.onOptionsItemSelected(item);
53      }
54  }
```

第 19 行代码表示获取布局文件中的帧布局管理器。第 20 行代码表示将自定义视图添加到帧布局管理器中。第 28 行代码表示定义一个画笔。第 29 行代码表示使用抗锯齿功能。第 31 行代码表示绘制背景。第 33 行代码表示绘制原图。第 34 行代码表示创建一个 Matrix 的对象。第 35 行代码表示将 matrix 旋转 30°。第 36 行代码表示将 matrix 平移到

(200,100)的位置。第 37 行代码表示绘制图像并应用 matrix 的变换。

(3) 部署 MatrixTranslateDemo 工程,程序运行结果如图 11-17 所示。

图 11-17　程序运行结果

11.5.5　使用 BitmapShader 渲染图像实例

Android 中提供的 BitmapShader 类主要用于渲染图像。如果需要将一张图片裁剪成椭圆或圆形等形状显示到屏幕上时,就可以使用 BitmapShader 类来实现。使用 BitmapShader 来渲染图像的基本步骤如下。

(1) 创建 BitmapShader 类的对象,可以通过以下的代码进行创建。

```
BitmapShader(Bitmap bitmap,Shader.TileMode tileX,Shader.ToleMode tileY)
```

其中的 bitmap 参数用于指定一个位图对象,通常是用于渲染原图像;tileX 参数用于指定在水平方向上图像的重复方式;tileY 参数用于指定在垂直方向上图像的重复方式。

(2) 通过 Paint 的 setShader()方法来设置渲染对象。

(3) 在绘制图像时,使用已经设置了 setShader()方法的画笔。

下面通过实例介绍如何渲染图像。

(1) 创建一个新的 Android 工程,工程名为 BitmapShaderDemo,应用程序名为 BitmapShaderDemo,包名为 hlju.edu.cn,创建的 Activity 的名字为 MainActivity,最小 SDK 版本根据选择的目标 API 会自动添加。

(2) 修改 src 目录中 hlju.edu.cn 包下的 MainActivity.java 文件,代码如下:

```
1    package hlju.edu.cn;
2    import android.app.Activity;
```

```java
3   import android.content.Context;
4   import android.graphics.Bitmap;
5   import android.graphics.BitmapFactory;
6   import android.graphics.BitmapShader;
7   import android.graphics.Canvas;
8   import android.graphics.Paint;
9   import android.graphics.RectF;
10  import android.graphics.Shader.TileMode;
11  import android.os.Bundle;
12  import android.view.Menu;
13  import android.view.MenuItem;
14  import android.view.View;
15  import android.widget.FrameLayout;
16  public class MainActivity extends Activity {
17      private int view_width;
18      private int view_height;
19      @Override
20      protected void onCreate(Bundle savedInstanceState) {
21          super.onCreate(savedInstanceState);
22          setContentView(R.layout.activity_main);
23          FrameLayout ll = (FrameLayout)findViewById(R.id.frameLayout1);
24          ll.addView(new MyView(this));
25      }
26      public class MyView extends View{
27          public MyView(Context context) {
28              super(context);
29              view_width = context.getResources().getDisplayMetrics().widthPixels;
30              view_height = context.getResources().getDisplayMetrics().heightPixels;
31          }
32          @Override
33          protected void onDraw(Canvas canvas) {
34              Paint paint = new Paint();
35              paint.setAntiAlias(true);
36              Bitmap bitmap_bg = BitmapFactory.decodeResource (MainActivity.this.getResources(), R.drawable.android);
37              BitmapShader bitmapshader = new BitmapShader (bitmap_bg, TileMode.REPEAT, TileMode.REPEAT);
38              paint.setShader(bitmapshader);
39              canvas.drawRect(0, 0, view_width, view_height, paint);
40              Bitmap bm = BitmapFactory.decodeResource(MainActivity.this.getResources(), R.drawable.img02);
41              BitmapShader bs = new BitmapShader(bm,TileMode.REPEAT, TileMode.MIRROR);
42              paint.setShader(bs);
43              RectF oval = new RectF(0,0,280,180);
44              canvas.translate(100, 20);
45              canvas.drawOval(oval, paint);
46              super.onDraw(canvas);
47          }
48      }
49      @Override
50      public boolean onCreateOptionsMenu(Menu menu) {
51          getMenuInflater().inflate(R.menu.main, menu);
```

```
52              return true;
53          }
54          @Override
55          public boolean onOptionsItemSelected(MenuItem item) {
56              int id = item.getItemId();
57              if (id == R.id.action_settings) {
58                  return true;
59              }
60              return super.onOptionsItemSelected(item);
61          }
```

第 23 行代码表示获取布局文件中的帧布局管理器。第 24 行代码表示将自定义视图添加到帧布局管理器中。第 29 行代码表示获取屏幕的宽度。第 30 行代码表示获取屏幕的高度。第 34 行代码表示定义一个画笔。第 35 行代码表示使用抗锯齿功能。第 38 行代码表示设置渲染对象。第 39 行代码表示绘制一个使用 BitmapShader 渲染的矩形。第 44 行代码表示将画面在 x 轴上平移 100 像素，在 y 轴上平移 20 像素。第 45 行代码表示绘制一个使用 BitmapShader 渲染的椭圆形。

（3）部署 BitmapShaderDemo 工程，程序运行结果如图 11-18 所示。

图 11-18　程序运行结果

习题

1. Android 访问图片的方式有哪几种。
2. 简述 Canvas 类的特点和作用。
3. 在 Android 中是否可以为图像添加特效，如何添加。
4. 在 Android 中如何渲染图像。

第12章 综合示例设计与开发

本章将以"理财系统"作为示例,综合运用以往章节所学习的知识和技巧,从需求分析、界面设计、模块设计和程序开发等几个方面,详细介绍Android应用程序的设计思路与开发方法。通过本章的学习可以让读者掌握Android应用程序的设计方法和多种组件应用的能力。

本章主要学习内容:
- 掌握Android应用程序的基本设计方法和思路;
- 掌握使用多种组件进行Android程序开发的方法。

12.1 需求分析

通过前面章节的学习,读者应该已经掌握了一些Android应用程序的开发知识和方法,但如何能够综合的运用这些知识和方法,解决实际开发中所遇到的问题,还是一个需要继续学习和探讨的问题。设计本章的初衷就是希望读者能够根据实际项目的需求,准确地分析出Android应用程序开发所可能涉及的知识点,通过分析软件的需求,快速设计出用户界面和模块结构,并最终完成应用程序的开发和调试。

本章提供的"理财系统"是一个略微复杂的示例。本系统主要是基于人们日常生活中对于个人理财具体情况的需求。研究设计出符合当今社会人们经济生活中出现的随时随地方便理财理念的软件系统。手机应用为人们的生活带来乐趣的同时也带来方便。本系统是基于当今社会流行的Android平台框架,通过总结PC理财软件设计理念和体系,为用户提供个人经典理财模式的手机应用版。系统需要密码登录,目的在于保护其个人财产情况,防止个人财务状况外流造成不良后果,通过本系统用户可对其日常的收入、日常的支出做出合理的预算和统计,对其他阶段的个人收入支出情况进行记录,利用对比可进行分析,并通过合理分配自己财务使其得到最大化的合理应用,完成个人财务积累并增值,有利于个人投资,满足人们"精打细算"的生活需求。

从上面的描述中可以基本了解软件的功能需求,系统能为用户提供基本的理财需求,包括数据插入、数据查询和数据删除等功能,用户可以通过系统对数据进行相应的操作,主要功能有以下几点:

- 用户登录,进入本软件需要密码验证登录以保护个人理财状况的隐私性;
- 辅助维护,用户可以通过交互界面操作相关项;

- 日常收入，用户可以将收入金额和日期等一起添加；
- 日常支出，用户可以将支出金额和日期等一起添加；
- 收入支出统计，按照用户的数据查询要求，用户可以对相关数据进行统计；
- 计算器，对用户设计出可计算银行中财务状况的特殊计算机；
- 收入查询，用户可以根据输入的内容通过数据进行查询，可以删除有误数据；
- 支出查询，支出查询功能与收入查询功能相似，同样根据条件对数据进行操作；
- 基本情况，用户可以增加自己的个人基本情况，修改个人信息及登录信息。

12.2 程序设计

12.2.1 系统功能模块设计

Android 手机理财软件系统有 6 个功能模块，这 6 个功能模块如图 12-1 所示。

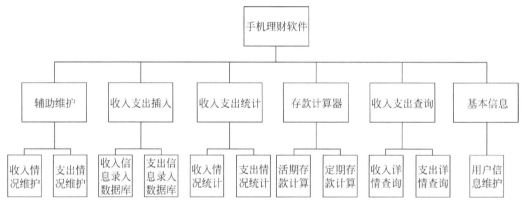

图 12-1 功能模块图

1．辅助维护模块

此模块会对用户每天的收入和支出进行详细的记录，会将数据进行分类收集。收集的数据主要是由用户在软件上手动输入的，确定数据后，会将数据保存到手机自带的数据库中，在保存时分析数据是否有效，如无效，会将其删除，保证数据库中数据不会出现重复。此模块被开发出于主要是用于记录人们生活中不同环境情况下的收入和支出。更加有助于人们记录和打理自己的财务情况。而移动端和台式计算机的输入方式不同，通过手机进行输入会减少人们实际输入的数据量。

2．收入支出插入模块

此模块主要是将个人数据插入到手机数据库中。Android 平台中自带的数据库 SQLite 中存储数据的形式主要是存储数据表，SQLite 中不需要独立的引擎，是由程序同过 API 直接对数据库进行操作，在数据库中主要是通过日期来确定时间段，对输入的数据进行相应的编辑和备注等，保存到手机自带的数据库中。

3. 收入支出统计模块

此模块主要功能是将数据库中存储的个人理财数据提取出来并进行相应的操作。通过此模块用户可将自己数据库中的数据提取出来，并进行不同类别的统计。如单日总结统计，按照一定的时间段进行统计，按照收入的来源进行统计，按照支出的来源进行统计，更加方便地进行自己的理财。

4. 存款计算器模块

此模块主要功能是针对用户在银行中个人存款变化进行计算，虽然银行的利率是较低的，但是人们还是将银行储蓄业务作为主要的理财手段。因为银行储蓄是最安全和稳定的理财方式，所以随时查看自己银行存款的变化是特别必要的，但是银行金额的计算复杂度不是可以口算的，这时人们会选计算器作为主要的工具来实现复杂的计算，但是传统的计算机只能实现简单的数字运算不能进行分类的数据运算，此时需要一款针对用户存款理财开发出的专用的计算机。使用户更加方便地进行理财。

5. 收入支出查询模块

此模块是方便用户对数据进行方便的查询操作。因为数据库中的信息属于后台数据，因此用户需要一个可操作的界面来辅助对数据的操作，在界面上用户可以自己设定不同的查询条件来进行查询，结果会显示到一个独立的窗口上，方便对数据的管理。

6. 基本信息模块

此模块是整个系统的基础也是最需要安全性的模块，因为用户所有的个人信息都会储存在此模块中，针对此模块的安全性，系统会由登录功能来防止别人对数据的访问和防止数据的泄露，而且此模块还是在后台运行，如果防止数据在手机丢失后泄露，之前可将所有数据备份到计算机上。这样就会保证理财数据的安全。系统 UML 用例图如图 12-2 所示。

12.2.2 系统流程设计

本手机理财软件主要有 3 个主界面，分别为欢迎动画界面、登录界面、理财详情管理界面，主要操作流程如图 12-3 所示。

12.2.3 数据库设计

对于本数据库中的数据中会设计 5 个数据表用于存储，命名为 Scy（日常收入类别表）、Zcy（日常支出类别表）、Income（日常收入表）、Spend（日常支出表）和 PersonDate（基本信息表），各个表之间会有关联，各个表之间的关系如图 12-4 所示。

具体数据库表字段设计如下。

（1）Scy（日常收入类别表）：本表记录平时的收入，并按照类别分开存储，主要字段有 Scy 名称、Scy 说明信息，具体如表 12-1 所示。

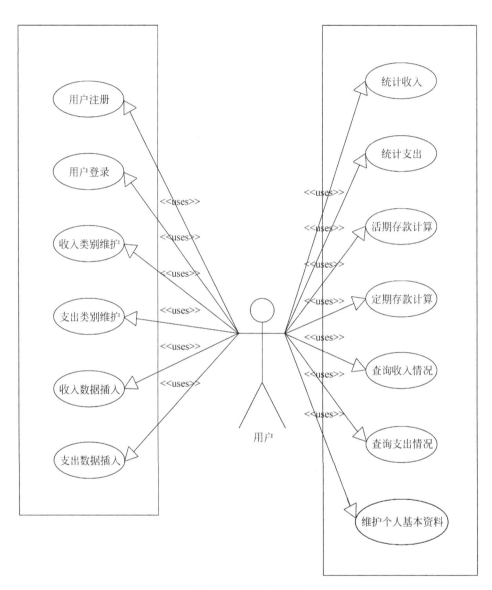

图 12-2　UML 用例图

第12章 综合示例设计与开发

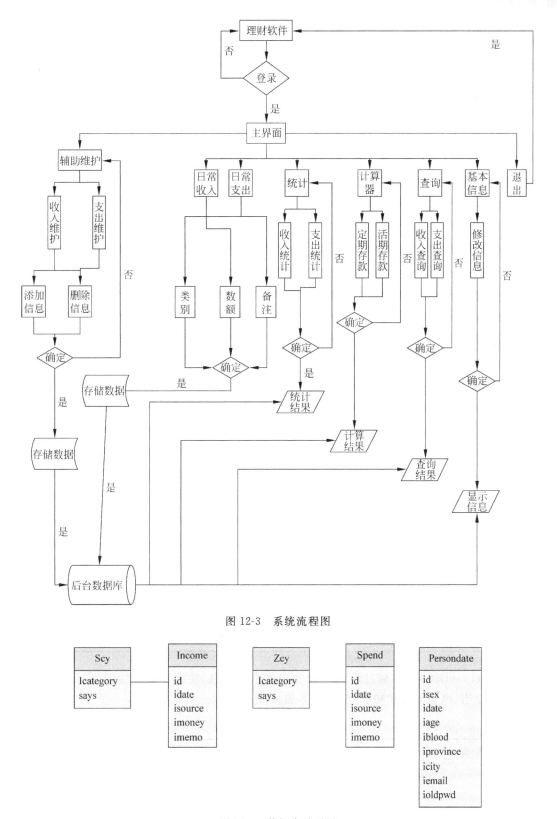

图 12-3 系统流程图

图 12-4 数据表关系图

表 12-1　日常收入类别表

字 段 名 称	数 据 类 型	字 段 大 小	是 否 主 键	说　　明
Icategory	Varchar	10	是	类别名称
says	Varchar	50	否	对类别的说明

(2) Zcy(日常支出类别表)：本表记录平时的支出，并按照类别分开存储，主要字段有 Zcy 名称、Zcy 说明信息，具体如表 12-2 所示。

表 12-2　日常支出类别表

字 段 名 称	数 据 类 型	字 段 大 小	是 否 主 键	说　　明
Icategory	Varchar	10	是	类别名称
says	Varchar	50	否	对类别的说明

(3) Income(日常收入表)：本表记录日常收入数据记录，主要字段有 Income 日期、Income 金额、Income 类别、Income 备注，具体如表 12-3 所示。

表 12-3　日常收入表

字 段 名 称	数 据 类 型	字 段 大 小	是 否 主 键	说　　明
id	Integer	8	是	收入 id
idate	Char	10	否	收入时间
isource	Varchar	8	否	收入类型
imoney	Integer	8	否	收入金额
imemo	Varchar	50	否	收入说明

(4) Spend(日常支出表)：本表记录日常支出数据记录，主要字段有 Spend 日期、Spend 金额、Spend 类别、Spend 备注，具体如表 12-4 所示。

表 12-4　日常支出表

字 段 名 称	数 据 类 型	字 段 大 小	是 否 主 键	说　　明
id	Integer	8	是	支出 id
idate	Char	10	否	支出时间
isource	Varchar	20	否	支出类型
imoney	Integer	8	否	支出金额
imemo	Varchar	50	否	支出说明

(5) PersonDate(基本信息表)：存储用户的个人信息，主要字段有性别、生日、年龄、血型、省市、城市、电子邮箱和初始密码，具体如表 12-5 所示。

表 12-5　基本信息表

字 段 名 称	数 据 类 型	字 段 大 小	是 否 主 键	说　　明
id	Integer	8	是	用户 id
isex	Char	1	否	用户性别
idate	Varchar	10	否	用户生日

续表

字段名称	数据类型	字段大小	是否主键	说　　明
iage	Varchar	2	否	用户年龄
iblood	Varchar	5	否	用户血型
iprovince	Varchar	5	否	用户省份
icity	Varchar	5	否	用户所在城市
iemail	Varchar	15	否	用户电子邮件
ioldpwd	Varchar	6	否	用户原先的密码

12.3　程序开发

12.3.1　工程结构

在程序开发阶段，首先确定"理财系统"的工程名称为 FinanceManager。然后根据程序模块设计的内容，建立 FinanceManager 示例。FinanceManager 示例源代码的文件结构如图 12-5 所示。

图 12-5　FinanceManager 示例的源代码文件

12.3.2　数据库操作类

数据库适配器是最底层的模块，主要用于封装用户界面和后台服务对 SQLite 数据库的操作。数据库适配器的核心代码主要在 DBUtil.java 文件中。

DBUtil.java 文件的核心代码如下：

```java
//创建或打开数据库
//创建或打开数据库的方法
    public static void createOrOpenDatabase()
    {
    try
        {
          sld = SQLiteDatabase.openDatabase
          (
             "/data/data/hlju.edu.FM/mydb",            //数据库所在路径
             null,                                    //CursorFactory
      SQLiteDatabase.OPEN_READWRITE|SQLiteDatabase.CREATE_IF_NECESSARY
//读、写若不存在则创建
          );

          String sql01 = "create table if not exists Income" +
                "(" +
                   "id INTEGER PRIMARY KEY AUTOINCREMENT," +
                   "idate char(10)," +
                   "isource Varchar(20)," +
                   "imoney Integer," +
                   "imemo Varchar(50)" +
                ")";
          String sql02 = "create table if not exists Scy" +     //收入类别
                "(" +
                   "icategory Varchar(10) PRIMARY KEY ," +
                   "says varchar(50)" +
                ")";
          String sql03 = "create table if not exists Spend" +
                "(" +
                   "id INTEGER PRIMARY KEY AUTOINCREMENT," +
                   "idate char(10)," +
                   "isource Varchar(20)," +
                   "imoney Integer," +
                   "imemo Varchar(50)" +
                ")";
          String sql04 = "create table if not exists Zcy" +     //支出类别
                "(" +
                   "icategory Varchar(10) PRIMARY KEY," +
                   "says Varchar(50)" +
                ")";
          String sql05 = "create table if not exists PersonDate" +
                "(" +
                   "id INTEGER PRIMARY KEY AUTOINCREMENT," +
                   "isex char(1)," +
                   "idate varchar(10)," +
                   "iage varchar(2)," +
                   "iblood varchar(5)," +
                   "iprovince varchar(5)," +
                   "icity varchar(5)," +
```

```
                "iemail varchar(15)," +
                "ioldpwd varchar(6)" +
            ")";
        sld.execSQL(sql01);
        sld.execSQL(sql02);
        sld.execSQL(sql03);
        sld.execSQL(sql04);
        sld.execSQL(sql05);
    }
    catch(Exception e)
    {
        e.printStackTrace();
    }
}

//类别维护插入方法
public static void insertCategory(String str,String str0,String str1)
{
    try
    {
        String sql = "insert into " + str0 + " values('" + str + "','" + str1 + "');";
        sld.execSQL(sql);
    }
    catch(Exception e)
    {
        e.printStackTrace();
    }
}
```

12.3.3 界面设计类

对于手机软件,界面是交互式最重要的根基,用户的设计对于第一印象特别重要,为了界面的简洁和美观,本软件采用了九宫格,和画布重绘来实现界面显示,对于动画显示是事先设定一定的时间来显示,设置透明度。下面是关于界面设计的核心代码:

```
//主界面的设计,九宫格布局
    public void initBitmap(Resources r) {}              //加载图片
    public void onDraw(Canvas canvas){}                 //绘制
    public boolean onTouchEvent(MotionEvent e)          //触屏
    {
        int x = (int)(e.getX());
        int y = (int)(e.getY());
        switch(e.getAction())
        {
            case MotionEvent.ACTION_DOWN:
    if(x > LEI_XOFFSET&&x < LEI_XOFFSET + PIC_WIDTH&&y > LEI_YOFFSET&&y < LEI_YOFFSET + PIC_HEIGHT)
```

```
                            {
                                activity.hd.sendEmptyMessage(0);
                            }
//动画视图的切换
    public void surfaceCreated(SurfaceHolder holder) {//创建时被调用
        new Thread()
        {
            public void run()
            {
                for(Bitmap bm:logos)
                {
                    currentLogo = bm;
                    //计算图片位置
                    currentX = screenWidth/2 - bm.getWidth()/2;
                    currentY = screenHeight/2 - bm.getHeight()/2;

                    for(int i = 255;i > -10;i = i - 10)
                    {//动态更改图片的透明度值并不断重绘
                        currentAlpha = i;
                        if(currentAlpha < 0)
                        {
                            currentAlpha = 0;
                        }
                        SurfaceHolder myholder = MySurfaceView.this.getHolder();
                        Canvas canvas = myholder.lockCanvas();    //获取画布
                        try{
                            synchronized(myholder){
                                onDraw(canvas);                    //绘制
                            }
                        }
                        catch(Exception e){
                            e.printStackTrace();
                        }
                        finally{
                            if(canvas != null){
                                myholder.unlockCanvasAndPost(canvas);
                            }
                        }
```

12.3.4 辅助工具类

本系统不仅会对数据库进行增、删、改和查的操作之外,还要对数据进行存储,为此需要一些工具类来帮助人们操作,还有利于界面更加美观,对于工具类的主要功能是日期对照和数据转换。下面是对于工具类的核心代码:

```
//日期工具类
    public static String getdate(String years,String monthes,String dates,int yearInterval)
```

```
        {
            int systemyears = dt.getYear() + 1900;    //得到系统日期
if((years.matches(str1))&&(monthes.matches(str2))&&(dates.matches(str3))&&(yearInterval>0))
            {//判断是否符合格式
                if(((Math.abs(insertyear - systemyears))<=yearInterval))

                    if((((insertmonth == 1)||(insertmonth == 3)||(insertmonth == 5)||
                        (insertmonth == 7)||(insertmonth == 8)||(insertmonth == 10)||
                        (insertmonth == 12))&&(date<32))

//数据转换类
public class FDSDUtil
{
    //保留两位小数
    public static String formatData(double d)
    {
        DecimalFormat myformat = new  DecimalFormat("0.00");
        return myformat.format(d);
    }
    //保留整数
    public static String formatDataInt(double d)
    {
        DecimalFormat myformat = new  DecimalFormat("0");
        return myformat.format(d);
    }
```

12.3.5 主控制类

此类是整个软件的核心类,此类会在软件初始阶段初始化界面,并根据用户的操作来切换到需要的界面,并且会监听事件和用户的操作。下面是主控制类的核心代码:

```
public class Pm_Activity extends Activity
{

    Handler hd = new Handler()                  //接收信息界面跳转
    {
        @Override
        public void handleMessage(Message msg)  //重写方法
        {
            switch(msg.what)
            {
            }
        }
    };

        //增加监听
    addbutton.setOnClickListener
```

```java
                (
                    new OnClickListener()
                    {
                        @Override
                        public void onClick(View v)
                        {
                            List<String> slist = null;
                            if(rb1.isChecked())
                            {
                                slist = DBUtil.queryCategory("Scy");
                                boolean flag = Pm_Activity.this.InOrNot(icategory, slist);
                                //测试通过后将 10 更改到 Constant 的类中
                if((icategory.length()!=0)&&(saytext.length()<textlength)&&(saytext.length()>0))
                                {
                                    if(flag)
                                    {
                                        DBUtil.insertCategory(icategory,"Scy",saytext);
                                        //插入类别
                                        ListView lv = (ListView)findViewById(R.id.ListView01);
                                        Pm_Activity.this.getDataToListView(lv,"Scy");
                                    }
                                    else
                                    {
                                        Toast.makeText(Pm_Activity.this, "不可以重复插
入!!!", Toast.LENGTH_SHORT).show();
                                    }
                                }
                                else if(icategory.length() == 0)
                                {
                                }
                                else
                                {
                                    Toast.makeText(Pm_Activity.this, "说明框不可以为空!",
Toast.LENGTH_SHORT).show();
                                }
                            }
                            if(rb2.isChecked())
                            {
                                slist = DBUtil.queryCategory("Zcy");
                                boolean flag = Pm_Activity.this.InOrNot(icategory, slist);
                                if((icategory.length()!=0)&&(saytext.length()<textlength)
&&(saytext.length()>0))
                                {
                                    if(flag)
                                    {
                                        DBUtil.insertCategory(icategory,"Zcy",saytext);
//添加适配器为不同行的 TextView 填充不同的值
```

```java
//GridView适配器
public void SelectBaseAdapter(ListView lv,final List<String> result,final int id,final int dbif)
{
    GridView gv = (GridView)this.findViewById(R.id.GridView01);
    SimpleAdapter sca = new SimpleAdapter
    (
        this,
        generateDataList(result),
        R.layout.gridview,
        new String[]{"col1","col2","col3"},
        new int[]{R.id.TextView01,R.id.TextView02,R.id.TextView03}
    );
    gv.setAdapter(sca);                                         //为GridView设置数据适配器
    gv.setOnItemClickListener
    (
        new OnItemClickListener()
        {
            @Override
            public void onItemClick(AdapterView<?> arg0, View arg1,
                                    int arg2, long arg3)
            {
                LinearLayout ll = (LinearLayout)arg1;
                TextView tvn1 = (TextView)ll.getChildAt(0);     //获取其中的TextView
                TextView tvn2 = (TextView)ll.getChildAt(1);     //获取其中的TextView
                TextView tvn3 = (TextView)ll.getChildAt(2);
                Pm_Activity.text01 = tvn1.getText().toString().trim();
                Pm_Activity.text02 = tvn2.getText().toString().trim();
                Pm_Activity.text03 = tvn3.getText().toString().trim();
                if(id == 1)
                {
                    goToInxxView(dbif);
                }
                else if(id == 2)
                {
                    goToSpxxView(dbif);
                }
            }
        }
    );
}

//GridView存储数据
public List<? extends Map<String, ?>> generateDataList(List<String> result)
{
ArrayList<Map<String,Object>> list = new ArrayList<Map<String,Object>>();;
int rowCounter = result.size()/4;
for(int i = 0;i<rowCounter;i++)
{
    HashMap<String,Object> hmap = new HashMap<String,Object>();
    hmap.put("col1",result.get(i*4));
```

```
            hmap.put("col2",result.get(i*4+1));
            hmap.put("col3",result.get(i*4+2));
            hmap.put("col4",result.get(i*4+3));
            list.add(hmap);
    }
    return list;
}

//返回按钮的实现,不关闭数据库
public void returnView(Button button)
{
    button.setOnClickListener
     (
        new OnClickListener()
        {
            @Override
            public void onClick(View v)
            {
                DBUtil.closeDatabase();
                goToLc_View();
            }
        }
```

12.3.6 用户界面

图 12-6 所显示的是应用软件通过验证登录后,进入程序主界面,此为主界面结构布局,采用九宫格布局方案,利用画布重绘技术,通过鼠标单击获取当前坐标,判断是否在规定的范围内,如果在某一规定范围内,则进入相关功能界面,否则不予响应。

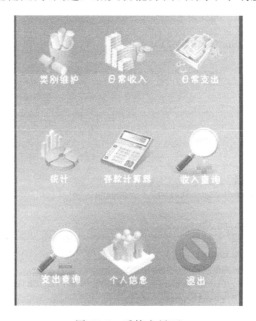

图 12-6 系统主界面

图12-7所示的界面为类别维护界面,主要功能是向数据库中添加收入支出类别,也就是详情表中的收支来源,通过添加类别,可以在以后的日常收入和日常支出界面进行相关数据的添加,类别带有说明栏,可以向其中写入添加此类别的具体情况,单击"添加"向数据库相应类别表中添加类别信息,单击"删除"则删除相应类别以及此类别下的所有数据。

图12-8所示的界面主要显示的是日常收入和日常支出状态下的日期插入情况,当单击插入日期的时候,会弹出日期插入对话框,用户可以通过单击上下按钮进行时间的调整,当确定时间后单击"确定"插入数据库。

图 12-7　辅助维护界面

图 12-8　日常收入界面

图12-9界面显示的是存款计算器,它的主要功能是对用户个人的银行储蓄业务进行相关的统计计算,通过调整不同的参数,可以得到个人存储业务的资金具体情况,达到更好理财的目标。

图12-10所示的界面是查询操作界面,它主要包括收入查询和支出查询,查询的方式为按照要求填入相关请求范围,单击数据框,弹出对话框插入日期范围,金额填写可以通过系统自带的模拟键盘手动输入,选择查询类别,单击"查询数据"就可以得到统计结果显示。

图 12-9　计算器界面

图 12-10　查询页面

图 12-11 所显示的界面是通过查询界面操作得到的数据列表,单击每一项数据,可以进入此详细情况界面,该界面详细记录了收入和支出的具体情况,包括类别、金额、备注信息,同时如果需要对数据库信息进行删除操作,也可以在此界面进行具体数据的删除。

图 12-11 详情页面

习题

编写完成本章综合示例,并详细说明 GGView 类、DBUtil 类、FM_Activity 类、FDSDUtil 类、MyDialog 类和 MySurfaceView 类的作用。

参 考 文 献

[1] 郑萌,赵常松,等.Android 应用程序开发与典型案例[M].北京：电子工业出版社,2012.
[2] 秦建平.Android 编程宝典[M].北京：北京航空航天大学出版社,2013.
[3] 陈会安.Android SDK 程序设计与开发范例[M].北京：清华大学出版社,2013.
[4] 王家林.Android 4.0 网络编程详解[M].北京：电子工业出版社,2012.
[5] 李文琴,李翠霞.Android 开发与实践[M].北京：人民邮电出版社,2014.

图书资源支持

感谢您一直以来对清华版图书的支持和爱护。为了配合本书的使用,本书提供配套的资源,有需求的读者请扫描下方的"书圈"微信公众号二维码,在图书专区下载,也可以拨打电话或发送电子邮件咨询。

如果您在使用本书的过程中遇到了什么问题,或者有相关图书出版计划,也请您发邮件告诉我们,以便我们更好地为您服务。

我们的联系方式:

地　　址:北京海淀区双清路学研大厦 A 座 707

邮　　编:100084

电　　话:010-62770175-4604

资源下载:http://www.tup.com.cn

电子邮件:weijj@tup.tsinghua.edu.cn

QQ:883604(请写明您的单位和姓名)

用微信扫一扫右边的二维码,即可关注清华大学出版社公众号"书圈"。

资源下载、样书申请

书圈